Lecture Notes in Physics

For information about Vols. 1–23, please contact your bookseller or Springer-Verlag.

Vol. 24: R. F. Snipes, Statistical Mechanical Theory of the Electrolytic Transport of Nonelectrolytes. V, 210 pages. 1973.

Vol. 25: Constructive Quantum Field Theory. The 1973 "Ettore Majorana" International School of Mathematical Physics. Edited by G. Velo and A. Wightman. III, 331 pages. 1973.

Vol. 26: A. Hubert, Theorie der Domänenwände in geordneten Medien. XII, 377 Seiten. 1974.

Vol. 27: R. K. Zeytounian, Notes sur les Ecoulements Rotationnels de Fluides Parfaits. XIII, 407 pages. 1974.

Vol. 28: Lectures in Statistical Physics. Edited by W. C. Schieve and J. S. Turner. V, 342 pages. 1974.

Vol. 29: Foundations of Quantum Mechanics and Ordered Linear Spaces. Advanced Study Institute, Marburg 1973. Edited by A. Hartkämper and H. Neumann. VI, 355 pages. 1974.

Vol. 30: Polarization Nuclear Physics. Proceedings 1973. Edited by D. Fick. IX, 292 pages. 1974.

Vol. 31: Transport Phenomena. Sitges International Schools of Statistical Mechanics, June 1974. Edited by G. Kirczenow and J. Marro. XIV, 517 pages. 1974.

Vol. 32: Particles, Quantum Fields and Statistical Mechanics. Proceedings 1973. Edited by M. Alexanian and A. Zepeda. V, 132 pages. 1975.

Vol. 33: Classical and Quantum Mechanical Aspects of Heavy Ion Collisions. Proceedings 1974. Edited by H. L. Harney, P. Braun-Munzinger, and C. K. Gelbke. VII, 311 pages. 1975.

Vol. 34: One-Dimensional Conductors GPS Summer School Proceedings, 1974. Edited by H. G. Schuster. VII, 371 pages. 1975.

Vol. 35: Proceedings of the Fourth International Conference on Numerical Methods in Fluid Dynamics, 1974. Edited by R. D. Richtmyer. V, 457 pages. 1975.

Vol. 36: R. Gatignol, Théorie Cinétique des Gaz à Répartition Discrète de Vitesses. II, 219 pages. 1975.

Vol. 37: Trends in Elementary Particle Theory. Proceedings 1974. Edited by H. Rollnik and K. Dietz. V, 472 pages. 1975.

Vol. 38: Dynamical Systems, Theory and Applications. Proceedings 1974. Edited by J. Moser. VI, 624 pages. 1975.

Vol. 39: International Symposium on Mathematical Problems in Theoretical Physics. Proceedings 1975. Edited by H. Araki. XII, 562 pages. 1975.

Vol. 40: Effective Interactions and Operators in Nuclei. Proceedings 1975. Edited by B. R. Barrett. XII, 339 pages. 1975.

Vol. 41: Progress in Numerical Fluid Dynamics. Proceedings 1974. Edited by H. J. Wirz. V, 471 pages. 1975.

Vol. 42: H II Regions and Related Topics. Proceedings 1975. Edited by D. Downes and T. L. Wilson. XII, 488 pages. 1975.

Vol. 43: Laser Spectroscopy. Proceedings 1975. Edited by S. Haroche, J. C. Pebay-Peyroula, T. W. Hänsch, and S. E. Harris. X, 466 pages. 1975.

Vol. 44: R. A. Breuer, Gravitational Perturbation Theory and Synchrotron Radiation. VI, 196 pages. 1975.

Vol. 45: Dynamical Concepts on Scaling Violation and the New Resonances in e^+e^- Annihilation. Edited by B. Humpert. VII, 248 pages. 1976.

Vol. 46: E. J. Flaherty, Hermitian and Kählerian Geometry in Relativity. VIII, 365 pages. 1976.

Vol. 47: Padé Approximants Method and Its Applications to Mechanics. Edited by H. Cabannes. XV, 267 pages. 1976.

Vol. 48: Interplanetary Dust and Zodiacal Light. Proceedings 1975. Edited by H. Elsässer and H. Fechtig. XII, 496 pages. 1976.

Vol. 49: W. G. Harter and C. W. Patterson, A Unitary Calculus for Electronic Orbitals. XII, 144 pages. 1976.

Vol. 50: Group Theoretical Methods in Physics. 4th International Colloquium. Nijmegen 1975. Edited by A. Janner, T. Janssen, and M. Boon. XIII, 629 pages. 1976.

Vol. 51: W. Nörenberg und H. A. Weidenmüller. Introduction to the Theory of Heavy-Ion Collisions. IX, 273 pages. 1976.

Vol. 52: M. Mladjenović, Development of Magnetic β-Ray Spectroscopy. X, 282 pages. 1976.

Vol. 53: D. J. Simms and N. M. J. Woodhouse, Lectures on Geometric Quantization. V, 166 pages. 1976.

Vol. 54: Critical Phenomena. Sitges International School on Statistical Mechanics, June 1976. Edited by J. Brey and R. B. Jones. XI, 383 pages. 1976.

Vol. 55: Nuclear Optical Model Potential. Proceedings 1976. Edited by S. Boffi and G. Passatore. VI, 221 pages. 1976.

Vol. 56: Current Induced Reactions. International Summer Institute, Hamburg 1975. Edited by J. G. Körner, G. Kramer, and D. Schildknecht. V, 553 pages. 1976.

Vol. 57: Physics of Highly Excited States in Solids. Proceedings 1975. Edited by M. Ueta and Y. Nishina. IX, 391 pages. 1976.

Vol. 58: Computing Methods in Applied Sciences. Proceedings 1975. Edited by R. Glowinski and J. L. Lions. VIII, 593 pages. 1976.

Vol. 59: Proceedings of the Fifth International Conference on Numerical Methods in Fluid Dynamics. 1976. Edited by A. I. van de Vooren and P. J. Zandbergen. VII, 459 pages. 1976.

Vol. 60: C. Gruber, A. Hintermann, and D. Merlini, Group Analysis of Classical Lattice Systems. XIV, 326 pages. 1977.

Vol. 61: International School on Electro and Photonuclear Reactions I. Edited by C. Schaerf. VIII, 650 pages. 1977.

Vol. 62: International School on Electro and Photonuclear Reactions II. Edited by C. Schaerf. VIII, 301 pages. 1977.

Vol. 63: V. K. Dobrev et al., Harmonic Analysis on the n-Dimensional Lorentz Group and Its Application to Conformal Quantum Field Theory. X, 280 pages. 1977.

Vol. 64: Waves on Water of Variable Depth. Edited by D. G. Provis and R. Radok. 231 pages. 1977.

Vol. 65: Organic Conductors and Semiconductors. Proceedings 1976. Edited by L. Pál, G. Grüner, A. Jánossy and J. Sólyom. 654 pages. 1977.

Vol. 66: A. H. Völkel, Fields, Particles and Currents. VI, 354 pages. 1977.

Vol. 67: W. Drechsler and M. E. Mayer, Fiber Bundle Techniques in Gauge Theories. X, 248 pages. 1977.

Lecture Notes in Physics

Edited by J. Ehlers, München, K. Hepp, Zürich
R. Kippenhahn, München, H. A. Weidenmüller, Heidelberg
and J. Zittartz, Köln
Managing Editor: W. Beiglböck, Heidelberg

93

Stochastic Behavior in Classical and Quantum Hamiltonian Systems

Volta Memorial Conference,
Como, 1977

Edited by G. Casati and J. Ford

Springer-Verlag
Berlin Heidelberg GmbH 1979

Editors
Giulio Casati
Istituto Scienze Fisiche
„Aldo Pontremoli"
Università Degli Studi di Milano
Milano
Italy

Joseph Ford
Georgia Institute of Technology
School of Physics
Atlanta, Georgia 30332/USA

Library of Congress Cataloging in Publication Data

Volta Memorial Conference, Como, Italy, 1977.
Stochastic behavior in classical and quantum Hamiltonian systems.

(Lecture notes in physics ; 93)
Includes bibliographies and index.
1. Hamiltonian systems--Congresses. 2. Stochastic processes--Congresses. I. Casati, Giulio, 1942- II. Ford, Joseph, 1927- III. Title. IV. Series.
QC174.17.H3V64 1977 530.1'2 79-12135

ISBN 978-3-540-09120-2 ISBN 978-3-540-35510-6 (eBook)
DOI 10.1007/978-3-540-35510-6

Originally published by Springer-Verlag Berlin Heidelberg New York in 1979

2153/3140-543210

PREFACE

The fact that completely deterministic, nonlinear systems can yield wildly chaotic solution behavior has, over the past two decades, been independently discovered and re-discovered by numerous scientists working in a host of distinct scientific disciplines. Separated from each other by thickets of specialized jargon and by specialized journals catering to mutually exclusive audiences, these workers had remained largely unaware of the communality of their work. In an effort to break down this scientific provincialism, the undersigned organized and held during the summer of 1977 a conference on stochastic behavior in classical and quantum Hamiltonian systems which, to our knowledge, brought together for the first time astronomers, biologists, economists, physicists, and mathematicians working in this common area. This volume includes some but far from all of the talks presented at that conference. Indeed, we deeply regret the fact that, for various reasons, many of the excellent presentations made during the conference do not appear in these pages. Despite this fact, we hope that the present volume will nonetheless focus the attention of a wider audience upon this subject area.

Historically, this is the second scientific conference to be sponsored by the city of Como as part of its traditional festivals honoring the memory of Alessandro Volta, a native son. The now-famous first Como Conference of 1927, which involved ten Nobel laureates among other notable physicists, was deeply concerned with the then young quantum mechanics and contributed significantly to its further development. The 1977 Como Conference therefore represents a continuation of this earlier conference in the sense that the 1977 Conference was also deeply concerned with quantum mechanics. In particular, much of its time was devoted to the problem of quantizing chaotic classical nonlinear systems, a difficulty anticipated by Einstein as early as 1917 but largely overlooked during the intervening decades. Here also, we hope that the beginning work described herein will be furthered by a larger audience.

To the Mayor of Como, Antonio Spallino, and to the Mayor of Campione, Felice De Baggis, we wish to relay the deep appreciation felt by all participants for the gracious hospitality provided by both their cities. Finally, to Dr. Barbara Giovannini, for service as that organizational mainspring without which a conference cannot succeed, we wish to express our own personal heartfelt gratitude.

Giulio Casati

Joseph Ford

TABLE OF CONTENTS

INTEGRABLE AND STOCHASTIC BEHAVIOUR IN DYNAMICAL ASTRONOMY

G. Contopoulos
European Southern Observatory
c/o CERN, Geneva, Switzerland

ABSTRACT

Some problems of Stellar Dynamics and Celestial Mechanics are presented, where integrability and stochasticity play a role. Such problems are: 1) the motions of stars in the meridian plane of an axisymmetric galaxy, 2) the motions in the plane of symmetry of a spiral galaxy, 3) the escapes of stars to infinity, 4) the bifurcations of families of periodic orbits, 5) Lynden-Bell's statistics in collapsing systems, 6) the general three-body problem, and 7) the applicability of Arnold's diffusion.

I. INTRODUCTION

Dynamical Astronomy is divided into Stellar Dynamics and Celestial Mechanics. The first deals with stellar systems composed of large numbers of stars, while the second deals with systems of few bodies, with the main emphasis on our planetary system, where the dominant role is played by the sun.

The notions of integrability and stochasticity play an important role in many problems of Stellar Dynamics and Celestial Mechanics and I will give some examples from both fields, mentioning also some of the most recent results.

A large part of Stellar Dynamics is devoted to collisionless problems, i.e. problems in which the time of relaxation is long. In fact in a galagy of 10^{11} stars the time of relaxation is of the order of 10^{13} years, and this is not only much larger than the period of galactic rotation, which is equal to a few 10^{8} years, but also larger than the age of the universe itself. In other words relaxation, due to the grainy character of the galaxy (as distinguished from a smooth average potential), is small.

The relaxation time of smaller systems (stellar clusters) is much shorter, but still longer than the orbital period. Therefore in many cases we can ignore the grainy character of a stellar system and consider the orbits of stars in the average, smooth, field.

II. STELLAR DYNAMICS

a) The third integral

The simplest type of problems in Stellar Dynamics deals with orbits of stars in a given potential that represents a model of a galaxy, or of a cluster. For example we may be interested to know the orbit of a star of known age in order to find its place of origin.

In general, however, we are not interested in particular orbits, but in the statistical behaviour of the stars in a stellar system. In particular we are interested in the distribution of stars in space and in the distribution of their velocities.

In collisionless dynamics the density of stars in phase space, f, (f is called also "distribution function"), is a constant of motion. In other words the stars move in phase space μ as an incompressible fluid. Thus if we have, say, dN stars in a phase element, these stars move around in time, but the volume, dV, that they occupy is always equal, and thus the density, f , remains the same.

However the shape of dV changes, and in general after some time this element becomes a very complicated long string that approaches every point in the available phase space. In such a case it would be useless to speak about the density f at a particular point, but we should consider only some average, or "coarse grained", density, $\bar{f}$, as meaningful.

But what is meant by "available phase space?" Of course if the field is time independent, the energy, E , of each star does not change. Also if the system has axial symmetry one component, J , of the angular momentum, does not change. Thus $\bar{f}$ is a function of the integrals E and J. That means that the stars do not go everywhere in phase space, but stay on the "surface" of constant E and J . The question now arises whether they cover that "surface" in an ergodic, or stochastic, way, or whether there are other, further restrictions.

Twenty years ago the answer would have been that no other useful integrals exist besides the energy and the angular momentum. This was based partly on some general ideas about ergodicity, but mainly on the well-known theorem of Poincaré about the non-existence of other analytic integrals of motion.

Thus it came as a surprise when some of the first applications of a computer in Stellar Dynamics in the 1950's indicated that, in general, the motion on the energy "surface" in phase space is not ergodic, and suggested the existence of a "third" integral of motion (Contopoulos 1958, 1960).

The existence of a "third" integral explained a curious phenomenon concerning the distribution of stellar velocities near the sun. If only two integrals exist, the energy

$$E=\frac{1}{2}\,(\dot{r}^2+\dot{\theta}^2+\dot{z}^2)+\ V(r,z), \tag{1}$$

and the angular momentum

$$J=r\theta, \tag{2}$$

(where z is the axis of symmetry, and r the radial direction, while $\dot{z}$ and $\dot{r}$ are the corresponding velocity components and θ is the transverse velocity), then the (coarse grained) distribution function depends only on E and J

$$\bar{f}=\bar{f}\ (E,J)\ . \tag{3}$$

If we write now $\bar{f}$ as a function of velocities, we see easily that it has the form

$$\bar{f}=F(\dot{r}^2+\ \dot{z}^2,\ \theta)\ . \tag{4}$$

Thus the surfaces $\bar{f}$ $=const$ (approximately ellipsoids) have the same axes along $\dot{r}$ and $\dot{z}$. But the observed distribution of stellar velocities shows that the $\dot{r}$-axis of the velocity ellipsoid is much larger that the $\dot{z}$-axis. This can be explained only if a third integral is operative in this case.

Many numerical and observational studies have been made to find when a third integral is applicable and when not. It is known now that even in cases where the stellar velocities are quite large (deviations from the circular velocity up to some hundreds of km/sec), approaching even the escape velocity, there seems to be an effective third integral of motion (Perek and Peterson 1966, Martinet and Mayer 1975).

On the other hand if the stellar orbits approach the central region of our Galaxy (i.e. if they have small angular momentum) the orbits become to a great extent stochastic (of course on the surface of constant E and J).

In such cases the galactic potential cannot be approximated at all by that of two uncoupled harmonic oscillators.

Since 1960 many studies have been made concerning the behaviour of the various integrals of motion in stellar systems that can be reduced to two-degrees of freedom. We now understand the general behaviour of the orbits, the role of resonances, and of the interactions between the various resonances, as well as the approach to ergodicity.

Most orbits on the meridian plane of a galaxy are boxes (Fig. 1), or tubes (Fig. 2) around stable periodic orbits.

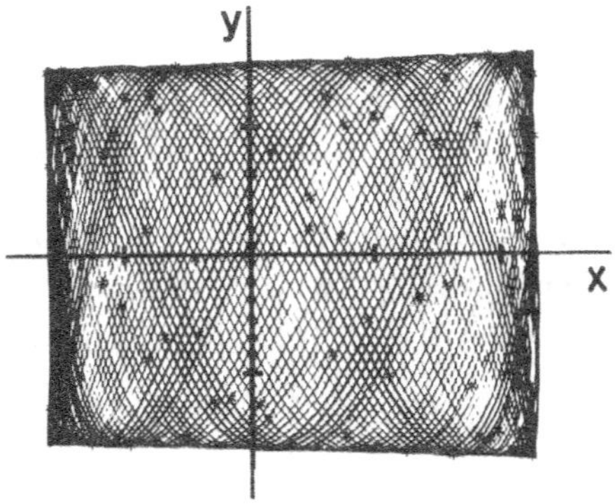

Figure 1. A box orbit. This is the general form of the (non-resonant) orbits in a hamiltonian of the form $H = \frac{1}{2}(\dot{x}^2+\dot{y}^2+\omega_1^2 x^2+\omega_2^2 y^2)$ *+ higher order terms.*

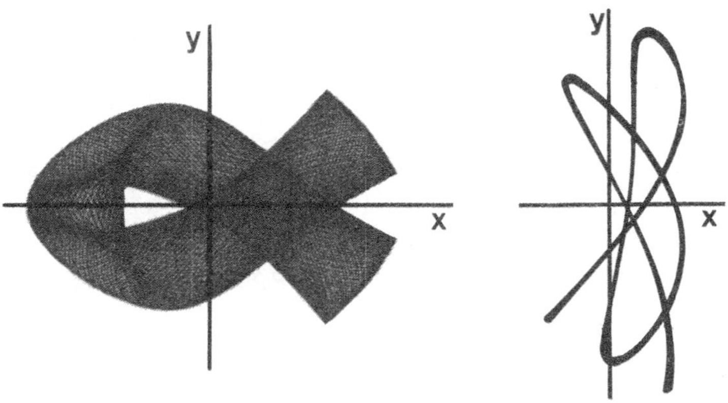

Figure 2. Two tube orbits. These are orbits near stable resonant periodic orbits in the hamiltonian of Figure 1.

On the other hand the unstable periodic orbits define two asymptotic surfaces (Fig. 3 represents sections of such asymptotic surfaces, called asymptotic curves), that intersect along homoclinic and heteroclinic points, and this kind of interaction between resonances produces partial

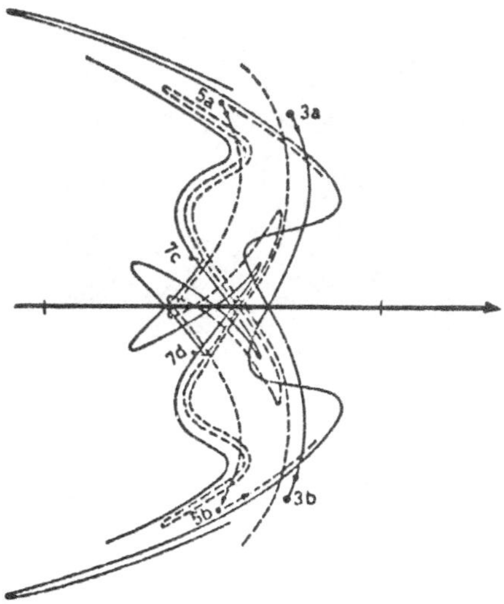

Figure 3. Asymptotic curves from various unstable invariant points in a particular galactic model (Martinet 1974). The asymptotic curves corresponding to the same periodic orbit intersect at homoclinic points, while the asymptotic curves corresponding to different periodic orbits intersect at heteroclinic points.

or almost complete stochasticity (Fig. 4). This phenomenon was found independently by several workers in many fields (Rosenbluth, Sagdeev and Taylor 1966, Contopoulos 1967, Chirikov 1969, Danby 1970, Bartlett and Wagner 1970, etc.).

The transition from ordered motion to stochastic seems to be rather abrupt (Hénon and Heiles 1964, Barbanis 1966). A similar abrupt transition was found also by Galgani and Scotti (1972) in systems of many interacting particles.

In galactic dynamics it is of interest to know when a third integral is applicable in order to use it in constructing models of galaxies. Such a study in the case of our Galaxy has been made by Martinet and Mayer (1975).

b) Spiral Galaxies

Up to now I have spoken about orbits in axisymmetric galaxies. Another topic of great current interest is the orbits and integrals of

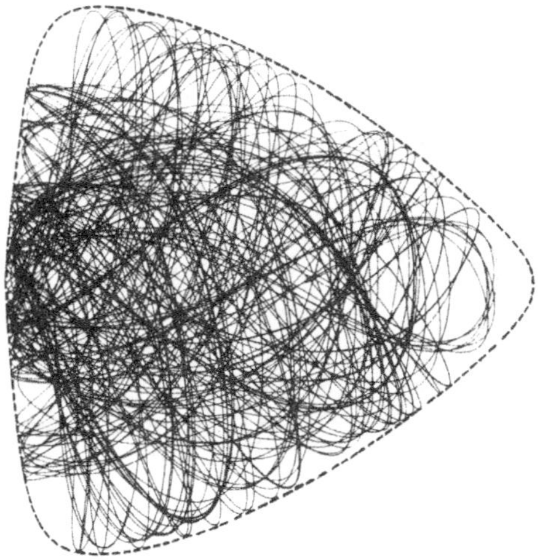

Figure 4. A stochastic orbit, filling the whole available space inside the curve of zero velocity. Such orbits appear in the hamiltonian of Fig. 1 for large enough energy, or large enough perturbation.

motion in spiral fields.

Let us consider, first, the orbits in the plane of symmetry of an axisymmetric galaxy. The distribution function, f_o, is specified as a given function of the energy and angular momentum. If a spiral density perturbation is introduced, σ_{1imp}, the distribution function is $f=f_o+f_1$, where f_1 is given by the linearized collisionless Boltzmann equation. By integrating f_1 over all velocities we find the response density

$$\sigma_{1resp}=\int f_1 d\bar{v}, \tag{5}$$

and by equating it with the imposed density

$$\sigma_{1resp}=\sigma_{1imp} \tag{6}$$

we have an equation that gives the possible spiral perturbations in a given axisymmetric galaxy. This is the basic idea of the current density wave theory of spiral structure (Lindblad 1950, Lin, Yuan and Shu 1969, Kalnajs 1971).

If we write this problem in action-angle variables, we find a hamiltonian of the form

$$H = \omega_1 I_1 + \omega_2 I_2 + \varepsilon_+ cos(\vartheta_1 - 2\vartheta_2 + const) + \varepsilon_o cos(2\vartheta_2 + const) + \varepsilon_- cos(\vartheta_1 + 2\vartheta_2 + const) + \text{higher order terms}, \tag{7}$$

where $\varepsilon_+, \varepsilon_o, \varepsilon_-$ are functions of the actions I_1, I_2. The linearized theory is applicable if we are away from the main galactic resonances, where $\omega_1 - 2\omega_2 = 0$, or $\omega_1 = 0$, or $\omega_1 + 2\omega_2 = 0$. However, the resonances play a very important role in galactic dynamics. It does not seem possible to have a global theory of spiral structure without a correct treatment of the resonances.

The study of the dynamics of spiral galaxies starts with a discussion of the orbits and integrals of motion in the case of the hamiltonian (7). The orbits in a spiral field away from resonances are rings (Fig. 5). However near the resonances the topology of the orbits is different (Figs. 6 and 7) (Contopoulos 1970a, 1973, 1975, 1978, Contopoulos and Mertzanides 1977, Vandervoort 1973, 1975, Vandervoort and Monet 1975).

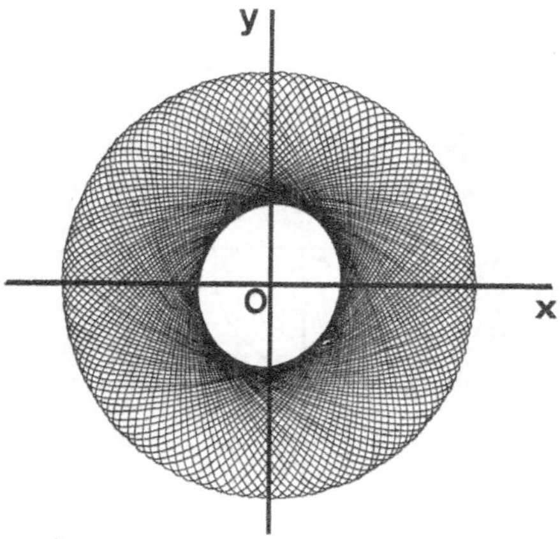

Figure 5. An epicyclic orbit in a spiral galaxy, which fills a distorted ring. The ring becomes exactly circular in the unperturbed (axisymmetric) case.

After the integrals of motion are found in each resonant or non-resonant case, the distribution function f can be derived as a function of the corresponding integrals of motion. Then one derives the total response density σ_{resp}, which has to be set equal to the original imposed density σ_{imp}:

$$\sigma_{resp} = \sigma_{imp} . \tag{8}$$

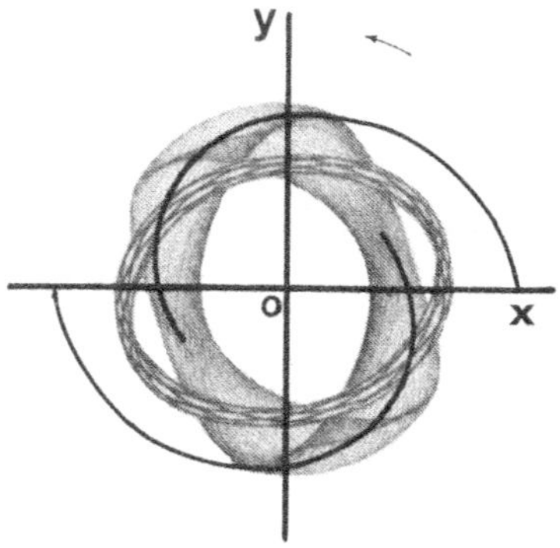

Figure 6. Two resonant orbits near the inner Lindblad resonance in a spiral galaxy (when $\omega_1 - 2\omega_2 = 0$).

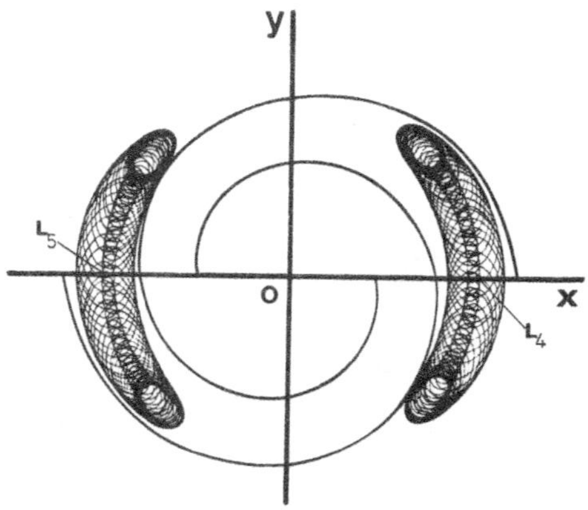

Figure 7. Two resonant orbits near the particle resonance in a spiral galaxy ($\omega_2 = 0$).

In order to solve this equation one has to adjust the parameters of the trigonometric terms of equation (7) so that self-consistency is secured.

This problem is a typical example of the collisionless problems encountered in Stellar Dynamics and Plasma Dynamics. One has to find approximate analytic expressions for the integrals of motion, in order to find the form of the distribution function, f , and the response density. Fortunately in realistic models of galaxies we do not have to worry very much about the role of stochasticity, because the spiral perturbation is small enough, so that the system is nearly integrable. It would be more difficult to consider cases with appreciable regions of stochasticity in a self-consistent way. On the other hand we know that for very large spiral perturbations the orbits are practically completely stochastic. Then the distribution function depends only on the hamiltonian and it seems that no self-consistent spiral solution exists at all (except barred solutions).

We have found that if we would double the observed amplitude of the spiral arms in our Galaxy, many orbits would become stochastic (Fig. 8). The question arises therefore what happens if during the evolution of a galaxy the amplitude of the spiral field goes above the critical

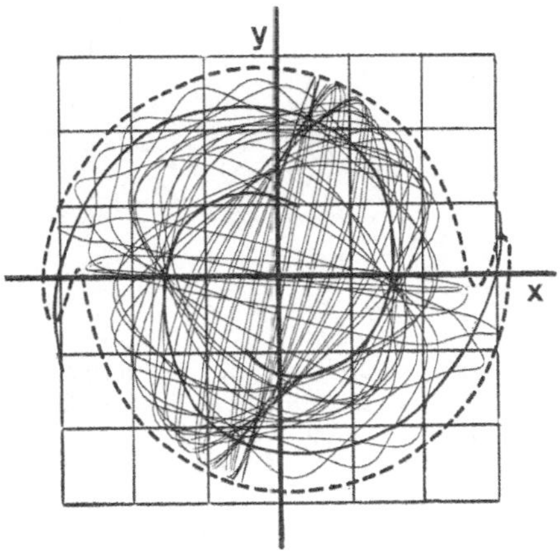

Figure 8. A stochastic orbit in a spiral galaxy with a strong spiral field (Barbanis).

value beyond which stochasticity becomes important. This is an unsolved problem of great theoretical interest.

c) Escapes of Stars

Another problem in Stellar Dynamics is the problem of escapes of stars. If a star has energy beyond the escape energy it will go to infinity unless there is some other restriction besides the energy, e.g. another integral of motion.

In order to study the problem of escapes we have explored numerically some strongly perturbed systems to find how many escapes occur when we go beyond the escape energy, and whether all stars escape at very large energies (a similar question, if there is a boundary forbidding escapes to infinity, is whether stochasticity becomes complete for very large perturbations).

Figure 9 shows a region beyond the escape energy in a particular case of the hamiltonian of Fig. 1. The main regions of initial conditions that lead to escape are shaded. It is seen that for energies

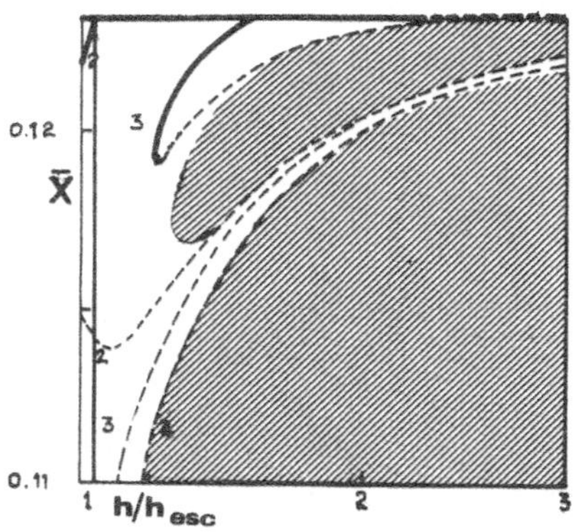

Figure 9. Characteristic curves of some families of periodic orbits in a particular case of the hamiltonian of Fig. 1 (Contopoulos 1970b); (——) stable, (------) unstable orbits; $\bar{x}=\omega_1 x$ is given as function of the energy, measured with the escape energy as unity, while $y=\dot{x}=0$. Each family is characterized by the number of crossings of the x-axis. The main escape regions are shaded.

larger that the escape energy there is a set of non-escaping orbits. The largest non-escaping regions are close to stable periodic orbits. We will see below (§III) that in Celestial Mechanics we also find cases where we have no escapes even if the energy surface extends to infinity. We conclude that the escapes do not depend only on the energy. In particular one can find, near the stable periodic orbits, new integrals of motion, that restrict the motion of the particles to a finite region of the energy surface, which may extend to infinity.

d) Periodic Orbits

The most important orbits in a dynamical system are the periodic orbits. On one hand the stable periodic orbits are followed by non-vanishing sets of quasi-periodic orbits, that have the same general characteristics. On the other hand the unstable periodic orbits produce (in non integrable systems) stochasticity in their vicinity. In general in a dynamical system we distinguish two types of periodic orbits:

a) The regular periodic orbits that can be derived from the unperturbed problem by continuous deformation. In Fig. 10 we see how the basic periodic orbit of the unperturbed problem (curve 1; this corresponds to oscillations up and down) produces by bifurcation a large number of families in the perturbed problem.

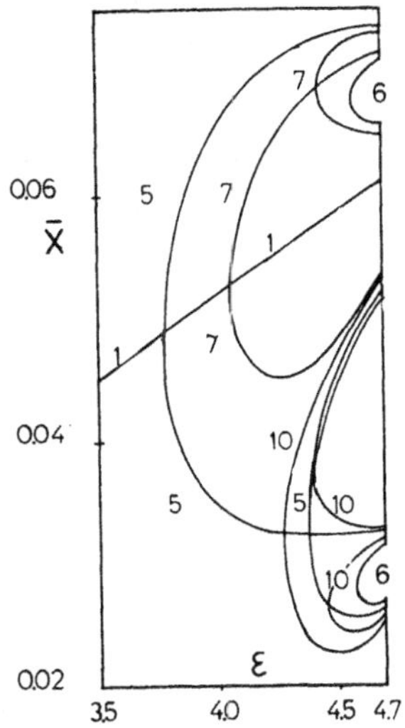

Figure 10. Characteristic curves of some families of periodic orbits, as in Fig. 9; $\bar{x}=\omega_1 x$ *is given as function of the perturbation* ε.

b) The irregular periodic orbits, which cannot be derived by continuous deformation from the unperturbed problem. The characteristic curves of these families of periodic orbits do not cross the characteristic curves of the regular families. Therefore they probably come from infinity and go to infinity, i.e. these orbits exist for arbitrarily large perturbations. They are important in producing stochasticity, because they join regions of phase space that are qualitatively quite different. E.g. the family 6 in Fig. 10 joins one region that contains orbits of multiplicity 5 (i.e. closing after 5, or 10, etc., oscillations along the y-axis) to a region of multiplicity 7 (where the orbits close after 7 oscillations).

I think that such are the periodic orbits that appear in the neighbourhood of heteroclinic points. There is a theorem, due to Birkhoff, that near homoclinic points there is an infinity of periodic orbits. Danby (1973) traced the periodic orbits near homoclinic points, by decreasing the energy, and he found that they are derived from the stable regular periodic orbits, which are represented by invariant points surrounded by islands on a surface of section.

Similarly one can find an infinity of periodic orbits near heteroclinic points. These periodic orbits are in general unstable. However, each family of such orbits has a region of stability near its minimum perturbation, as it was proved already by Poincaré (1899). What is new in our numerical investigations is an indication that new types of irregular orbits appear for larger and larger perturbations. Thus it seems that by increasing the perturbation we find further and further types of heteroclinic points. If this is the case for arbitrarily large perturbations then we will never have complete stochasticity but there will always be minute islands of stability in the stochastic region.

e) Lynden-Bell Statistics

Before leaving the area of Stellar Dynamics I would like to mention briefly another type of stochasticity, namely stochasticity in a varying potential.

If the potential changes considerably and irregularly in time, e.g. during the collapse of a stellar system, the individual energy and angular momentum of each star changes in an apparently stochastic way. What is preserved is only the total energy and angular momentum. One can then use the usual arguments about maximum entropy to find the final distribution of energy and angular momentum among the stars. However, even in this case, the density f of the various elements dV in phase space remains constant, and this fact acts as an "exclusion principle". Thus Lynden-Bell (1967) could derive a distribution function for systems that went through a collapse which is similar to Fermi-Dirac's.

Numerical experiments have verified the Lynden-Bell statistics to a certain degree, but in other cases the results agree less with the theoretical predictions. It seems that in many cases of collapsing systems the stochasticity that is produced is not enough. It is not yet clear what factors determine the non-stochastic behaviour, but of course this is a subject of great theoretical and practical interest.

III. CELESTIAL MECHANICS

I come now to some examples of integrability and stochasticity from Celestial Mechanics, namely the restricted 3-body problem, the general 3-body problem and, even further, the general N-body problem.

A very thorough study of the circular restricted 3-body problem, when the masses of the two primaries are equal, was made by Hénon (1965, 1966). Hénon found the regions where we have good invariant curves, other regions where we have stochasticity, and regions of escape.

The topology in phase space is extremely complicated. There are regions where most of the orbits are on integral surfaces, while other regions contain only small islands of stability in an otherwise "ergodic sea". Finally in other regions most orbits extend to infinity.

A similar systematic exploration has started in recent years in the case of the general 3-body problem. This study, which is done to a certain degree analytically, but mainly numerically, aims at finding the various families of periodic orbits, whether these periodic orbits are stable or unstable, and the extent of the regions of non-periodic orbits that follow the stable periodic orbits.

This problem is much more difficult than the restricted 3-body problem because even the planar 3-body problem is a problem of three-degrees of freedom.

If this problem were integrable then there would be three integrals of motion $H_1=h_1$, $H_2=h_2$, $H_3=h_3$, where H_1, H_2, H_3 are functions of the variables x_1, $\dot{x}_1$, x_3, $\dot{x}_3$, y_3, $\dot{y}_3$ and h_1, h_2, h_3 are constants. These would allow the elimination of two variables, say $x_3, \dot{y}_3$, therefore if we then used a "surface of section", say $y_3=0$, we would have an invariant surface $f(x_1, \dot{x}_1, \dot{x}_3)=const.$ that would contain the points of intersection of an orbit by the plane $y_3=0$. Such invariant surfaces were found in the neighbourhood of stable periodic orbits of the general 3-body problem that were discovered by Hadjidemetriou (1975). Fig. 11 shows such an invariant surface. We notice that successive points of intersection of an orbit by the plane $y_3=0$ are close to each other and can be joined by a curve lying on the invariant surface. After about 50 points of intersection we come close to the initial point, but following a new loop on the invariant surface. Successive loops deviate further and further from the initial loop, and about 250 loops are needed to cover the whole invariant surface.

If the problem were integrable every orbit would lie on an integral surface. However, one does not expect the general 3-body problem to be exactly integrable. One can only use the Kolmogorof-Arnold-Moser theorem to prove the existence, near the stable periodic orbits, of a non-vanishing set of non-periodic orbits that lie on invariant tori. However, while such tori in the case of two degrees of freedom separate the interior from the exterior phase space,this is not the case with three or more degrees of freedom. Thus if the problem is not integrable, but allows for some degree of stochasticity, an orbit, which is not exactly on an integral surface, may diffuse through the stochastic regions and escape to infinity. This phenomenon is the well known "Arnold diffusion" (Arnold and Avez 1967, Chirikov 1969).

One particular case where the Arnold diffusion is expected to play an important role is the basic problem of Celestial Mechanics, namely the stability of the solar system. Perhaps the strongest theorem in this connection up to now is one proved by Arnold, which states that if the masses, the eccentricities and the inclinations of a set of planets are sufficiently small, then for most of the initial conditions the planetary system is quasi-periodic. However arbitrarily close to every set of initial conditions there are small regions of stochasticity. Thus if some irregular force, like friction with the interplanetary gas, can slightly change the orbits in a random way, the system may find its way, through diffusion, to an orbit that escapes to infinity. Thus Arnold reaches the somewhat paradoxical conclusion: The solar system is stable for the large majority of initial conditions; nevertheless it is, in fact, unstable (Arnold 1962).

In order to find the Arnold diffusion we have done several numerical experiments with Galgani and Giorgilli in the case of three coupled oscillators and with Hadjidemetriou in the case of the general 3-body problem. The calculations were made for times much longer than those that one would expect by using a rough estimate of the Arnold diffusion time (Nehoroshev 1971), but no diffusion was found.

In the case of the three coupled oscillators we found large regions of phase space where the motion is stochastic ("ergodic seas"), which do not communicate with each other. In fact we calculated the average values of various quantities and we found quite different values in the different ergodic seas, while if the ergodic regions communicated, one would expect these average values to be the same. This fact indicates that there is a kind of barrier that does not allow the communication between the various ergodic regions.

On the other hand we know that stochasticity is produced by the interaction of resonances.

For this reason we are presently making some exploratory studies to find the behaviour of the orbits near unstable periodic orbits in systems of three degrees of freedom. Because of the analogy with problems of two degrees of freedom we expect that the stochasticity is largest in these regions, therefore any diffusion should be found there first.

In this way we expect to get a better understanding of the role of stochasticity in systems of three or more degrees of freedom. The importance of such studies for a better understanding of the behaviour of systems of many bodies in Celestial Mechanics, Molecular Physics, Solid State Physics, etc., is obvious.

Much further research is required in systems of many degrees of freedom. In such systems people usuually expect complete stochasticity. However we have found indications that non-stochastic behaviour is also quite general in certain cases. Such indications are the discovery of stable periodic orbits in the N-body problem (Hadjidemetriou) and the non-stochastic behaviour of systems of N-coupled oscillators with small enough energy (Galgani).

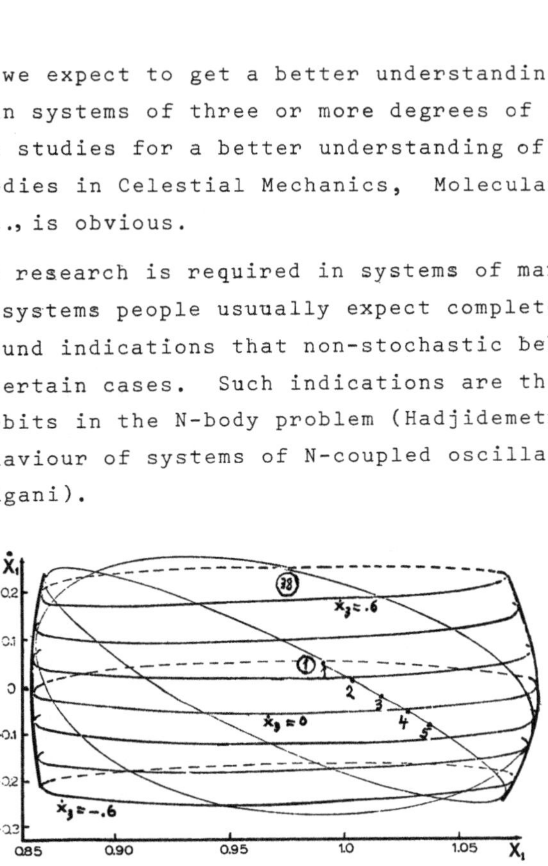

Figure 11. An "invariant surface" in the space (x_1 $\dot{x}_1$ $\dot{x}_3$), containing the intersections of an orbit by the plane $y_3=0$. The successive points of intersection 1, 2, 3 ... are joined by a continuous line; this line forms loops like ① or ㊳. The average period between two intersections is P_1. A period $P_2 \simeq 50P_1$ is required to describe a loop, while a period $P_3 \simeq 250P_2$ is needed for the successive loops to cover the whole invariant surface, which has the form of an open barrel.

We have given some examples of integrable and stochastic behaviour in various areas of Dynamical Astronomy. This is not an exhaustive list of problems. (Furthermore the list of references given here is quite incomplete and should be used only as a starting point). However this review allows one to draw two conclusions: a) That there are many interesting problems in Astronomy, where integrability and stochasticity play an important role, which await further research, and b) that these problems are very similar, in structure, to many non-linear problems from other disciplines. Therefore an interaction of people working in non-linear problems in general is highly desirable.

This is why I consider the present Symposium extremely useful.

REFERENCES

Arnold, V.I. 1962, Dokl. Akad. Nauk. 145, 487.

Arnold, V.I. and Avez, A. 1967, Problẽmes Ergodiques de la Mécanique Classique, Gauthier Villars, Paris.

Barbanis, B. 1966, Astron. J. 71, 415.

Bartlett, J.M. and Wagner, C.A. 1970, Celes. Mech. 2, 228.

Chirikov, B.V. 1969, Nucl. Phys. Inst. Siberian Section USSR Acad. Sciences Rep. 267; CERN Trans. 71-40.

Contopoulos, G. 1958, Stockholm Obs. Ann. 20, No. 5.

Contopoulos, G. 1960, Z. Astrophys. 49, 275.

Contopoulos, G. 1967, Bull. Astron. Ser. 3, 2, 223.

Contopoulos, G. 1970a, Astrophys. J. 160, 113.

Contopoulos, G. 1970b, Astron. J. 75, 96.

Contopoulos, G. 1973, Astrophys. J. 181, 657.

Contopoulos, G. 1975, Astrophys. J. 201, 566.

Contopoulos, G. 1978, Astron. Astrophys. (in press).

Contopoulos, G. and Mertzanides, C. 1977, Astron. Astrophys. 61, 477.

Danby, J.M.A. 1970, in G.E.O. Giacaglia (ed.) Periodic Orbits Stability and Resonances, Reidel, Dordrecht, p. 272.

Danby, J.M.A. 1973, Celes. Mech. 8, 273.

Galgani, L. and Scotti, A. 1972, Rev. Nuovo Cimento, 2, 189.

Hadjidemetriou, J. 1975, Celes. Mech. 12, 155, 255.

Hénon, M. 1965, Ann. Astrophys. 28, 499, 922.

Hénon, M. 1966, Bull. Astron. Ser. 3, 1, fasc. 1, 57; fasc. 2. 49.

Hénon, M. and Heiles, C. 1964, Astron. J. 69, 73.

Kalnajs, A.J. 1971, Astrophys. J. 166, 275.

Lin, C.C., Yuan, C. and Shu, F.H. 1969, Astrophys. J. 155, 721.

Lindblad, B. 1950, Stockholm Obs. Ann. 16, No. 1.

Lynden-Bell, D. 1967, Monthly Notices Roy. Astron. Soc. 136, 101.

Martinet, L. 1974, Astron. Astrophys. 32, 329.

Martinet, L. and Mayer, F. 1975, Astron. Astrophys. 44, 45.
Nehoroshev, N.N. 1971, Funksionalnii Analyz 5, 81.
Perek, L. and Peterson, D.M. 1966, IAU Symposium No. 25, Academic Press, p. 113.
Poincaré, H. 1899, Les Méthodes Nouvelles de la Mécanique Céleste, Gauthier Villars, Paris.
Rosenbluth, M.N., Sagdeev, R.A. and Taylor, J.B. 1966, Nuclear Fusion 6, 297.
Vandervoort, P.O. 1973, Astrophys. J. 180, 739.
Vandervoort, P.O. 1975, Astrophys. J. 201, 50.
Vandervoort, P.O. and Monet, D.G. 1975, Astrophys. J. 201, 311.

ADIABATIC and STOCHASTIC MOTION OF CHARGED PARTICLES in the FIELD of a SINGLE WAVE

by
Allan J. Lichtenberg
University of California

ABSTRACT

A unified treatment of particle motion in a wave field is presented both for propagation oblique to a magnetic field and propagation across a magnetic field. It is shown that both cases are related to the dynamical motion of two nonlinear coupled oscillators whose frequencies are harmonically related at some values of their actions. The oblique propagation corresponds to accidental degeneracy of the coupled oscillators for which the oscillator frequencies are functions of the action in the absence of the perturbation. The motion with perpendicular propagation corresponds to intrinsic degeneracy for which the nonlinearity occurs only in the coupling term. For the former case islands in the phase space trajectories are formed around the resonant actions with the ratio of the bounce to cyclotron frequency $\Omega_B/\Omega \propto \varepsilon^{1/2}$ and the island amplitude $\Delta p_M \propto \varepsilon^{1/2}$, where ε is the coupling parameter. In the latter case $\Omega_B/\Omega \propto \varepsilon$ and $\Delta p_M \propto 1$. For both cases coupled oscillator theory[1] predicts overlap ($2\Delta p_M/\delta p = 1$, where δp is the resonance separation) at $\Omega_B/\Omega = 1/4$. With similar parameters this implies overlap at smaller coupling for accidental degeneracy, although the use of practical parameters, corresponding to real plasma waves, may reverse this situation. The importance of 2nd order resonances in leading to stochasticity is demonstrated by showing that the 2nd order islands are exponentially small for small Ω_B/Ω, but that the ratio of the 2nd order island width to their separation becomes comparable to the 1st order island ratio near overlap. The result is a nearly complete randomization of the trajectories in the phase plane near $\Omega_B/\Omega = 1/4$ leading to stochasticity and heating.

I. Introduction

Recently there has been renewed interest in the conditions under which the dynamical motion of a phase trajectory of two nonlinear coupled oscillators can lead to stochastic motion in the phase plane, that is, motion which tends to fill a three dimensional phase volume, rather than being restricted by a constant of the motion to a two dimensional torus within that volume. This renewed interest has arisen from numerical observations that waves propagating either obliquely[1] or perpendicularly[2,3] to a magnetic field can give rise to stochastic heating of particles gyrating in that field, providing appropriate field and dynamical variables are chosen. In both cases these results have been interpreted in terms of the breakdown of the invariants arising from system resonances[1,2,3].

The purpose of this paper is twofold. First, we show the relation between particle motion in a wave field in which the wave is propagating obliquely to a d.c. magnetic field[1] and particle motion for a wave propagating perpendicularly to the magnetic field.[2,3] In both cases we consider the wave field as a perturbation on the particle motion during a single gyroperiod rather than the dominant behavior.[4] Second, we demonstrate, in a more systematic way than previously[5] the mechanism by which the second and higher order islands lead to ergodic motion over portions of the available phase space. We shall show that the resonant forms of the basic Hamiltonian for the oblique and perpendicular wave propagation, correspond to the two basic forms of the Hamiltonian for two nonlinearly coupled oscillators, which we have treated previously,[5] that of accidental and intrinsic degeneracy.

II. First Order Resonances

For an oblique wave, Smith and Kaufman[1] find the Hamiltonian for a particle as measured in the wave frame $v_z = \frac{\omega}{k_z}$

$$H = \frac{p_z^2}{2M} + \Omega p_\phi + e\Phi_o \sum_m J_m(k_\perp \rho) \sin(k_z z - m\phi) \tag{1}$$

where p_z is the axial momentum, p_ϕ the magnetic moment, Ω the gyrofrequency, ϕ the gyrophase, k_z and $k_\perp$ the componants of wave vector along and perpendicular to the static field B_o, ρ the gyroradius, $e\Phi_o$ the magnitude of the perturbing potential, and J_m the Bessel functions of the 1st kind. The Bessel function summation arises from the nonlinear phase shift resulting from the Larmor orbit extending over a spatially varying wave phase. For $\Phi_o = 0$, p_z and p_ϕ are constants of the motion.

For $k \perp B_o$ Karney and Bers[2] and Fukuyama et al[3] obtain the Hamiltonian

$$H = \Omega p_\phi + e\Phi_o \sum_m J_m(k\rho) \sin(\omega t - m\phi) \tag{2}$$

One difference between the Hamiltonians in Eqs. (1) and (2) is that the 2nd is explicitly time dependent. However, introducing the near-identity canonical transformation with the generating function

$$F_2 = \phi \tilde{p}_\phi + \omega t \; \tilde{p}_\psi \tag{3}$$

the new Hamiltonian given by

$$\tilde{H} = H + \frac{\partial F_2}{\partial t} \tag{4}$$

becomes

$$\tilde{H} = \Omega\tilde{p}_\phi + \omega\tilde{p}_\psi + e \; \Phi_o \sum J_m(k\rho) \sin(\psi - m\phi) \tag{5}$$

which does not contain time explicitly.

Eqs. (1) and (5) can now be considered as the Hamiltonians of two-dimensional oscillators which are coupled through a perturbation term of coupling strength $e\Phi_o$, considered small. There remains, however, a fundamental difference between the forms of the two equations. In (5) both momenta in the unperturbed terms are linear and in action-angle form, while the z-momentum in (1) is quadratic. These two cases have been previously treated for a pair of weakly coupled oscillators in which a resonance existed between some harmonics of the two degrees of freedom.[5] A resonance in Eq. (1) represents an "accidental degeneracy" for a given value of m=ℓ and

$$p_z = M\ell\ \Omega/k_z \tag{6}$$

The existance of the perturbation causes p_z to vary, moving the particles' momentum away from resonance, and thus limiting the effect of the resonant perturbation. A resonance in Eq. (5) represents an "intrinsic degeneracy" for which the frequency shift with momentum occurs only due to the non-linearity within the purturbation itself, thus allowing much larger variations in momentum before shifting the particle away from resonance. We calculate these effects, explicitly, below.

Assume Eq. (6) is satisfied for some ℓ we transform Eq. (1) with a generating function

$$F_2 = (k_z z - \ell\phi)\ \hat{p}_z + \phi\hat{p}_\phi \tag{8}$$

to obtain the Hamiltonian in terms of the new canonical variables

$$\hat{z} = \frac{\partial F_2}{\partial \hat{p}_z} = k_z z - \ell\phi \qquad p_z = \frac{\partial F_2}{\partial z} = k_z \hat{p}_z \tag{9}$$

$$\hat{\phi} = \frac{\partial F_2}{\partial \hat{p}_\phi} = \phi \qquad p_\phi = \frac{\partial F_2}{\partial \phi} = (\hat{p}_\phi - \ell\hat{p}_z) \tag{10}$$

as

$$\hat{H} = \frac{k_z^2}{2M}\hat{p}_z^{\,2} + \Omega(\hat{p}_\phi - \ell\hat{p}_z) + e\Phi_o \sum_m J_m(k_\perp\rho)\ \sin[\hat{z} - (m-\ell)\hat{\phi}] \tag{11}$$

where ρ is implicitly a function of the actions. Sufficiently close to a resonance $\hat{z}$ is slowly varying; we can average over a period of the $(m-\ell)\hat{\phi}$ coordinate, obtaining zero for the perturbation term except for $m = \ell$, for which Eq. (11) becomes

$$\hat{H} = \frac{k_z^{\,2}}{2M}\hat{p}_z^{\,2} + \Omega(\hat{p}_\phi - \ell\hat{p}_z) + \varepsilon M v_z^{\,2}\ J_\ell(k_\perp\rho)\ \sin\hat{z}. \tag{12}$$

Here we have written $\varepsilon = \dfrac{e\Phi_o}{Mv_z^{\,2}}$ which is assumed to be a small quantity.

The motion is singular at $\hat{p}_{zo}$, $\hat{z}_o$ obtained as in Eq. (13):

$$\frac{\partial H}{\partial \hat{p}_z} = 0 \quad \Rightarrow \quad \hat{p}_{zo} = \frac{\ell\Omega}{k_z^2/M} + O(\varepsilon)$$

$$\frac{\partial H}{\partial \hat{z}} = 0 \quad \Rightarrow \quad \hat{z}_o = \pm\frac{\pi}{2}\,. \tag{13}$$

Expanding around the singularity as $\hat{p}_z = \hat{p}_{zo} + \Delta\hat{p}_z$, $\hat{z} = \hat{z}_o + \Delta\hat{z}$, we obtain the linearized Hamiltonian for the perturbed motion

$$\Delta\hat{H} = g^{(a)}\,\frac{(\Delta\hat{p}_z)^2}{2} + f\,\frac{(\Delta\hat{z})^2}{2} = \text{Const.} \tag{14}$$

where the superscript (a) refers to the accidentally degenerate case. Here

$$g^{(a)} = \frac{\partial^2 H}{\partial \hat{p}_z^2} = \frac{k_z^2}{M} + O(\varepsilon) \tag{15}$$

and

$$f = \frac{\partial^2 H}{\partial \hat{z}^2} = \pm\,\varepsilon M v_z^2\, J_\ell(k_\perp\rho)\,. \tag{16}$$

To lowest order in ε, the frequency near the elliptic singular point for the perturbed oscillation, corresponding to the Hamiltonian of Eq. (14), is [6]

$$\hat{\Omega}_{Bo}^{(a)} = (fg^{(a)})^{1/2} = \left[\varepsilon\,\ell^2\, J_\ell(k_\perp\rho)\right]^{1/2}\Omega \tag{18}$$

and the peak amplitude at the separatrix, as obtained from Eq. (12), is [6]

$$\Delta\hat{p}_{z_{Max}} = 2\left(\frac{f}{g^{(a)}}\right)^{1/2} = 2\left[\frac{\varepsilon M v_z^2\, J_\ell(k_\perp\rho)M}{k_z^2}\right]^{1/2} = \frac{2\,\hat{\Omega}_{Bo}^{(a)}}{g^{(a)}}\,. \tag{19}$$

Both $\Omega_B^{(a)}$ and $\Delta\hat{p}_z$ are proportional to the square root of the small perturbation. The separation of adjacent resonances is given from Eqs. (6) and (9) as

$$\delta\hat{p}_z = \frac{M\Omega}{k_z^2} \tag{20}$$

such that the ratio of the momentum oscillation to momentum separation is, from Eqs. (15) and (19) and (20),

$$\frac{2\,\Delta\hat{p}_{zmax}}{\delta\hat{p}_z} = \frac{4\,\hat{\Omega}^{(a)}_{Bo}}{\Omega}\,. \tag{21}$$

A simple overlap condition, $\Delta\hat{p}/\delta\hat{p} \geq 1$ is then

$$\frac{\hat{\Omega}^{(a)}_{Bo}}{\Omega} = \frac{1}{4} \tag{22}$$

i.e. the frequency of the perturbed resonant oscillation is one fourth of the lowest fundamental frequency. These results, although not presented in the same way, have been obtained by Smith and Kaufman.

We now compare the results obtained from Eq. (1) with those for the intrinsicly degenerate system given by Eq. (5). Applying the generating function $F_2 = (\psi - \ell\phi)\,\hat{p}_\psi + \phi\hat{p}_\phi$, assuming sufficient closeness to resonance, as previously, to keep a single term in Eq. (5) after averaging, transformation equations analogous to Eqs. (9) and (10) then transform Eq. (5) to

$$\hat{H} = \Omega(\hat{p}_\phi - \ell\,\hat{p}_\psi) + \omega\,\hat{p}_\psi + \varepsilon M v_\perp^2 J_\ell(k\rho)\,\sin\hat{\psi} \tag{23}$$

where $\varepsilon = \dfrac{e\Phi_o}{Mv_\perp^2}$. The gyroradius ρ is a function of $\hat{p}_\psi$ and $\hat{p}_\phi$ through the transformation Eq. $p_\phi = \hat{p}_\phi - \ell\,\hat{p}_\psi)$ and the definition of ρ

$$\rho = [2\,p_\phi/M\Omega]^{\frac{1}{2}}\,.$$

Expanding around the singular point as previously, we obtain

$$\Delta\hat{H} = g^{(i)}(\Delta\hat{p}_\psi)^2 + f(\Delta\hat{\psi})^2 = \text{Const.}, \tag{24}$$

$$g^{(i)} = \varepsilon M v_-^2\,\frac{d^2 J_\ell(k\rho)}{d\hat{p}_\psi^{\,2}}\,, \tag{25}$$

and

$$f = \varepsilon M v_{\perp}^{2} J_{\ell}, \tag{25}$$

where the superscript (i) is for the intrinsicly degenerate case. To lowest order in ε the frequency and momentum excursion are

$$\hat{\Omega}_{Bo}^{(i)} = (fg^{(i)})^{\frac{1}{2}} = \varepsilon M v_{\perp}^{2}\left(J_{\ell}\ \frac{d^{2}J_{\ell}}{d\hat{p}_{\psi}^{2}}\right)^{\frac{1}{2}} \tag{26}$$

and

$$\Delta\hat{p}_{max}^{(i)} = 2\ \frac{f}{g^{(i)}} = 2\left(J_{\ell}\ /\ \frac{d^{2}J_{\ell}}{d\hat{p}_{\psi}^{2}}\right)^{\frac{1}{2}} = \frac{2\hat{\Omega}_{Bo}^{(i)}}{g^{(i)}}\ . \tag{27}$$

Comparing Eqs. (26) and (27) with Eqs. (18) and (19) we observe that for intrinsic degeneracy the frequency of the beat oscillation is of order ε, $\varepsilon^{\frac{1}{2}}$ slower than for accidental degeneracy, while the excursion in momentum of order unity, $\varepsilon^{-\frac{1}{2}}$ larger than for accidental degeneracy.

Unlike the situation for wave propagation at an angle to the magnetic field, the higher freqeuncy is fixed at ω, and thus there are no resonances at higher harmonics of Ω. For the Hamiltonian of Eq. (23), it is still possible to obtain resonance at a succession of values of $v_{\perp}$. This can be seen by setting the derivative of (23) equal to zero

$$\frac{\partial\hat{H}}{\partial\hat{p}_{\theta}} = \omega - \ell\Omega + \varepsilon M v_{\perp}^{2}\ \frac{dJ_{\ell}(k\rho)}{d\hat{p}_{\psi}} = 0 \tag{28}$$

which give the values of $v_{\perp}$ at the singularities of the motion. We note that these zeros can occur over a range of values of $k\rho$. In particular, for $\omega - \ell\Omega = 0$ they occur for

$$J_{\ell}'(k_{\rho}) = 0 \tag{29}$$

Although no first order resonant overlap exists, to lead to stochastic motion, resonances can occur between the island frequency and the fundamental cyclotron frequency, leading to 2nd order island formation and stochasticity. We shall show below, that this behavior if similar in character to the second order island formation for the accidentally degenerate motion.

II. Second Order Resonances

Second order islands play an important part in the development of the ergodic motion. Considering the linearized perturbed motion about the resonance given either by Eq. (14) or Eq. (24) the generating function

$$F_1 = \frac{1}{2} R\,(\Delta q)^2 \cot\theta, \quad R = (f/g)^{\frac{1}{2}} \tag{31}$$

generates the canonical transformation to new variables

$$\begin{aligned} \Delta p &= (2\,I\,R)^{\frac{1}{2}} \cos\theta \\ \Delta q &= (2\,I/R)^{\frac{1}{2}} \sin\theta \end{aligned} \tag{32}$$

where Δp and Δq represent the perturbation momentum and position coordinates for either problem. The transformed Hamiltonian is

$$K_o = \hat{\Omega}_{Bo}\, I \tag{33}$$

which is in action-angle form. We can extend this result to the nonlinear region by expanding $\Delta\hat{H}$ to fourth order in $\Delta\hat{p}_z$ and $\Delta\hat{z}$ to obtain

$$K = K_o + K_2 \tag{34}$$

and

$$\hat{\Omega}_B = \hat{\Omega}_{Bo} + \frac{\partial K_2}{\partial I}\,. \tag{35}$$

We have suppressed the rather lengthy details of the calculation of the K_2's, from 4th order perturbation theory, as they are not required for the following argument.

In localized regions of the phase space harmonics of the slow oscillation $\hat{\Omega}_B$ resonate with the slowest fundamental frequency Ω to produce new local distortions of the phase plane. We exhibit these resonances by explicitly reintroducing the oscillatory terms into the perturbation Hamiltonian

$$K = K_o + K_2 + \tilde{\Lambda} \; . \tag{36}$$

Here

$$\tilde{\Lambda} = \varepsilon M v^2 \sum_{m(m\neq\ell)} J_m(k_\perp\rho)\,\sin\left[\hat{q}_o + (2I/R)^{\frac{1}{2}}\sin\theta - (m-\ell)\phi\right]$$

$$= \varepsilon M v^2 \sum_{m(m\neq\ell)} J_m(k_\perp\rho) \sum_n J_n\left[(2I/R)^{\frac{1}{2}}\right] e^{i(\hat{q}_o - (m-\ell)\phi + in\theta)} \tag{37}$$

and v^2 is either v_z^2 or $v_\perp^2$ for accidental or intrinsic resonance, respectively. Taking only the lowest order resonant term, as previously,

$$\tilde{\Lambda} \cong \Lambda_n \sin(n\theta-\phi) = \varepsilon\, J_{\ell+1}(k_\perp\rho)\, J_n\left[(2I/R)^{\frac{1}{2}}\right] \sin(n\theta-\phi) , \tag{38}$$

transforming to a new locally slow variable

$$\hat{\theta} = n\theta - \phi \quad , \tag{39}$$

by a generating function as in Eq. (8), and expanding around the singularity, as in (19), we obtain the Hamiltonian for the secondary islands

$$\Delta K = g_s \frac{(\Delta\hat{I})^2}{2} + f_s \frac{(\Delta\hat{\theta})^2}{2} \tag{40}$$

where

$$g_s = \frac{\partial^2 \hat{K}_2}{\partial \hat{I}^2} , \quad f_s = \Lambda_n , \quad \text{and} \quad \hat{I} = I/n. \tag{41}$$

The frequency and peak to peak momentum excursion of the second order islands are then given, as previously, by

$$\hat{\Omega}_{Bs} = (f_s g_s)^{\frac{1}{2}} \tag{42}$$

and

$$\Delta\hat{I}_m = \frac{2\,\hat{\Omega}_{Bs}}{g_s} \tag{43}$$

Substituting for f_s and g_s in Eqs. (42) and (43) from Eqs. (41) and (38), we find that $\hat{\Omega}_{Bs}$ and ΔI_m are proportional through f_s to

$$\left\{J_n\left[(2I/R)^{\frac{1}{2}}\right]\right\}^{\frac{1}{2}} = 0\left[(1/n!)^{\frac{1}{2}}\right] \tag{44}$$

where the last relationship can be shown by expanding the Bessel function for large n.[5] Thus for large n (small ε) the factorial dominates and the islands become vanishingly small. However, the secondary resonances are also close together. Without calculating the perturbation Hamiltonian for the secondary resonances in detail we can compare the island width to the distance between islands, as previously. We calculate

$$\delta\hat{\Omega}_B = \frac{\partial^2 K_2}{\partial I^2}\,\delta I = \frac{\hat{\Omega}_B}{n} \tag{45}$$

for the distance between adjacent resonances. In terms of the hat variables the distance between resonances is given by

$$\delta\hat{\Omega}_B = \frac{\partial^2 K_2}{\partial(n\hat{I})^2}\,\delta\,(n\hat{I}) = \frac{g_s}{n}\,\delta\,\hat{I} \tag{46}$$

or

$$\delta\hat{I} = \frac{\hat{\Omega}_B}{g_s} \tag{47}$$

Substituting for ΔI_m from Eq. (43) we obtain, for overlap,

$$\frac{2\Delta\hat{I}_m}{\delta\hat{I}} = \frac{4\hat{\Omega}_{Bs}}{\hat{\Omega}_B} \geq 1 \tag{48}$$

which is identical in form to the overlap condition obtained for the primary resonances in Eq. (21) and (22). By induction, higher order resonances would also have the same form. Note that the secondary and higher order resonances are always accidentally degenerate. To determine when second order overlap occurs for increasing size of the perturbation, we must explicitly calculate the secondary bounce frequency $\hat{\Omega}_{Bs}$ in terms of the perturbation amplitude. We calculate g_s in Appendix B, obtaining to lowest order in ε

$$g_s \simeq g/8 \tag{49}$$

Substituting g_s from Eq. (49) and f_s from Eq. (41) into Eq. (42), we obtain

$$\hat{\Omega}_{Bs} \simeq \left[\varepsilon M v^2 J_{\ell+1}(k_\perp\rho) \, J_n\left[(2I/R)^{\frac{1}{2}}\right] g/8\right]^{\frac{1}{2}} \tag{50}$$

For simplicity we take $J_{\ell+1} \simeq J_\ell$, and taking J_n at its maximum value at the separatrix, $J_n(\pi)$, we obtain using either Eqs. (16) and (18) or Eqs. (25) and (26)

$$\frac{\hat{\Omega}_{Bs}}{\Omega_B} = \frac{1}{n} \simeq \left(\frac{J_n(\pi)}{8}\right)^{\frac{1}{2}} . \tag{51}$$

For accidental degeneracy the secondary islands closest to the fixed point would overlap first if Eq. (48) is satisfied before (22); i.e., with $n = 4$

$$\left(\frac{J_4(\pi)}{8}\right)^{\frac{1}{2}} \gtrsim \frac{1}{4} \tag{52}$$

which is marginally not satisfied. The important point, however, is that for either type of primary degeneracy the second order islands rapidly become important as the first order islands become large. For smaller values of perturbation it has been shown[3,8] that overlap exists near the island separatrices, giving bands of ergodicity in the phase plane that grow in area with the strength of the perturbation. The regions of ergodicity near secondary island resonances are very small until the first order resonances become large; they then increase rapidly leading to an ergodic phase plane with isolated adiabatic islands. This behavior has been confirmed numerically, for both oblique[1] and perpendicular[2] waves.

The basic results obtained here for perpendicular propagation, including the calculation of second order island amplitudes, have also been obtained by Fukuyama et al.,[3] but in a form emphasizing the stochastic regions near separatrices. It should also be noted that second order island calculations, in the neighborhood of an elliptic singularity, are in the same form for all coupled oscillator problems[5,9].

III. Discussion of Numerical Examples

In particular numerical examples for the two cases, Smith and Kaufman[1] found that stochasticity occurred for a perturbation amplitude considerably larger than that found by Karney and Bers.[2] This appears to be a surprising result in that $\hat{\Omega}_B/\Omega \propto \varepsilon^{1/2}$ for accidental degeneracy as seen from Eq. (18) while $\hat{\Omega}_B/\Omega \propto \varepsilon$ for intrinsic degeneracy as seen from Eq. (26). We would therefore expect the opposite result, that for ε small, the accidentally degenerate case should exhibit stochasticity first. To resolve this problem we analytically estimate the perturbation amplitude for overlap for the two cases in which comparable parameters are used. In Appendix A we calculate an approximate value of g in Eq. (25) for the intrinsic resonance problem, for the parameters that give maximum perturbation, as

$$g \cong \varepsilon M v_\perp^2 \, J_\ell(k\rho) \, \frac{\ell^4 \Omega^2}{M^2 v_\perp^4} \tag{53}$$

Substituting this result and f, from Eq. (25) into the expression for $\hat{\Omega}_B$ we obtain

$$\hat{\Omega}_B^{(i)} = \varepsilon \, \ell^2 \, J_\ell(k_\perp\rho) \, \Omega \tag{54}$$

Comparing this result with Eq. (18) for accidental degeneracy we find that the resonance conditions are, approximately,

$$\varepsilon \, \ell^2 \, J_\ell(k_\perp \rho) = \begin{cases} \dfrac{1}{n^2} & \text{accidental degeneracy} \\[2ex] \dfrac{1}{n} & \text{intrinsic degeneracy} \end{cases} \tag{55}$$

Thus, as expected, for identical ℓ and $k_\perp\rho$, and assuming $v_z = v_\perp$, the value of perturbation field for a given harmonic resonance is smaller in the case of accidental resonance. This situation reversed in the numerical examples studied, because a large value of $\ell=30$ was used for the intrinsic resonance[2] (wave propagation perpendicular to B) while $\ell=1$ was used to study the accidental resonance[2] (wave propagation at an angle to B) which corresponded to the physical plasma waves being studied. In both cases $k_\perp \rho$ was chosen to put J_ℓ near its maximum value; otherwise n remains large for all reasonable values of the perturbing field. In the intrinsicly degenerate case the stochasticity occurred due to the large amplitude of the 2nd and higher order islands, while interaction of both first and second order islands were important for accidental degeneracy.

IV. Transition Between Accidental and Intrinsic Degeneracy

There remains the question of the transition from accidental to intrinsic degeneracy as the wave direction approaches a normal to the magnetic field. The transition can be found by keeping the $O(\varepsilon)$ term in Eq. (15) for g

$$g = \frac{k_z^2}{M} + \varepsilon M v_z^2 \frac{\partial^2 J_\ell(k_\perp \rho)}{d^2 \hat{p}_z} \tag{56}$$

Approximating the Bessel function derivative as in Appendix A, we obtain

$$g \simeq \frac{k_z^2}{M} + \frac{\varepsilon M v_z^2 J_\ell \ell^2 (k_\perp \rho)^2 \Omega^2}{(M v_\perp^2)}$$

which, after substituting $k_z v_z = \ell\Omega$ and $k_\perp \rho = \ell$, becomes

$$g \simeq \frac{\ell^2 \Omega^2}{M v_z^2} \left[1 + \varepsilon \left(\frac{M v_z^2}{M v_\perp^2} \right)^2 \ell^2 J_\ell \right] . \tag{57}$$

The two terms are equal for

$$\frac{v_z^2}{v_\perp^2} = 1/(\varepsilon \ell^2 J_\ell)^{\frac{1}{2}} = n^{(a)} . \tag{58}$$

Assuming that $v_\perp$ is the characteristic velocity, we are considering resonance with an axial velocity class on the wings of the distribution function. In terms of the propagation vector

$$\frac{k_z}{k_\perp} = \left(\frac{1}{n^{(a)}}\right)^{1/2} . \tag{59}$$

For overlap we would put $n^{(a)} = 4$ in Eqs. (58) and (59).

Appendix A - Evaluation of $g^{(i)}$

We wish to evaluate

$$g^{(i)} = \varepsilon M v_\perp^2 \frac{d^2 J_\ell(k_\perp \rho)}{d\hat{p}_\psi^2} \qquad \text{A-1}$$

near its maximum value. Carying out the derivative implicitly we obtain

$$\frac{d^2 J_\ell(k\rho)}{d\hat{p}_\psi} = \frac{d^2 J_\ell(x)}{dx^2}\left(\frac{d(k\rho)}{d\hat{p}_\psi}\right)^2 + \frac{dJ_\ell(x)}{dx}\frac{d^2(k\rho)}{d\hat{p}_\psi^2} . \qquad \text{A-2}$$

To order the terms we assume that the Bessel function derivatives are chosen near their maximum values such that we can approximately set

$$\frac{d^2 J_\ell(x)}{dx^2} = \frac{dJ_\ell(x)}{dx} = J_\ell(x) \qquad \text{A-3}$$

obtaining for A-2

$$\frac{d^2 J_\ell(k\rho)}{d\hat{p}_\psi^2} \cong J_\ell(x)\left[\left(\frac{d(k\rho)}{d\hat{p}_\psi}\right)^2 + \frac{d^2(k\rho)}{d\hat{p}_\psi^2}\right] . \qquad \text{A-4}$$

We also have

$$\frac{d(k\rho)}{d\hat{p}_\psi} = \frac{k}{\rho}\frac{\ell}{M\Omega} .$$

Taking $k\rho \simeq \ell$ at the maximum of the Bessel Function we find the first term in A-4 is larger than the 2nd by ℓ, and assuming $\ell >> 1$ keep only the 1st term to obtain

$$\frac{d^2\, J_\ell(k\rho)}{d\hat{p}_\psi^{\,2}} = J_\ell \frac{\ell^4\, \Omega^2}{M^2 v_\perp^4} \quad . \qquad \text{A-5}$$

Substituting A-5 in A-1, Eq. (53) is obtained.

Appendix B - Evaluation of g_s

We wish to calculate g_s. From the Hamiltonian in Eq. (12) or Eq. (23) we calculate the fourth order terms in the expansion about the elliptic singularity

$$\hat{H}_2 = \varepsilon M v^2 \left[\frac{1}{4!} J_\ell (\Delta\hat{z})^4 - \frac{1}{4} \frac{\partial^2 J_\ell}{\partial \hat{p}_z^{\,2}} (\Delta\hat{p}_z)^2 (\Delta\hat{z})^2 + \frac{1}{4!} \frac{\partial^4 J_\ell}{\partial \hat{p}_z^{\,4}} (\Delta\hat{p}_z)^4 \right]$$

and transforming to action-angle form by using Eq. (32) and averaging, we have

$$K_2 = \langle \hat{H}_2 \rangle = \varepsilon M v^2 \left[\frac{1}{64} J_\ell \left(\frac{2I}{R} \right)^2 - \frac{1}{32} \frac{\partial^2 J_\ell}{\partial \hat{p}_z^{\,2}} (2I)^2 + \frac{1}{64} \frac{\partial^4 J_\ell}{\partial \hat{p}_z^{\,4}} (2RI)^2 \right] .$$

By differentiating twice with respect to I we obtain

$$g_s = \varepsilon M v^2 \left[\frac{1}{8} J_\ell \frac{1}{R^2} - \frac{1}{4} \frac{\partial^2 J_\ell}{\partial \hat{p}_z^{\,2}} + \frac{1}{8} \frac{\partial^4 J_\ell}{\partial \hat{p}_z^{\,4}} R^4 \right] .$$

For accidental degeneracy $R^2 = f/g = O(\varepsilon)$, then to lowest order in ε, $O(\varepsilon^\circ)$, we keep only the leading term giving

$$g_s^{(a)} = \frac{\varepsilon M v^2}{8} \frac{J_\ell}{f} g^{(a)} = \frac{g^{(a)}}{8}$$

where the superscript (a) stands for accidental degeneracy and the last equality was obtained by substituting for f from Eq. (16) or (25). For intrinsic degeneracy $f/g = O(1)$, and all of the terms in g_s must be kept. Using the results from Appendix A we find that all terms scale in the same manner, such that for intrinsic degeneracy

$$g_s^{(i)} = O\left[\frac{g^{(i)}}{8} \right]$$

Appendix C - Regions of validity for resonant and non-resonant transformations

In Karney and Bers numerical example, treating particles moving in a perpendicularly propagating wave, a slightly off harmonic resonance frequency was chosen. This was required to compare the results with their analytic trajectories that were calculated from a first order invariant obtained from a non-resonant transformation.[7] Their transformation generated an invariant whose maximum rate of change can be estimated as

$$\tilde{p}_\phi = \frac{\varepsilon \ell \ J_\ell(\ell)}{\ell-\nu} \dot{p}_\phi \tag{56}$$

where $\nu = \omega/\Omega$ and the argument of the Bessel function was taken equal to its index. Thus, for near resonant fields ε must be at its maximum sufficiently small that

$$\left| \frac{\varepsilon \ell J_\ell(\ell)}{\ell-\nu} \right| << 1 \tag{57}$$

This restricts the region of validity of their transformation near a resonance. The resonant transformation, on the other hand, requires that the applied frequency be sufficiently close to a resonance that the resonant term dominates the summation. The perturbation term can then shift the harmonic of the gyration frequency sufficiently to bring it into resonance with the applied frequency. This value can be estimated from Eq. (23) by noting that the singularity (resonance) occurs at

$$\frac{\partial \hat{H}}{\partial \hat{p}_\theta} = \omega - \ell\Omega \pm \varepsilon M v_\perp^2 \frac{\partial J_\ell(k\rho)}{\partial \hat{p}_\theta} = 0 \tag{58}$$

where the plus and minus correspond to the stable and unstable singular points, respectively. The condition for resonance is then that Eq. (29a) can be satisfied for some value of $\hat{p}_\theta$, and using the same approximations as in Eq. (29) we obtain

$$\left| \frac{\varepsilon \ell^2 J_\ell(\ell)}{\ell-\nu} \right| > 1 \ . \tag{59}$$

For $\ell=1$ the regions of validity of the two transformations are mutually exclusive. For large ℓ there can be considerable overlap between them. For an example in Karney and Bers $\ell-\nu \cong .1$ $\ell = 30$ and $\varepsilon\ell^2 = 1-3$ we find that the left hand side of Eq. (57) lies between 0.1 and 0.3 and thus their transformation is well satisfied for the smaller perturbation and marginally for the larger. For the resonant transformation the inequality in Eq. (59) is well satisfied, the left hand side lying between 3 and 10.

Acknowledgement

The author would like to acknowledge helpful conversations with Gary Smith and William Nevins. The work was supported in part by the National Science Foundation under Grant ENG-75-02709, and by ERDA under contract E(04-3)-34 PP215.

References

1. G. R. Smith and A.N. Kaufman, Phys. Rev. Lett. 34 1613 (1975).
2. C. F. Karney and A. Bers, "Stochastic Ion Heating by a Perpendicularly Propagating Ion Wave" RLE MIT report PRR 76/24 (1976).
3. A. Fukuyama, H. Momota, and R. Itatani, Stochastic Acceleration by an Electrostatic Wave near Ion Cyclotron Harmonics" Institute of Plasma Physics, Nogoya IPPJ 259 (1976).
4. R. Sugihara and Y. Midzuno, "Trapping and Detrapping in a Single Wave etc." Institute of Plasma Physics, Nagoya Report IPPJ-263 (1976).
5. E. F. Jaeger and A. J. Lichtenberg, Annals of Phys. 71 319 (1972).
6. A. J. Lichtenberg, Phase Space Dynamics of Particles, Wiley (1969) Chapt. 2.
7. G. H. Walker and J. Ford, Phys. Rev. 188, 416 (1969).
8. B. V. Chirikov, "Research Concerning the Theory of Non-Linear Resonance and Stochasticity," CERN Translation 71-40 of Nuclear Physics Institute of the Siberian Academy of Sciences Report 267 (1971).
9. F. Jaeger, A. J. Lichtenberg and M. A. Lieberman, Plasma Phys. 14, 317 (1972).

NUMERICAL STUDY OF PARTICLE MOTION IN TWO WAVES

Gary R. Smith

Lawrence Livermore Laboratory
University of California
Livermore, CA 94550/USA

ABSTRACT

We are studying the dynamical system described by the Hamiltonian $H = H_0 + \varepsilon V$, where

$$H_0 = \tfrac{1}{2} p^2 - \cos x, \quad V = - \cos(\lambda x - \Omega t).$$

We have encountered this system in a number of problems of practical importance. In addition, the system has intrinsic interest for the theory of adiabaticity and stochasticity. The invariant action J of the unperturbed Hamiltonian H_0 is subject to strong modification or destruction because of the perturbation εV. Absence of an invariant (i.e., stochasticity) occurs in a phase space region whose size and shape vary with the three parameters ε, λ, Ω. Previous studies have varied the amplitude of a perturbation (our ε); we emphasize here the strong dependences on the space (λ) and time (Ω) scales of the perturbation. Our results show that a perturbation is most effective at causing stochastic motion if its space and time scales are comparable ($\lambda \sim 1$, $\Omega \sim 1$) to those in the unperturbed Hamiltonian H_0.

1. INTRODUCTION

Much study has been devoted to the Hénon - Heiles Hamiltonian, which we write in the form

$$H = \left(p_x^2 + p_y^2\right)/2m + \frac{1}{2} k(x^2 + y^2) + \varepsilon\left(x^2 y - \frac{1}{3} y^3\right). \tag{1}$$

It is customary to eliminate the constants m, k, and ε by using a set of units in which

unit of mass $= m$
unit of length $= k/\varepsilon$
unit of time $= (m/k)^{1/2}$
unit of energy $= k^3/\varepsilon^2$.

A particle of given dimensional energy C, moving in a well with a given k, has a large value of the dimensionless energy $E = C\varepsilon^2/k^3$ if ε is large. The onset of stochasticity[1] as E is increased is thus caused by the increasing amplitude ε of the perturbation in (1).

More generally, the space and time scales of a perturbation, in addition to its amplitude, can be important in determining the onset of stochasticity. A dynamical system which displays these properties is given by

$$H(x,p,t) = H_0(x,p) + \varepsilon V(x,t), \qquad (2)$$

$$H_0 = \tfrac{1}{2} p^2 - \cos x, \qquad V = - \cos(\lambda x - \Omega t).$$

This system occurs in several problems in plasma physics, ranging from purely theoretical problems[2-4] through a modest laboratory experiment[5] to studies of interest to the important fusion energy program[6-8]. Hamiltonian (2) occurs frequently because many problems can be reduced to the motion of a particle in a one-dimensional potential which varies periodically in space and time. Fourier analysis of the potential yields, in general, a large number of sinusoidal terms. Sometimes all but two of these terms can be ignored, resulting in (2). We view this dynamical system as an elementary problem in stochasticity theory which exhibits features seen in many, more complicated problems. We use the phrase "particle motion in two waves" to describe (2).

2. HAMILTONIAN IN ACTION-ANGLE VARIABLES

In preparation for the presentation below of our numerical results, we introduce the variables which are most convenient in theoretical work. These are the action and angle variables of the unperturbed Hamiltonian H_0. Since H_0 describes the well-known (nonlinear) pendulum problem, one expects elliptic integrals and Jacobian elliptic functions to appear. In the notation of Ref. 9 the definitions of the action J and of the angle ϕ are

$$J(H_0) = \frac{8}{\pi} [E(\kappa) - (1- \kappa^2)\, K(\kappa)]$$

$$\phi(x,H_0) = \pi\, F(\xi,\kappa)/2\, K(\kappa),$$

where $\kappa \sin \xi = \sin \tfrac{1}{2}x$, and where the modulus κ is given by

$$\kappa^2 \equiv \tfrac{1}{2}(1 + H_0).$$

These definitions are valid only for a particle which has $H_0 < 1$ and is therefore trapped in the unperturbed potential well. Our definitions for an untrapped particle are available in Ref. 10.

The action-angle variables are convenient because the unperturbed motion is described so simply in terms of them:

$$\dot{J} = - \partial H_0/\partial\phi = 0 \Rightarrow J = J(t = 0)$$

$$\dot{\phi} = \partial H_0/\partial J \equiv \omega_b(J) = \text{const} \Rightarrow \phi = \phi_0 + \omega_b t.$$

In terms of J and ϕ, the Hamiltonian appears

$$H(\phi,J,t) = H_0(J) + \varepsilon V(\phi,J,t) \qquad (3a)$$

$$V = - \sum_{n=-\infty}^{\infty} V_n(J,\lambda) \cos(n\phi - \Omega t). \qquad (3b)$$

The Fourier coefficients are given by

$$V_n = \frac{1}{2\pi} \int_{-\pi}^{\pi} d\phi \cos[\lambda x(\phi,J) - n\phi].$$

Expressions for $V_n(J,\lambda = \frac{1}{2})$ are available in standard references, including Ref. 9. For half-integral values of λ (like 1, 3/2, 2, ...), the coefficients V_n can be computed by contour integration. The expressions obtained from reference works or by contour integration are valid for all J but only certain λ. On the other hand, it is very simple to find the approximate formula

$$V_n[J(\kappa), \lambda] \approx J_n(2\kappa\lambda), \tag{4}$$

valid for any λ, but accurate only for small values of κ. The Bessel function J_n on the right-hand side of (4) should not be confused with the action J on the left-hand side.

Our numerical results will show the presence of certain resonances (called bounce resonances) which are evident in (3). The nth term in (3b) has a large effect on the motion when it varies slowly in time:

$$n\dot{\phi} - \Omega \approx n\omega_b - \Omega \approx 0.$$

Chains of islands will be observed near the resonant values of the action J, which are defined by

$$n\omega_b(J) = \Omega . \tag{5}$$

3. NUMERICAL METHODS

Our study of the dynamical system (2) proceeds by numerical integration of the Hamiltonian equations of motion

$$\dot{x} = p, \ \dot{p} = - \sin x - \varepsilon \lambda \sin(\lambda x - \Omega t) .$$

We make stroboscopic plots, analogous to Poincaré surface of section maps, by recording a point whenever $\Omega t = 2\pi N$, N = 0, 1, 2, ... The recorded points are plotted not in the xp-plane, but in the ϕJ-plane. Since the unperturbed ($\varepsilon = 0$) motion results in points on a horizontal line in the ϕJ-plane, our plots are particularly useful for seeing the effects of the perturbation.

4. DISCUSSION OF RESULTS

We keep the amplitude of the perturbation fixed at $\varepsilon = 0.2$ and study the motion for various λ and Ω. As a reference case we choose $\lambda = \Omega = 1$.

First, we consider the results of Fig. 1, for which λ is varied while Ω is kept equal to unity. In each part of the figure we show trajectories commencing at several different initial points in the ϕJ-plane. Initial points lying in the stochastic region of the plane lead to a set of trajectory points which cannot be connected by a smooth curve. Such trajectories indicate the absence of an isolating constant of the motion. Initial points lying in the nonstochastic parts of the ϕJ-plane lead to sets of trajectory points which have been connected by smooth curves.

The four plots in Fig. 1 dramatically show the growth of the stochastic region as λ is increased from much less than unity to the reference case of $\lambda = 1$. The prominent island in each plot is due to the bounce resonance (5) with $n = 1$. Notice how the features on the plots change size and shape as λ varies but do not move around much.

In Fig. 2 we keep λ equal to unity and choose four different values of Ω. The four plots in Fig. 2, together with the reference case ($\lambda = \Omega = 1$) in Fig. 1, demonstrate that the largest stochastic region occurs for $\Omega = 1$. In contrast to Fig. 1, the features on the plots of Fig. 2 move around rapidly as Ω is varied. As a result, different features are seen in each part of Fig. 2. As indicated in the figure caption, we attribute the chains of islands to particular resonances of the form $n\omega_b = m\Omega$, where n and m are positive integers. The appearance of such prominent islands with $m \neq 1$ was unexpected. Description of these islands will require a different theoretical formulation than that given in (3).

To graphically display the variations in size of the stochastic part of phase space we have devised a quantitative measure of the size. Our measure is definite, but admittedly arbitrary. For Figs. 1 and 2 we attempted to locate nonstochastic trajectories as close as possible to the separatrix, the horizontal line at $J = 8/\pi \approx 2.55$. The area of the ϕJ-plane between the separatrix and the "closest" trajectory is divided by the area of the plane bounded by $J = 0$ and $8/\pi$ and by $\phi = 0$ and 2π. We refer to the ratio as the "stochastic fraction". In Figs. 3 and 4 we plot our measurements of the stochastic fraction as a function of λ and Ω. The curves connecting the measured points are merely aids to the eye; a theoretical expression for stochastic fraction as a function of ε, λ, and Ω is not available.

Besides being visually pleasing, the plots in Figs. 1 and 2 suggest that the simple dynamical system (2) has many intriguing properties. Interesting results should be obtained by theoretical and numerical work extending that begun here.

Work performed under the auspices of the U.S. Energy Research and Development Administration under contract No. W-7405-Eng-48.

REFERENCES

[1]M. Hénon and C. Heiles, Astron. J. 69, 73 (1964).

[2]G. M. Zaslavskii and N. N. Filonenko, Zh. Eksp. Teor. Fiz. 54, 1590 (1968) [Sov. Phys. - JETP 27, 851 (1968)]. Particle in two waves.

[3]P. K. Kaw and W. L. Kruer, Phys. Fluids 14, 190 (1971). Particle in two waves.

[4]G. R. Smith and A. N. Kaufman, Phys. Rev. Lett. 34, 1613 (1975). Particle in a uniform magnetic field and a single, oblique wave.

[5]G. Dimonte and J. H. Malmberg, Phys. Rev. Lett. 38, 401 (1977). Electron in two waves propagating on a traveling wave tube.

[6]A. B. Rechester and T. H. Stix, Phys. Rev. Lett. 36, 587 (1976). Magnetic braiding in a tokamak.

[7]M. Dobrowolny, A. Orefice, and R. Pozzoli, Plasma Phys. 16, 479 (1974). Particle in a tokamak magnetic field and a single wave.

[8]G. R. Smith, Phys. Rev. Lett. 38, 970 (1977). Particle in a tokamak magnetic field and a single wave.

[9]P. F. Byrd and M. D. Friedman, Handbook of Elliptic Integrals for Engineers and Scientists (Springer-Verlag, New York, 1971).

[10]G. R. Smith, Ph.D. Thesis, University of California, Berkeley, Sept. 1977, LBL-6824 (unpublished).

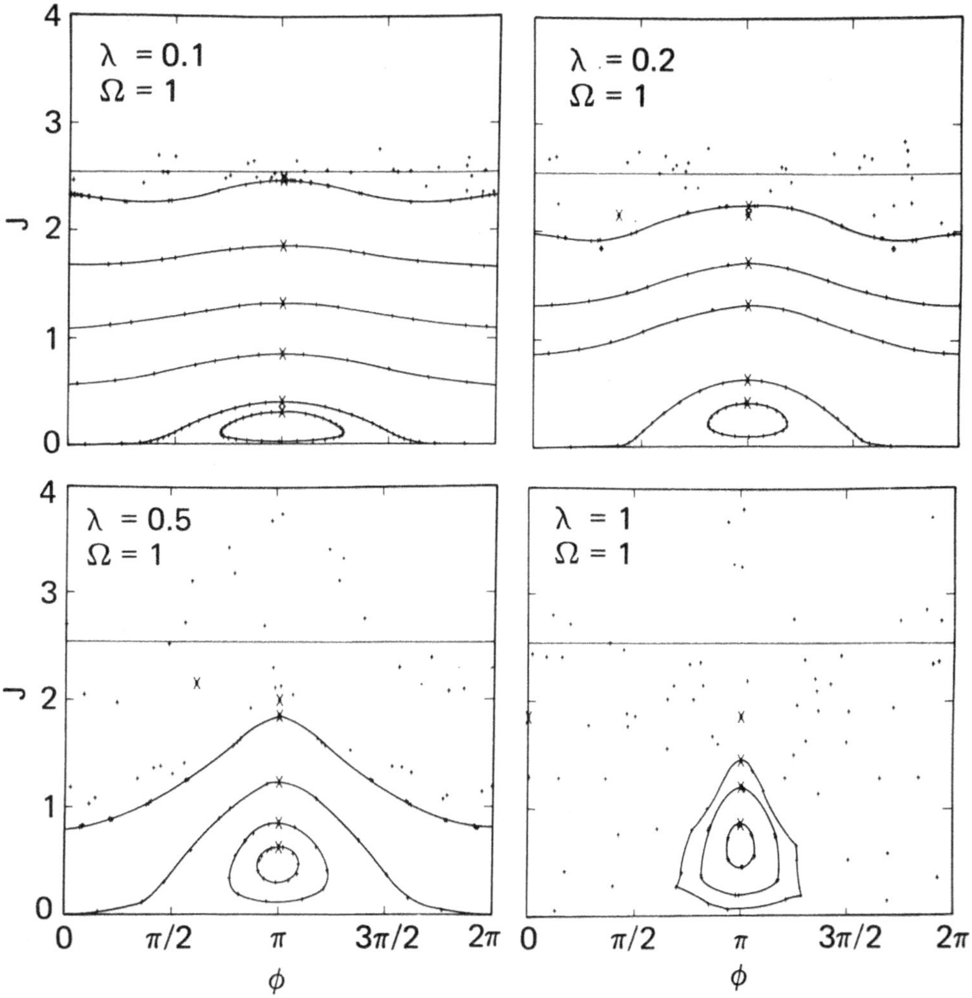

Fig. 1. Stroboscopic plots of particle trajectories, illustrating changes in the motion as λ is varied. The parameters $\varepsilon = 0.2$ and $\Omega = 1$ are kept fixed. The initial points for the trajectories are marked by X's, and each initial point represents a particle trapped in the unperturbed potential at $t = 0$. The horizontal line at $J \approx 2.55$ is the separatrix which divides trapped particles from untrapped ones.

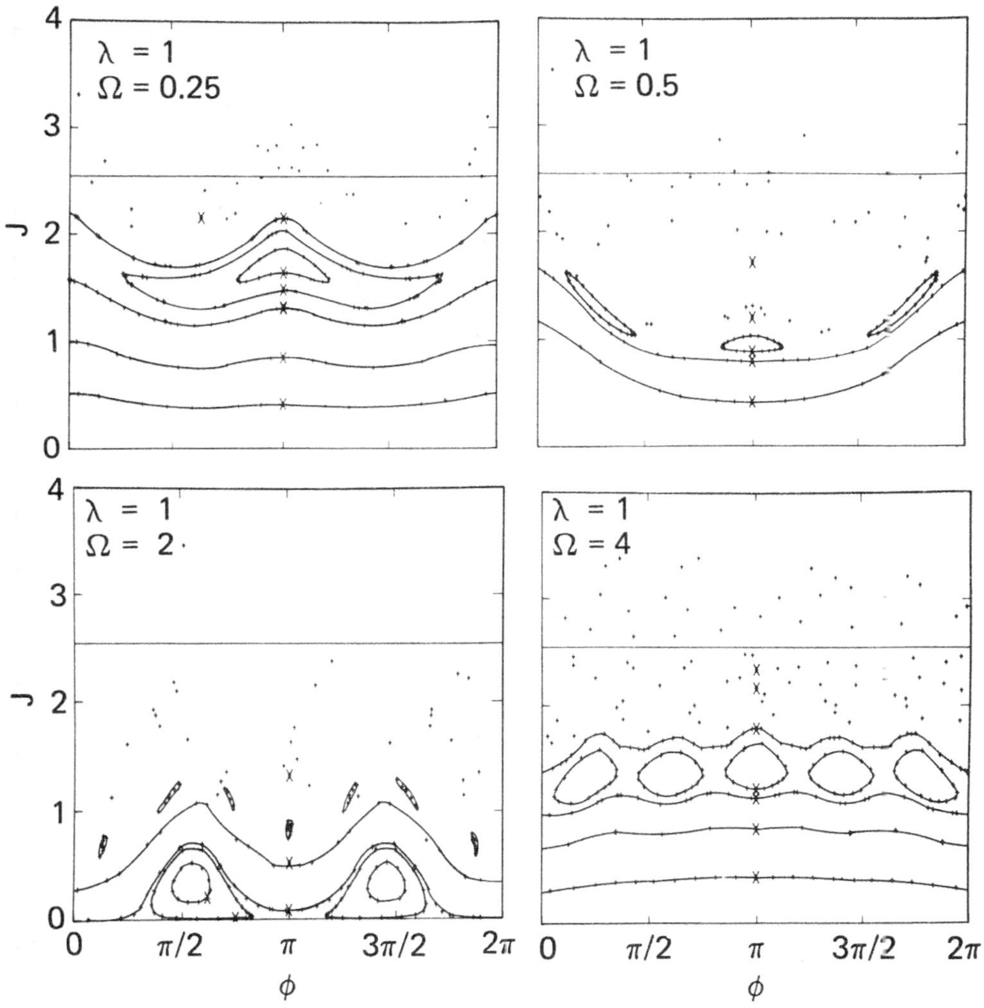

Fig. 2. Stroboscopic plots illustrating changes in the particle motion as Ω is varied. The fixed parameters are $\varepsilon = 0.2$ and $\lambda = 1$. The chains of islands seem to be due to the following resonances: the plot for $\Omega = 0.25$ shows $\omega_b = 3\Omega$, $\Omega = 0.5$ shows $3\omega_b = 5\Omega$, $\Omega = 2$ shows $2\omega_b = \Omega$ and $7\omega_b = 3\Omega$, and $\Omega = 4$ shows $5\omega_b = \Omega$.

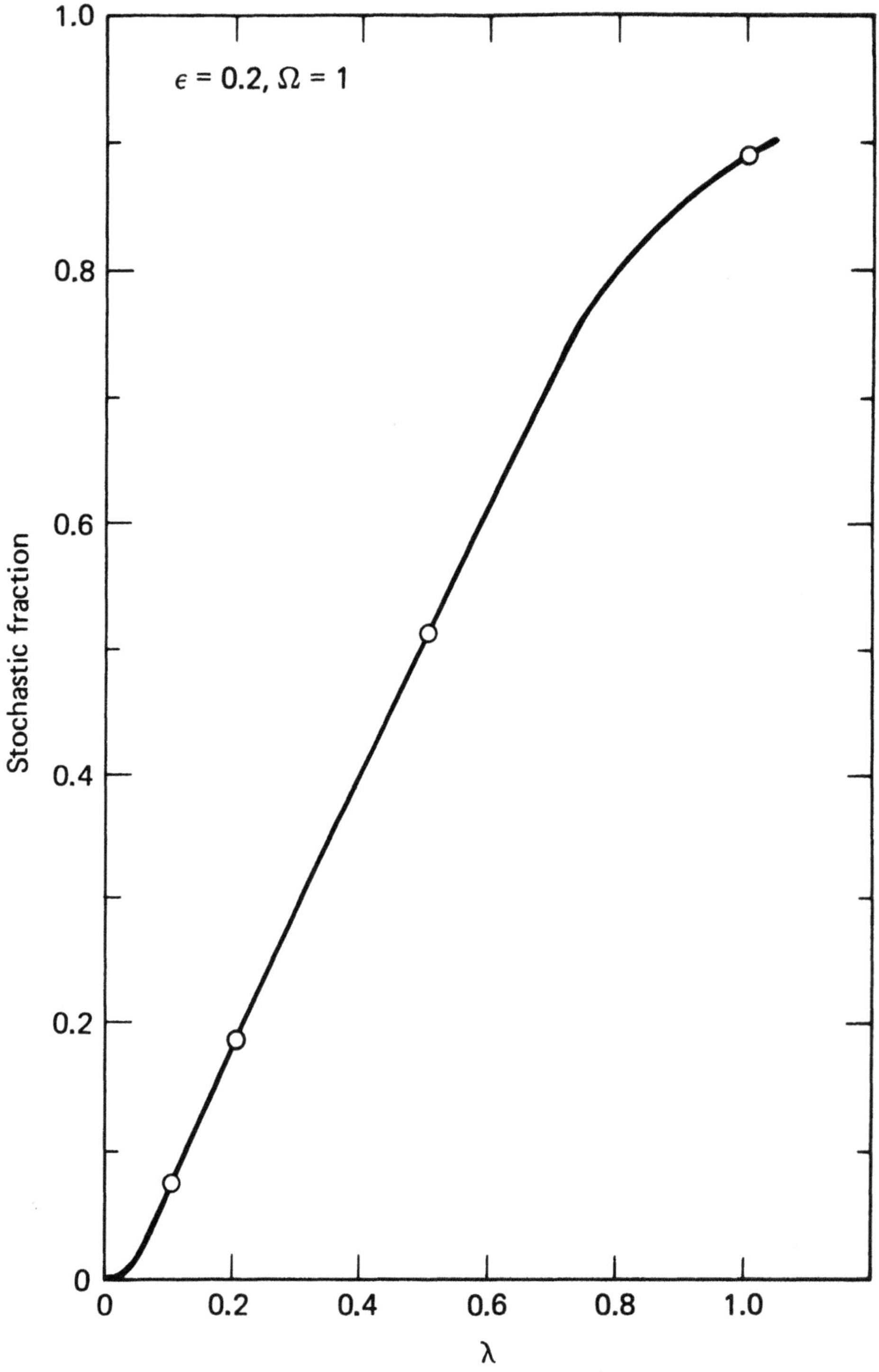

Fig. 3. The stochastic fraction as a function of λ for $\varepsilon = 0.2$ and $\Omega = 1$.

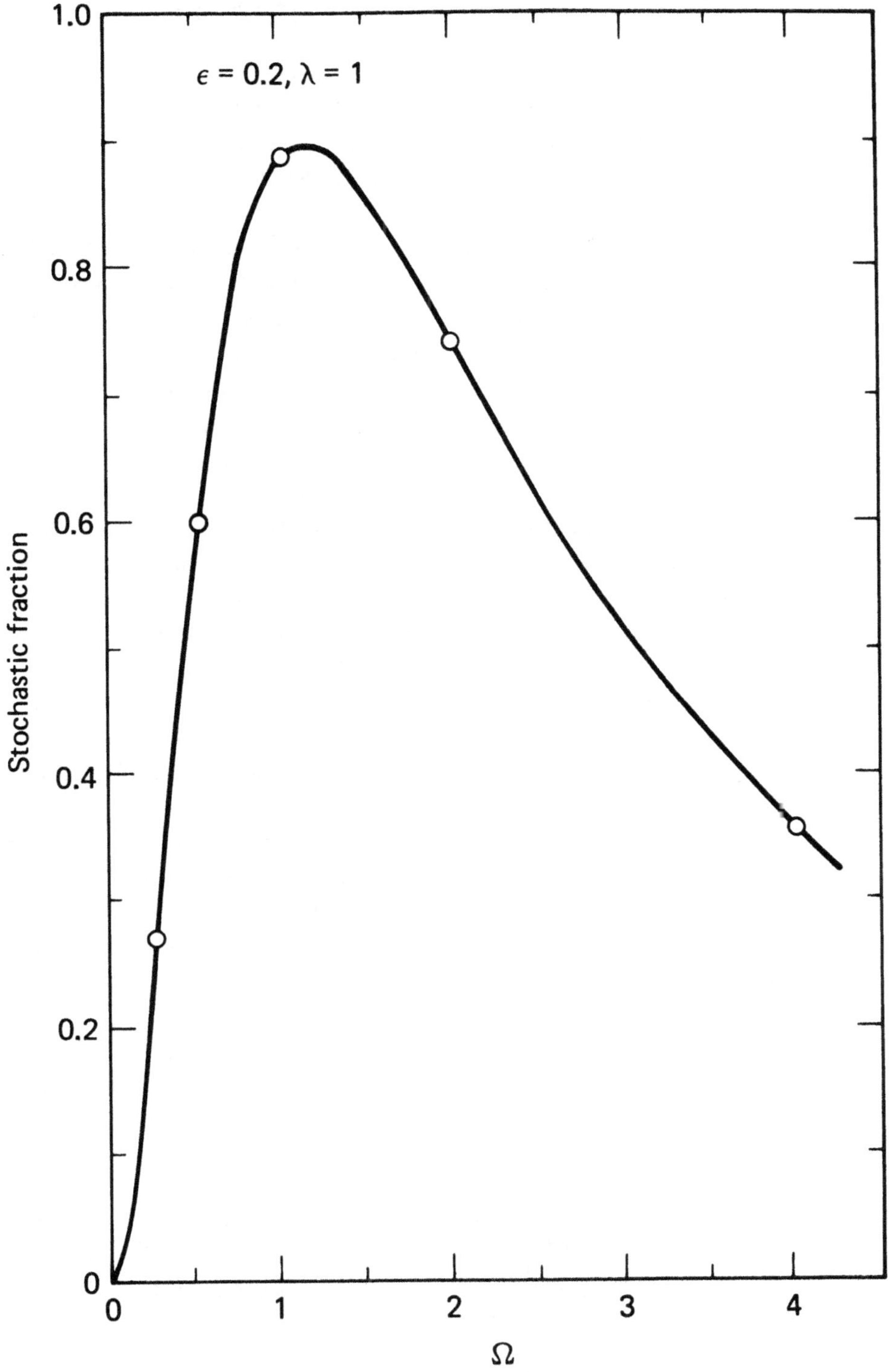

Fig. 4. The stochastic fraction as a function of Ω for $\varepsilon = 0.2$ and $\lambda = 1$.

STOCHASTIC ION HEATING BY A PERPENDICULARLY PROPAGATING ELECTROSTATIC WAVE*

C. F. F. Karney† and A. Bers
Research Laboratory of Electronics and Plasma Fusion Center
Massachusetts Institute of Technology
Cambridge, Massachusetts 02139

The motion of an ion in the presence of a constant magnetic field and a perpendicularly propagating electrostatic wave with frequency several times the ion cyclotron frequency is shown to become stochastic for fields satisfying $E/B_0 > 1/4(\Omega/\omega)^{1/3}$ (ω/k). This stochasticity condition is independent of how close ω is to a cyclotron harmonic. Applications of current interest in supplementary heating of plasmas with rf power near the lower-hybrid frequency are suggested.

It is well known that a small-(infinitesimal-) amplitude electrostatic wave traveling across a constant magnetic field suffers linear damping on the ions only if its frequency ω is an exact multiple of the ion cyclotron frequency, Ω. At finite amplitudes of the wave, nonlinear effects become important and we can expect that this resonance is broadened. We show that a single wave leads to stochastic ion motion at an amplitude which is independent of how close ω is to a cyclotron harmonic, $n\Omega$. This provides a mechanism whereby the ions in a plasma can be heated by such coherent waves.

There are two, physically distinct, nonlinear mechanisms operative in this interaction. The first consists of a transient trapping of the ions in the potential of the wave; this leads to a rapid heating of the ions near the lower boundary of the stochastic region.[1] The second is due to nonlinear resonances that arise from the perturbed cyclotron motion of the ions in the wave field, and which produce stochastic ion motion in a region of ion velocity space extending from the trapping region to an upper bound determined by the field amplitude; this leads to a slower heating of the ions up to the upper boundary of the stochastic region.

In current schemes for the supplementary heating of tokamak plasmas with microwave power, wave-packet fields that propagate nearly across the magnetic field (e.g., lower-hybrid-resonance cones and some of their parametrically excited decay waves) are

*Reprinted from Phys. Rev. Lett. 39, 550-4 (1977) by kind permission of the American Physical Society. Fuller accounts of the material presented here may be found in: C. F. F. Karney, "Stochastic Heating of Ions in a Tokamak by RF Power," Ph.D. Thesis, Department of Electrical Engineering and Computer Science, Massachusetts Institute of Technology (May 1977). C. F. F. Karney, "Stochastic Heating by a Lower Hybrid Wave," PPPL-1429, Princeton University (April 1978).
†Address from September 1977: Plasma Physics Laboratory, Princeton University, Princeton, NJ 08540.

generated. The stochastic ion heating mechanism we present can play an important role in sucn heating schemes.

Stochastic ion motion can also be caused by an obliquely propagating wave that traps ions in the potential wells parallel to the magnetic field.[2] This mechanism is inapplicable for the wave of interest here, which we take to propagate exactly across the magnetic field.

We consider an ion in a constant magnetic field $B_o\hat{z}$ and a perpendicularly propagating electrostatic wave $\hat{y}E\cos(ky - \omega t - \phi)$. The equations of motion are

$$\ddot{y} + \dot{x} = \alpha\cos(y - \nu t - \phi),\ \ddot{x} - \dot{y} = 0, \tag{1}$$

where lengths are normalized to k^{-1} and times to Ω^{-1}, $\alpha = qkE/(m\Omega^2)$, and $\nu = \omega/\Omega$. The Hamiltonian for this system is

$$H = I_1 + \nu I_2 - \alpha\sin[(2I_1)^{1/2}\sin w_1 - w_2], \tag{2}$$

where (I_1,w_1) and (I_2,w_2) are conjugate action-angle variables of the unperturbed system ($\alpha \to 0$), $y = (2I_1)^{1/2}\sin w_1$, and $x = -I_2 - (2I_1)^{1/2}\cos w_1$ (without loss of generality, we take the constant of the motion, $\dot{x} - y$, to be zero). In (2) w_2 is the wave phase and so H describes two harmonic oscillators, the ion in a magnetic field described by (I_1, w_1) and the wave described by (I_2,w_2), coupled by the last term in (2).

In order to find the condition for the onset of stochasticity we use the surface-of-section method.[3] We choose the cross section defined by $w_1 = \pi$ and numerically compute $r \equiv (2I_1)^{1/2}$, the normalized velocity, against w_2 for each crossing of the $w_1 = \pi$ plane. Figure 1 shows the results of such computations for $\nu = 30.23$, $\alpha = 2.2$ and 4, and in a region of velocity space with $r > \nu$. With $\alpha = 0$ the trajectories would all lie on horizontal straight lines, since I_1 is a conserved quantity. At $\alpha = 2.2$ the trajectories have become quite complex, many lying on chains of "islands." the order of the islands is given by the number of cyclotron orbits it takes for the ion to return to the island it started on. Clearly visible are chains of fourth- and fifth-order islands. Outside the islands the motion appears stochastic, but the stochastic regions are separated from one another by coherent regions, preventing the movement of particles from one stochastic region to another and limiting the amount of energy the particles can gain. At $\alpha = 4$, the motion is stochastic over nearly all of the region of velocity space shown in the figure, allowing particles to be heated through the region.

From the study of these figures and others at different frequencies, we form the following picture of how the motion becomes stochastic: The presence of the wave causes a nonlinear change in the cyclotron frequency, $\langle\dot{w}_1\rangle$. This in turn causes the ratio of the wave frequency to the cyclotron frequency to become a "simple' rational number in certain regions of velocity space, leading to fixed points. [In Fig. 1(a), for instance, the cyclotron frequency has become $\nu/30\frac{1}{4}$ and $\nu/30\frac{1}{5}$ at the fourth- and

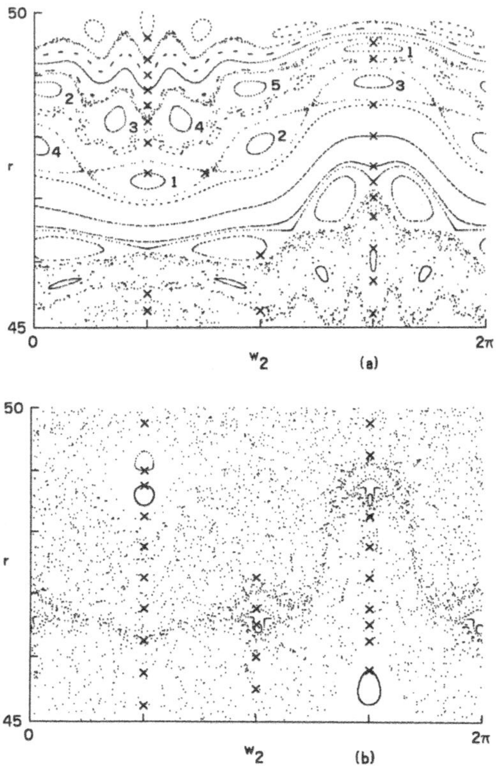

FIG. 1. The $w_1 = \pi$ cross sections of phase space for $\nu = 30.23$ and (a) $\alpha = 2.2$, (b) $\alpha = 4$. The trajectories of 24 particles are followed for 300 orbits in each case. Crosses indicate initial conditions; dots, subsequent crossings. In (a) the numbers indicate the position of fourth- and fifth-order islands and the order in which they are visited. The initial conditions were picked to illustrate the dominant features of the motion.

fifth-order islands, respectively.] Around half of these fixed points, a chain of islands forms; the other half are hyperbolic fixed points at the separatrices between the islands. As is well known,[3,4] the addition of any small perturbation to the system causes the motion in the vicinity of the separatix to become stochastic [see Fig. 1(a)]. At what we define to be the "stochasticity threshold" the chains of islands of various orders overlap, causing the stochasticity to become much more widespread [see Fig. 1(b)]; above this threshold appreciable heating is possible. As the field is increased two factors contribute to the eventual overlap of the islands. Firstly, the islands grow in size (although this effect is not always visible in surface-of-section plots because the size of the stochastic regions surrounding the islands also increases, eating away at the islands). Secondly, and perhaps more significantly for our problem, islands of different orders appear.

To find the stochasticity threshold in terms of island overlap, we should therefore do two things: to find out at what field the islands of various orders appear, and to find out their size. As we will see, all but first-order islands appear as the result of an $O(\alpha^2)$ shift in the frequency. The size of the corresponding islands is determined by higher-order terms. The following analysis gives the threshold for islands of various orders, but not their size. This together with the numerical simulations gives us the stochasticity threshold.

The location of pth-order islands may be derived analytically by finding pth-order fixed points. These in turn are found by setting

$$\langle\dot{w}_1\rangle/\dot{w}_2 = \langle\dot{w}_1\rangle/\nu = p/s, \tag{3}$$

where s and p are integers and the (time) average is performed over p cyclotron orbits (or, equivalently, s wave periods). For first-order islands ($p = 1$, $s = n$), $\langle\dot{w}_1\rangle$ is readily found by a Fourier transformation[5] of (2) with respect to w_1 and w_2. Then to order α, $nw_1 - w_2$ is constant, and so only one term in the Fourier sum contributes to give

$$\langle\dot{w}_1\rangle = 1 - \alpha(\partial/\partial I_1)J_n[(2I_1)^{1/2}]\sin(nw_1 - w_2). \tag{4}$$

To find the threshold for first-order islands we substitute (4) in (3) and take $\sin(nw_1 - w_2) = \pm 1$. If we also asymptotically expand the Bessel function in the limit $r >> \nu$, we obtain

$$\alpha = (\tfrac{1}{2}\pi r^3)^{1/2}|\delta|/n, \tag{5}$$

where $\nu = n + \delta$. For $p \neq 1$, we transform[5] H to a set of action-angle variables for which the angles are cyclic to order α. Then, if we are not too close to the threshold for first-order islands, there are contributions to H of order α^2 varying as $\sin(mw_1 - kw_2)$, where k = 0 or 2. if $p \neq 2$, then on averaging only the angle-independent term ($m = k = 0$) contributes, giving

$$\langle\dot{w}_1\rangle = 1 - \frac{\alpha^2}{4} \sum_{m=-\infty}^{\infty} \frac{\partial^2 J_m^2[(2I_1)^{1/2}]}{\partial I_1^2} \frac{m}{m-\nu} . \tag{6}$$

For $p = 2$, there is an additional contribution when $m = s$ and $k = 2$; however, with $r > \nu >> 1$ and $\nu \approx n + \frac{1}{2}$, this term is negligible, so that we use (6) in this case also. In Fig. 2(a) we plot $\nu/\langle\dot{w}_1\rangle$ as given by (6). As α is increased, the island condition, (3), is satisfied for more and more values of s and p. The threshold for a particular chain of higher-order islands is found by substituting (6) in (3). Expanding in the limit $r >> \nu >> 1$ gives

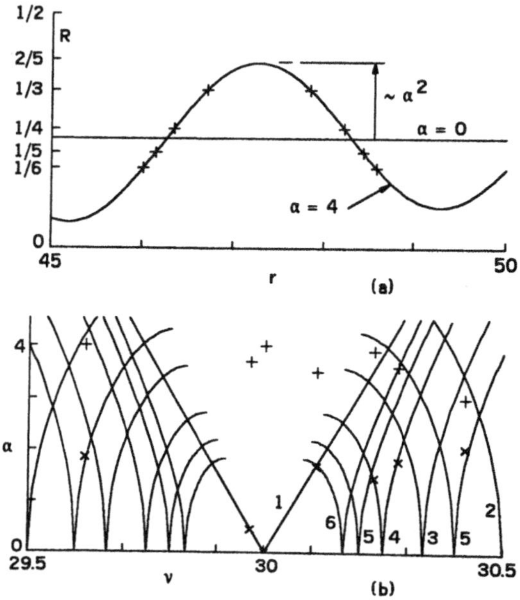

FIG. 2. (a) Plot of $R \equiv \nu/\langle\dot{w}\rangle - 30$ as given by (6) for $\nu = 30.23$. The plusses indicate the values of r for which $\langle\dot{w}_1\rangle$ is a simple rational multiple of ν. (b) The threshold α for the formation of islands of various orders (the numbers by the curves) for $r \simeq 47.5$ as a function of ν. The lines give the analytically predicted values and the crosses give the numerically observed values. The plusses give the numerically observed stochasticity thresholds.

$$\alpha = |\sin(\pi\delta)\varepsilon r^3|^{1/2}/n, \tag{7}$$

where $\nu = s/p + \varepsilon$.

In Fig. 2(b) we plot the predicted values of the thresholds for island formation as given by (5) and (7) for $29\frac{1}{2} < \nu < 30\frac{1}{2}$, and $r \simeq 47.5$ (in this plot we have allowed for the fact that $r \gg \nu$ is not satisfied), together with the numerically observed thresholds for various ν. Also given in Fig. 2(b) are some numerically observed stochasticity thresholds. We observe that this threshold is insensitive to δ, varying by about 20% for δ between $-\frac{1}{2}$ and $\frac{1}{2}$. Since the overlap of islands causes the motion to become stochastic, an accurate expression for the stochasticity threshold is obtained merely by replacing δ and ε in (5) or (7) by a constant [which we determine from Fig 2(b)], while retaining the slow dependence on r and n, giving, for $r \gg \nu \gg 1$,

$$\alpha = (\tfrac{1}{2}\pi r^3)^{1/2}/4n. \tag{8}$$

For $r \simeq \nu$, we must go back to (4) and (6) and expand in the limit $\nu \gg 1$ to obtain[6]

$$\alpha = \nu^{2/3}/4, \tag{9}$$

or, in unnormalized terms,

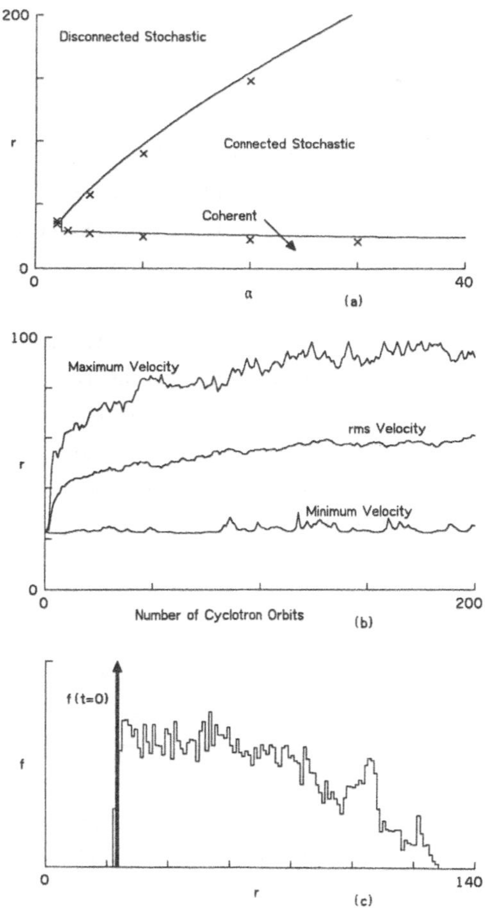

FIG. 3. (a) The limits of the stochastic region as given by Eqs. (9), (11), and (12). The crosses give the numerically observed values. The terms "disconnected stochastic" and "connected stochastic" are used to distinguish motion of the type shown in Figs. 1(a) and 1(b), respectively. (b) Heating of a group of 50 particles with $\alpha = 20$, initial velocity $r = 23$, and evenly distributed phases. (c) Velocity (r) distribution function for the particles in (b) averaged over orbits 800-1100. Normalization is such that $\int 2\pi r f\, dr = 1$.

$$E/B_o = \frac{1}{4}(\Omega/\omega)^{1/3}(\omega/k). \tag{10}$$

Our results can be used to determine what region of velocity space is stochastic for a given α. For α much larger than the threshold given by (9), the upper limit may be found by inverting (8) to give

$$r = (4\alpha n)^{2/3}(2/\pi)^{1/3}. \tag{11}$$

The lower limit is determined by trapping[1] and so is given by

$$r = \nu - \sqrt{\alpha} \quad . \tag{12}$$

Figure 3(a) shows these limits for $\nu = 30.23$, together with the numerically observed values. How an ion gains energy within the stochastic region may be illustrated by integrating the equations for a number of ions with velocities just above the lower limit of the stochastic region and with evenly distributed phases. In Fig. 3(b) we plot the rms speed of 50 particles as a function of cyclotron-orbit number (which is nearly proportional to time) for $\alpha = 20$ and $\nu = 30.23$. We note that the fairly rapid initial energy gain is followed by a slower heating. We can attribute the initial energy gain to the effects of trapping. Indeed, the rms velocity after the first few cyclotron orbits, is close to $\nu + 2\sqrt{\alpha}$. The slow heating continues for longer than 1000 cyclotron orbits, at which time the rms velocity is about 90, and the velocity distribution function has spread over most of the stochastic region, Fig. 3(c).

As an example, consider a wave near the lower-hybrid frequency in a plasma with $n_o = 10^{14}$ cm^{-3}, $B_o = 50$ kG, and $T_i = 1$ keV. Taking $\omega/\Omega_i \equiv \nu = 24$, and $k_\perp v_{Ti}/\Omega_i = 6$, we find from (10) that the threshold field for stochasticity is about 5.5 keV/cm (at this field $\varepsilon_o E^2/4n_o T_i$ is about 4×10^{-5}). The motion becomes stochastic for particles traveling at the wave phase velocity which is $4v_{Ti}$ so that an appreciable number of tail particles is heated. In tokamaks the lower-hybrid wave is of finite extent, determined by the excitation structure. In this example, if the extent along B_o is 10 cm, then the ions spend about 20 cyclotron periods in this field and can increase their energy by an amount corresponding to the trapping width of the fields. At much higher amplitudes of the lower-hybrid fields, the conditions for parametric decay may also be satisfied. For decay waves which are of lower frequency and shorter wavelength, the conditions for stochastic ion heating (10) and (12) may be more readily satisfied even for ions with velocities nearer to v_{Ti} so that bulk ion heating can result.

This work was supported by the U.S. Energy Research and Development Administration under Contract No. E(11-1)-3070.

REFERENCES

[1] C. F. F. Karney, A. Bers, and D. C. Watson, Bull. Am. Phys. Soc. 20, 1313 (1975); C. F. F. Karney and A. Bers, MIT Plasma Research Reports No. 76/7, 1976, and No. 76/24, 1976 (unpublished).

[2] G. R. Smith and A. N. Kaufman, Phys. Rev. Lett. 34, 1613 (1975).

[3] G. M. Zaslavskii and B. V. Chirikov, Usp. Fiz. Nauk 105, 3 (1971)[Sov. Phys. Usp. 14, 549 (1972)].

[4] J. Ford, in Fundamental Problems in Statistical Mechanics, edited by E. D. G. Cohen (North-Holland, Amsterdam, 1975), Vol. 3.

[5] G. H. Walker and J. Ford, Phys. Rev. 188, 416 (1969).

[6] This result may also be obtained (to within a factor of order unity) by demanding that trapping be effective, in the sense that an ion in its orbit spends at least a bounce period ($2\pi\alpha^{-1/2}$) within the trapping region given by $|\dot{y} - \nu| < \sqrt{\alpha}$.

PRESERVATION OF CONDITIONALLY PERIODIC MOVEMENTS WITH SMALL CHANGE IN THE HAMILTON FUNCTION*

Academician A. N. Kolmogorov
Department of Mathematics
Moscow State University
117234 Moscow, B-234
U.S.S.R.

ABSTRACT

This paper is a translation of Kolmogorov's original article announcing the theorem now known as the KAM theorem.

THEOREM AND DISCUSSION OF PROOF

Let us consider in the 2s-dimensional phase space of a dynamic system with s degrees of freedom the region G, represented as the product of an s-dimensional torus, T, and a region S, of a Euclidian s-dimensional space. We will designate the points of the torus, T, by the circular coordinates $q_1, \ldots, q_s$ (replacing q_α with $q'_\alpha = q_\alpha + 2\pi$ does not change points q), and the coordinates of the points, p, of S we will designate as $p_1, \ldots, p_s$. We will assume that in region G, in the coordinates $(q_1, \ldots, q_s, p_1, \ldots, p_s)$ the equations of motion have the canonical form

$$\frac{dq_\alpha}{dt} = \frac{\partial}{\partial p_\alpha} H(q,p), \quad \frac{dp_\alpha}{dt} = - \frac{\partial}{\partial q_\alpha} H(q,p). \tag{1}$$

The Hamilton function, H, is further assumed as dependent on the parameter θ and determined for all $(q,p) \in G$, $\theta \in (-c; +c)$, but not time-dependent. Moreover, further considerations require fairly significant restrictions on the smoothness of the function $H(q, p, \theta)$, stronger than infinite differentiability. For simplicity, in the following it is assumed that the function $H(q, p, \theta)$ is analytic over the set of variables (q, p, θ).

Summation over the Greek indices is further assumed to be from 1 to s. The usual vector designations $(x,y) = \sum_\alpha x_\alpha y_\alpha$, $|x| = -\sqrt{(x,x)}$ are used. A whole number vector indicates a vector for which all the components are whole numbers. The set of points (q, p) of G with

---

* Los Alamos Scientific Laboratory translation LA-TR-71-67 by Helen Dahlby of Akad. Nauk. S.S.S.R., Doklady 98, 527 (1954).

$p = c$ are designated by T_c. In theorem 1 it is assumed that S contains the point $p = 0$, i.e., $T_0 \subseteq G$.

Theorem 1. Let

$$H(q,p,0) = m + \sum_{\alpha} \lambda_\alpha p_\alpha + \frac{1}{2} \sum_{\alpha\beta} \Phi_{\alpha\beta}(q) p_\alpha p_\beta + O(|p|^3), \qquad (2)$$

where m and λ_α are constants and for a certain choice of constants $c > 0$ and $\eta > 0$ for all whole-number vectors, n, the inequality

$$(n,\lambda) \geq \frac{c}{|n|^\eta} . \qquad (3)$$

is satisfied.

Let, moreover, the determinant composed of the average values

$$\varphi_{\alpha\beta}(0) = \frac{1}{(2\pi)^s} \int_0^{2\pi} \cdots \int_0^{2\pi} \Phi_{\alpha\beta}(q) dq_1 \ldots dq_s$$

of the functions

$$\Phi_{\alpha\beta}(q) = \frac{\partial^2}{\partial p_\alpha \partial p_\beta} H(q,0,0)$$

be different from zero:

$$|\varphi_{\alpha\beta}(0)| \neq 0. \qquad (4)$$

Then there exist analytical functions $F_\alpha(Q, P, \theta)$ and $G\ (Q, P, \theta)$ which are determined for all sufficiently small θ and for all points (Q, P) of some neighborhood, V, of the set T_0, which bring about a contact transformation

$$q_\alpha = Q_\alpha + \theta F_\alpha(Q,P,\theta), \quad p_\alpha = P_\alpha + \theta G\ (Q,P,\theta)$$

of V into $V' \subseteq G$, which reduced H to the form

$$H = M(\theta) + \sum_\alpha \lambda_\alpha P_\alpha + O(|P|^2) \qquad (5)$$

($M(\theta)$ does not depend on Q and P).

It is easy to grasp the meaning of Theorem 1 for mechanics. It indicates that an s-parametric family of conditionally periodic motions

$$q_\alpha = \lambda_\alpha t + q_\alpha^{(o)}, \quad p_\alpha = 0,$$

which exists at $\theta = 0$ cannot, under conditions (3) and (4), disappear as a result of a small change in the Hamilton function H: there occurs only a displacement of the s-dimensional torus, T_0, around which the trajectories of these motions run, into the torus $P = 0$, which remains filled by the trajectories of conditionally periodic motions with the same frequencies $\lambda_1, \ldots, \lambda_s$.

The transformation

$$(Q, P) = K_\theta(q, p),$$

the existence of which is confirmed in Theorem 1, can be constructed in the form of the limit of the transformations

$$(Q^{(k)}, P^{(k)}) = K_\theta^{(k)}(q, p),$$

where the transformations

$$(Q^{(1)}, P^{(1)}) = L_\theta^{(1)}(q, p), \quad (Q^{(k+1)}, P^{(k+1)}) = L_\theta^{(k+1)}(Q^{(k)}, P^{(k)})$$

are found by the "generalized Newton method" (see Ref. 1). In this note we confine ourselves to the construction of the transformation: $K_\theta^{(1)} = L_\theta^{(1)}$, which itself permits grasping the role of conditions (3) and (4) of Theorem 1. Let us apply the transformation $L_\theta^{(1)}$ to the equations

$$Q_\alpha^{(1)} = q_\alpha + \theta Y_\alpha(q),$$

$$p_\alpha = P_\alpha^{(1)} = \theta \left\{ \sum_\beta P_\beta^{(1)} \frac{\partial Y_\beta}{\partial q_\alpha} + \xi_\alpha + \frac{\partial}{\partial q_\alpha} X(q) \right\} \tag{6}$$

(it is easy to verify that this is a contact transformation) and seek the constants ξ_α and ζ and the functions $X(q)$ and $Y_\beta(q)$, starting from the requirement that

$$H = m + \sum_\alpha \lambda_\alpha p_\alpha + \frac{1}{2} \sum_{\alpha\beta} \Phi_{\alpha\beta}(q) p_\alpha p_\beta + \\ + \theta \left\{ A(q) + \sum_\alpha B_\alpha(q) p_\alpha \right\} + O(|p|^3 + \theta|p|^2 + \theta^2) \tag{7}$$

take the form

$$H = m + \theta\zeta + \sum \lambda_\alpha P_\alpha^{(1)} + O(|P^{(1)}|^2 + \theta^2). \tag{8}$$

Substituting (6) into (7), we get

$$H = m + \sum_{\alpha} \lambda_\alpha P_\alpha^{(1)} + \theta \left\{ A + \sum_{\alpha} \lambda_\alpha \left(\xi_\alpha + \frac{\partial X}{\partial q_\alpha} \right) \right\} +$$

$$+ \theta \sum_{\alpha} P_\alpha^{(1)} \left\{ B_\alpha + \sum_{\beta} \Phi_{\alpha\beta}(q) \left(\xi_\beta + \frac{\partial X}{\partial q_\beta} \right) + \sum_{\beta} \lambda_\beta \frac{\partial Y_\beta}{\partial q_\beta} \right\} + O(|P^{(1)}|^2 + \theta^2).$$

Thus, our requirement (8) reduces to the equations

$$A + \sum_{\alpha} \lambda_\alpha \left(\xi_\alpha + \frac{\partial X}{\partial q_\alpha} \right) = \zeta \tag{9}$$

$$B_\alpha + \sum_{\beta} \Phi_{\alpha\beta} \left(\xi_\beta + \frac{\partial X}{\partial q_\beta} \right) + \sum_{\beta} \lambda_\beta \frac{\partial Y_\alpha}{\partial q_\beta} = 0. \tag{10}$$

being fulfilled.

Let us introduce the functions

$$Z_\alpha(q) = \sum_{\beta} \Phi_{\alpha\beta}(q) \frac{\partial}{\partial q_\beta} X(q). \tag{11}$$

Expanding the functions $\Phi_{\alpha\beta}$, A, B_α, X, Y_α, Z_α into a Fourier series of the type

$$X(q) = \sum x(n) e^{i(n,q)}$$

and assuming for definiteness that

$$x(0) = 0, \quad y(0) = 0, \tag{12}$$

we get for the remaining Fourier coefficients $x(n)$, $y_\alpha(n)$, and $z_\alpha(n)$ and constants ξ_α and ζ of the equation which are relevant to the determination

$$a(0) + \sum \lambda_\alpha \xi_\alpha = \zeta, \tag{13}$$

$$a(n) + (n,\lambda)\, x(n) = 0 \quad \text{for } n \neq 0, \tag{14}$$

$$b_\alpha(0) + \sum_{\beta} \varphi_\alpha(0) \xi_\beta + z_\alpha(0) = 0, \tag{15}$$

$$b_\alpha(n) + \sum_{\beta} \varphi_{\alpha\beta}(n) \xi_\beta + z_\alpha(n) + (n,\lambda)\, y_\alpha(n) = 0 \quad \text{for } n \neq 0. \tag{16}$$

It is easy to see that the system (11) - (16) is unambiguously solved under conditions (3) and (4). Condition (3) is important in the determination of $x(n)$ from (14), and in the determination of $y_\alpha(n)$ from (16). Condition (4) is important in the determination of ξ_β from (15). Since, as $|n|$ increases, the coefficients of the Fourier series of the analytical functions $\varphi_{\alpha\beta}$, A, and B_α have an order of decrease not less than $\rho^{|h|}, \rho < 1$, then from condition (3) there results not only the formal solvability of equations (13) - (16) but also the convergence of the Fourier series for the functions X, Y_α, and Z_α and the analyticity of these functions. The construction of further approximations is not associated with new difficulties. Only the use of condition (3) for proving the convergence of the recursions, $K_\theta^{(k)}$, to the analytical limit for the recursion K_θ is somewhat more subtle.

The condition of the absence of "small denominators" (3) should be considered, "generally speaking," as fulfilled, since for any $\eta > s - 1$ for all points of an s-dimensional space $\lambda = (\lambda_1, \ldots, \lambda_s)$ except the set of Lebesque measure zero it is possible to find $c(\lambda)$, for which

$$(n, \lambda) \geqslant \frac{c(\lambda)}{|n|^\eta}$$

whatever the integers $n_1, n_2, \ldots, n_s$ were[2]. It is also natural to consider condition (4) as, "generally speaking," fulfilled. Since

$$\varphi_{\alpha\beta}(0) = \frac{\partial}{\partial p_\alpha} \lambda_\beta(0),$$

where

$$\lambda_\beta(p) = \frac{1}{(2\pi)^s} \int_0^{2\pi} \cdots \int_0^{2\pi} \frac{dq_\beta}{dt} \, dq_1 \ldots dq_s$$

is the frequency averaged over the coordinate q_β with fixed momenta $p_1, \ldots, p_s$, condition (3) means that the Jacobian of the average frequencies over the momenta is different from zero.

Let us turn now to a consideration of the special case where $H(q, p, 0)$ depends only on p, i.e., $H(q, p, 0) = W(p)$. In this case, for $\theta = 0$ each torus, T_p, consists of the complete trajectories of the conditionally periodic movements with frequencies

$$\lambda_\alpha(p) = \frac{\partial W}{\partial p_\alpha} .$$

If the Jacobian

$$J = \left| \frac{\partial \lambda_\alpha}{\partial p_\beta} \right| = \left| \frac{\partial^2 W}{\partial p_\alpha \partial p_\beta} \right| \qquad (17)$$

is different from zero, then it is possible to apply Theorem 1 to almost all tori, T_p. There arises the natural hypothesis that at small θ the "displaced tori" obtained in accordance with Theorem 1 fill a larger part of region G. This is also confirmed by Theorem 2, pointed out later. In the formulation of this theorem we will consider the region S to be bounded and will introduce into the consideration the set, M_θ, of those points $(q^{(0)}, p^{(0)}) \in G$ for which the solution

$$q_\alpha(t) = f_\alpha(t; q^{(0)}, p^{(0)}, \theta), \quad p_\alpha(t) = G_\alpha(t; q^{(0)}, p^{(0)}, \theta)$$

of the system of equations (1) with initial conditions

$$q_\alpha(0) = q_\alpha^{(0)}, \quad p_\alpha(0) = p_\alpha^{(0)}$$

leads to trajectories not moving out of region G with change in t from $-\infty$ to $+\infty$, and conditionally periodic with periods $\lambda_\alpha = \lambda_\alpha(q^{(0)}, p^{(0)}, \theta)$, i.e., it has the form

$$f_\alpha(t) = \varphi_\alpha(e^{i\lambda_1 t}, \ldots, e^{i\lambda_s t}), \quad g_\alpha(t) = \psi_\alpha(e^{i\lambda_1 t}, \ldots, e^{i\lambda_s t}).$$

Theorem 2. If $H(q, p, 0) = W(p)$ and determinant (17) is not equal to zero in region S, then for $\theta \to 0$ the Lebesque degree of the set M_θ converges to the complete degree of region S.

Apparently, in the usual sense of the phrase, "general case" is when the set M_θ at all positive θ is everywhere dense. In such a case the complications arising in the theory of analytical dynamic systems are indicated more specifically in my note.[3]

REFERENCES

1. L. V. Kantorovich. Uspekhimatem. Nauk 3, 163 (1948).

2. J. F. Koksma. Diophantische Approximationen, Chelsea 1936. 157p.

3. A. N. Kolmogorov. Doklady Akad. Nauk 93, 763 (1953).

ON RESONANT HAMILTONIANS WITH TWO DEGREES OF FREEDOM NEAR AN EQUILIBRIUM POINT

Martin Kummer
University of Toledo, Toledo, Ohio 43606

ABSTRACT

This paper discusses the flow of a classical Hamiltonian with two degrees of freedom near an equilibrium point under the assumption that its quadratic part assumes the form

$$L_{\pm} = \frac{1}{2}(x_1^2 + y_1^2) \pm \frac{1}{2}(x_2^2 + y_2^2) .$$

1. INTRODUCTION

The purpose of the present paper is to review certain aspects of the flow that a classical Hamiltonian of the form

$$H_{\pm}(x_1,y_1,x_2,y_2) = \frac{1}{2}(x_1^2+y_1^2) \pm \frac{1}{2}(x_2^2+y_2^2) + \sum_{k=3}^{\infty} H_k(x_1,y_1,x_2,y_2) \quad (1)_{\pm}$$

induces in a sufficiently small neighborhood of the origin of its phase space R^4. Here, the vector field associated with $H_{\pm}$ is defined by the property that it is mapped onto the exact differential $-dH_{\pm}$ by the (symplectic) 2-form:

$$\omega = \sum_{k=1}^{2} (dy_k \wedge dx_k) . \quad (2)$$

In particular, we are interested in finding families of periodic orbits that are located close to the origin of R^4. It is clear that in this region the quadratic term of $H_{\pm}$ dominates the remainder and therefore it is natural to view the convergent sum of higher order polynomials $H_k(x_1,y_1,x_2,y_2)$ as a perturbation. Because all solutions of the unperturbed Hamiltonian are periodic with period 2π the task of finding periodic solutions of the Hamiltonian $H_{\pm}$ is intimately related to finding critical points of the averaged perturbation viewed as a function on the orbit manifold of the unperturbed Hamiltonian. A clear exposition of this method of finding periodic solutions is found in Ref. [1].

On account of the definiteness of the quadratic term in H_+ the origin of R^4 is a stable equilibrium point of H_+. However, this need no longer be the case for the Hamiltonian H_- and as a consequence a special investigation of the nature of this equilibrium point becomes necessary. It will be based on previous work by Sokol'skii [2].

A more detailed discussion together with proofs of most statements made in the present exposition can be found in Refs. [3] and [4].

2. THE HAMILTONIAN H_+

In this section we study the Hamiltonian H_+ given in $(1.1)_+$. We introduce complex variables

$$z_k = 2^{-1/2}(x_k + iy_k) \tag{1}$$

$$Z = \begin{pmatrix} z_1 \\ z_2 \end{pmatrix}, \qquad Z^\dagger = (\bar{z}_1, \bar{z}_2) \tag{2}$$

and the sesquilinear forms

$$M_k = \frac{1}{2} Z^\dagger \sigma_k Z, \qquad k = 0,1,2,3, \tag{3}$$

where

$$\sigma_0 = \begin{pmatrix} 1 & 0 \\ 0 & 1 \end{pmatrix}, \quad \sigma_1 = \begin{pmatrix} 0 & 1 \\ 1 & 0 \end{pmatrix}, \quad \sigma_2 = \begin{pmatrix} 0 & -i \\ i & 0 \end{pmatrix}, \quad \sigma_3 = \begin{pmatrix} 1 & 0 \\ 0 & -1 \end{pmatrix} \tag{4}$$

are the Pauli-matrices.

The quadratic term in $(1.1)_+$ becomes $2M_0$ and the fundamental 2-form becomes

$$\omega = \frac{1}{i} \sum_{k=1}^{2} (dz_k \wedge d\bar{z}_k) = id(Z^\dagger \sigma_0 dZ) . \tag{5}$$

The group of linear transformations leaving ω and M_0 invariant is U(2). Correspondingly, the M_k's (k = 0,1,2,3) satisfy the bracket-relations

$$\begin{aligned} [M_1,M_2] &= iM_3, \quad [M_2,M_3] = iM_1, \\ [M_3,M_1] &= iM_2, \quad [M_k,M_0] = 0, \quad k=1,2,3 \end{aligned} \tag{6}$$

with respect to the Poisson-bracket

$$[f,g] = \sum_{k=1}^{2} \left(\frac{\partial f}{\partial z_k} \frac{\partial g}{\partial \bar{z}_k} - \frac{\partial g}{\partial z_k} \frac{\partial f}{\partial \bar{z}_k} \right) \tag{7}$$

that is associated with the 2-form $i\omega$. Moreover, the following relation exists between the M_k's

$$M_1^2 + M_2^2 + M_3^2 = M_0^2 . \tag{8}$$

Dropping the perturbation term in $(1.1)_+$ leaves us with the Hamiltonian $2M_0$. Its flow decomposes R^4 into a family of concentric 3-spheres

$$S_m^3 = \left\{(x_1,y_1,x_2,y_2) \mid M_0 = m\right\}, \qquad m \geq 0 , \tag{9}$$

each of which is filled with periodic orbits of period 2π.

The map Π: $R^4 - \{0\} \to R^3$ defined by

$$\Pi(x_1,y_1,x_2,y_2) = \frac{1}{M_0}(M_1,M_2,M_3) \equiv (x,y,z) \tag{10}$$

identifies the points of each periodic orbit of the Hamiltonian $2M_0$ and relation (8) shows that the new manifold obtained in this way is nothing but the 2-sphere S^2. According to a theorem of J. Moser (Ref. 1, Theorem 4) it is possible to find periodic orbits of the full Hamiltonian H_+ (for small enough energies) by looking for the critical points (c.p.) of the averaged perturbation viewed as a function over the orbit space of the unperturbed Hamiltonian, i.e. in our case over S^2. Before we initiate this program we find it convenient to introduce in the region $x_1^2 + y_1^2 > 0$ of R^4 the symplectic chart (α_1,L,ξ,η) by means of the transformation:

$$\begin{pmatrix} x_1 \\ y_1 \end{pmatrix} = \left(2L - \xi^2 - \eta^2\right)^{1/2} \begin{pmatrix} \cos\alpha_1 \\ -\sin\alpha_1 \end{pmatrix}$$

$$\begin{pmatrix} x_2 \\ y_2 \end{pmatrix} = \begin{pmatrix} \cos\alpha_1 & \sin\alpha_1 \\ -\sin\alpha_1 & \cos\alpha_1 \end{pmatrix} \begin{pmatrix} \xi \\ \eta \end{pmatrix} . \tag{11}$$

In the new chart the map Π has the following expression

$$\Pi(\alpha_1,L,\xi,\eta) = \vec{x}(L,\xi,\eta) ,$$

where

$$\vec{x}(L,\xi,\eta) = \frac{1}{L}\left\{\left(2L - \xi^2 - \eta^2\right)^{1/2}\xi , \left(2L-\xi^2-\eta^2\right)^{1/2}\eta, L-\xi^2-\eta^2\right\}. \tag{12}$$

Its restriction to the sphere S_m^3 is obtained by setting $L = 2m$. Finally, if Π is restricted to the following transversal section of the unperturbed flow

$$S_m^3 \wedge \{\alpha_1 = \alpha_1^0 = \text{const}\}$$

it becomes a diffeomorphism with inverse

$$\xi = x\left(\frac{2m}{z+1}\right)^{1/2}, \qquad \eta = y\left(\frac{2m}{z+1}\right)^{1/2}, \tag{13}$$

i.e. the transversal section to the flow of the Hamiltonian $L = 2M_0$ on S_m^3 is diffeomorphic to S^2. Obviously, the last statement will also hold for the flow that H^+ induces on a surface of constant, sufficiently small energy.

Applying Gustavson averaging to our Hamiltonian H_+ we recognize that it can be brought into the following form

$$H_+ = P(M_0) + K_n(M_0,\vec{M}) + \mathcal{O}_{2n+1} \tag{14}$$

by means of a convergent canonical transformation. Here, $P(M_0)$ is a polynomial of degree $\leq$ n-1 in the variable M_0 with lowest order term $2M_0$ and $K_n(M_0,\vec{M})$ is a homogeneous polynomial in the variables M_0, $\vec{M} = (M_1,M_2,M_3)$ that, on account of relation (8), is only determined up to an additive polynomial belonging to the ideal generated in the algebra of all polynomials in the variables $(M_0,\vec{M})$ by the quadratic polynomial

$$M_1^2 + M_2^2 + M_3^2 - M_0^2 .$$

The integer n is determined by the condition that (modulo a polynomial in this ideal) $K_n(M_0,\vec{M})$ is not just a polynomial of M_0 alone. Thus, $K_n(M_0,\vec{M})$ can be assumed to possess a representation

$$K_n(M_0,\vec{M}) = \sum_{j=0}^{n} M_0^j \, K_{nj}(\vec{M}) \tag{15}$$

in which at least one of the polynomials $K_{nj}(\vec{M})$ of degree n-j in the variables $\vec{M}$ (j=0,1,...,n-1) is neither zero nor divisible by $M_1^2+M_2^2+M_3^2$.

Remark: The coefficients $K_{nj}(\vec{M})$ are determined by $K_n(M_0,\vec{M})$ uniquely only if we require that they be harmonic polynomials.

The symbol $\mathcal{O}_{2n+1}$ in (14) represents a convergent power series in the variables x_1,y_1,x_2,y_2 that begins with a term of order 2n+1. Dropping it, we are left with a Hamiltonian which we shall refer to as truncated Hamiltonian $\tilde{H}_+$. Our strategy is first to discuss the flow of the truncated Hamiltonian and only afterwards to study the effect of the inclusion of the term O_{2n+1}.

It follows from the bracket relations (6) that the flow induced by the truncated Hamiltonian on S^2 is governed by the differential equations

$$\frac{d\vec{x}}{d\tau} = \nabla K^{(n)}(\vec{x}) \times \vec{x} , \tag{16}$$

where $\tau = M_0^{n-1}t$, $\times$ denotes cross product and $K^{(n)}(x)$ is related to the function defined in (15) by $K^{(n)}(\vec{x}) = K_n(1,\vec{x})$.

The flow on S^2 uniquely determines the flow of $\tilde{H}_+$ in R^4. Indeed, if $x(\tau)$ is a solution of (16), then ξ and η are obtained as functions of time from (13) and α_1 by means of a simple integration of the equation

$$\frac{d\alpha_1}{dt} = \left.\frac{\partial H_+}{\partial L}\right|_{L=2m} = \frac{1}{2} P'(m) + \frac{n}{2} m^{n-1}[K^{(n)}(\vec{x}) + \frac{1}{z+1} \nabla K^{(n)}(\vec{x})\cdot(\vec{k}-z\vec{x})] , \tag{17}$$

where $\vec{k} = (0,0,1)$ is the unit vector in the positive z-direction.

A critical point (c.p.) of $K^{(n)}(\vec{x})$ on S^2 is a unit vector $\vec{e}$ with the property that there exists a real number λ, called the multiplier of $\vec{e}$, such that

$$\nabla K^{(n)}(\vec{e}) = \lambda\vec{e} . \tag{18}$$

From the equations (16) and (17) we easily deduce: Each c.p. of $K^{(n)}$ on S^2 gives rise to a family of periodic orbits of the truncated Hamiltonian $\tilde{H}_+$. The family parameter is m and the periodic orbit lying on S_m^3 has frequency

$$\frac{1}{2} P'(m) + \frac{n}{2} m^{n-1} K^{(n)}(\vec{e}) . \tag{19}$$

Because the flow of $\tilde{H}_+$ (at least for m small enough) possesses transversal sections that are diffeomorphic to our sphere S^2 , it is clear that the family of periodic orbits arising from a c.p. $\vec{e}$ of $K^{(n)}$ on S^2 is stable/unstable precisely if the c.p. $\vec{e}$ is stable/ unstable (c.p. in the present paper will always mean nondegenerate c.p. unless stated otherwise). Using the Hessian $A(\vec{e})$ of the function $K^{(n)}$ at the point $\vec{e}$, we can easily obtain the following formula for the index of a c.p. $\vec{e}$

$$\text{ind } \vec{e} = \text{sgn } \Delta(\vec{e}) \ ,$$

where

$$\Delta(\vec{e}) = - \det \begin{pmatrix} \lambda - A(\vec{e}) & \vec{e} \\ \vec{e}^T & 0 \end{pmatrix} = \left(\vec{e}, \text{ adj } (\lambda - A(\vec{e}))\ \vec{e}\right). \tag{20}$$

In the last expression the parentheses denote the usual inner product of R^3 and adj$(\cdot)$ means "algebraic adjoint of $(\cdot)$".

The question to be considered is whether the inclusion of the term $\mathcal{O}_{2n+1}$ in (14) preserves the stability of a stable periodic orbit of $\tilde{H}_+$. In order to investigate this question, we make use of the fact that a transformation

$$Z = U\hat{Z} \ , \qquad U \in SU(2) \tag{21}$$

is symplectic and that it induces a rotation $O(U)$ of the vector $\vec{x}$ by means of the formula

$$O(U)\ \vec{\sigma} = U^\dagger\ \vec{\sigma}\ U, \text{ i.e. } \vec{x} = O(U)\hat{\vec{x}} \ . \tag{22}$$

Hence it is no restriction to assume that our stable c.p. has position vector $\vec{k}$ and that the Hessian $A(\vec{k})$ has vanishing 12-entry (i.e. $A_{12} = A_{21} = 0$). It follows that

$$\lambda - A(\vec{k}) = \begin{pmatrix} A & 0 & -A_{13} \\ 0 & B & -A_{23} \\ -A_{13} & -A_{23} & C \end{pmatrix}, \quad \Delta(\vec{k}) = A \cdot B > 0,$$

where $A(\vec{k}) = (A_{ik})_{1 \leq i,k \leq 3}$, $A = \lambda - A_{11}$, $B = \lambda - A_{22}$, and $C = \lambda - A_{33}$. In order to formulate a sufficient criterion which guarantees that a stable periodic orbit of $\tilde{H}_+$ can (for sufficiently small energy) be continued to a stable orbit of the full Hamiltonian H_+ we need an expansion of the function

$$F(\xi,\eta) = K^{(n)}\left(\vec{x}(2,\xi,\eta)\right) - K^{(n)}(\vec{k}) \tag{23}$$

about the point $\xi = \eta = 0$. Here $\vec{x}(2,\xi,\eta)$ is obtained from the expression (12) by substituting $L = 2$. We find

$$- F(\xi,\eta) = \frac{1}{2}(A\xi^2 + B\eta^2) + \sum_{k=0}^{3} B_{k3-k}\,\xi^k\eta^{3-k} + \sum_{k=0}^{4} C_{k4-k}\xi^k\eta^{4-k} + \mathcal{O}_5 , \tag{24}$$

where $AB > 0$ and

$$\begin{aligned} B_{30} &= \frac{1}{2}A_{13} - \frac{1}{6}A_{111}, \qquad & B_{03} &= \frac{1}{2}A_{23} - \frac{1}{6}A_{222} \\ B_{12} &= \frac{1}{2}(A_{13} - A_{122}), \qquad & B_{21} &= \frac{1}{2}(A_{23} - A_{112}) \\ C_{40} &= \frac{1}{8}(C - A) + \frac{1}{4}A_{113} - \frac{1}{24}A_{1111} \\ C_{04} &= \frac{1}{8}(C - A) + \frac{1}{4}A_{223} - \frac{1}{24}A_{2222} \\ C_{22} &= \frac{1}{8}(2C - A - B) + \frac{1}{4}(A_{113} + A_{223} - A_{1122}) \end{aligned} \tag{25}$$

with

$$A_{113} = \frac{\partial^3 K^{(n)}}{\partial x^2\,\partial z}(\vec{k}) , \quad \text{etc.}$$

(C_{13} and C_{31} are not needed in the sequel.)

A suitable application of Moser's twist theorem yields the following result:

THEOREM 1. The family of stable periodic orbits of the truncated Hamiltonian H_+ that is associated with the stable critical point $\vec{k}$ of the function $K^{(n)}$ on S^2 can be continued to a family of stable periodic orbits of the full Hamiltonian H_+ provided that the energy is kept sufficiently small and provided that the following expression is nonzero:

$$\frac{3}{2}A[4A^2B_{03}^2 + (B_{03}A + B_{21}B)^2] + \frac{3}{2}B[4B^2B_{30}^2 + (B_{30}B + B_{12}A)^2] - AB(3C_{40}B^2 + C_{22}AB + 3C_{04}A^2) \tag{26}$$

In the case $n = 2$ this expression becomes (after multiplication by the factor 8)

$$12(A^3A_{13}^2 + B^3A_{23}^2) + 4(A)B^2T - 3(A+B)^2D \tag{27}$$

where $T = \mathrm{Tr}\,(\lambda - A(k))$ and $D = \det\,(\lambda - A(k))$. In the case $n = 2$ expression (27) is always nonzero provided that the multiplier λ (defined in (18)) is an eigenvalue of the matrix $A(\vec{k})$. Indeed, under this assumption we find

$$D = ABC - A_{13}^2 B - A_{23}^2 A = 0 .$$

Hence, because $AB > 0$ it follows that $C \neq 0$ and $\operatorname{sgn} A = \operatorname{sgn} B = \operatorname{sgn} C$ in case $A_{13}^2 + A_{23}^2 \neq 0$, and $C = 0$ in case $A_{13} = A_{23} = 0$. In both cases these conclusions imply that the expression (27) is nonzero.

Calling a c.p. $\vec{e}$ of the second or first kind, depending on whether its multiplier λ (defined in (18)) is an eigenvalue of $A(\vec{e})$ or not, we have:

THEOREM 2. If $n = 2$, then to every stable c.p. of the second kind which the function $K^{(2)}$ possesses on S^2, there corresponds a one parametric family of stable periodic orbits of the Hamiltonian H_+, the family parameter being the energy (which we keep sufficiently small).

In the following we study the case $n = 2$ in more detail. In view of the possibility of inducing rotations in $\vec{x}$-space by means of symplectic transformations of the phase space R^4 (see formulas (21), (22)) it is no restriction to assume that $K^{(2)}(\vec{x})$ has the form

$$K^{(2)}(x) = b_1 x + b_2 y + b_3 z + \frac{1}{2}(a_1 x^2 + a_2 y^2 + a_3 z^2) . \tag{28}$$

In the special case $\vec{b} = (b_1, b_2, b_3) = 0$, the differential equations (16) associated with the function $K^{(2)}$ become the Euler equations:

$$\begin{aligned} \frac{dx}{d\tau} &= (a_2 - a_3)\, yz \\ \frac{dy}{d\tau} &= (a_3 - a_1)\, zx \\ \frac{dz}{d\tau} &= (a_1 - a_2)\, xy , \end{aligned} \tag{29}$$

i.e. the flow of the truncated Hamiltonian on S^2 is the Euler flow. In particular it possesses four stable and two unstable c.p., all of the second kind. Returning now to the case of nonzero vector $\vec{b}$ we easily obtain:

THEOREM 3. The truncated Hamiltonian

$$\tilde{H}_+ = 2M_0 + M_0^2 \left[\frac{a_0}{2} + K^{(2)}\left(\frac{\vec{M}}{M_0}\right) \right]$$

where a_0 is some real number and the function $K^{(2)}$ defined in (28) induces on S^2 a flow that to an observer rotating with angular velocity $\vec{b}$ with respect to S^2 appears to be the Euler flow.

In the following we will look for critical points of this flow. For simplicity we make the additional assumption that the a_i's in (28) are pairwise distinct. Specializing the equations (16) to the case of the function (28) we obtain

$$(\lambda - a_i)\, e_i = b_i \qquad (i = 1,2,3). \tag{30}$$

If λ_0 is a (simple) zero of the function

$$g(\lambda) = 1 - \frac{b_1^2}{(\lambda - a_1)^2} - \frac{b_2^2}{(\lambda - a_2)^2} - \frac{b_3^2}{(\lambda - a_3)^2}$$

then

$$\vec{e}(\lambda_0) = \left(\frac{b_1}{\lambda_0 - a_1}, \frac{b_2}{\lambda_0 - a_2}, \frac{b_3}{\lambda_0 - a_3}\right) \tag{31}$$

is a c.p. of the first kind with index

$$\text{ind}\, \vec{e}(\lambda_0) = \text{sgn}\, (\lambda_0 - a_1)(\lambda_0 - a_2)(\lambda_0 - a_3)\, g'(\lambda_0)\ . \tag{32}$$

A pair of c.p. of the second kind

$$\vec{f}_1^{\pm} = \left(\pm\, [g(a_1)]^{1/2}, \frac{b_2}{a_1 - a_2}, \frac{b_3}{a_1 - a_3}\right)$$

with total index $2\varepsilon_1$ $(\varepsilon_1 = \text{sgn}\,(a_1 - a_2)(a_1 - a_3))$ arises if $g(a_1) > 0$. Of course, this can only happen if $b_1 = 0$. By a cyclic permutation of all subscripts in the last statement we obtain analogous statements for the pairs of c.p. of the second kind: $\vec{f}_2^{\pm}$, $\vec{f}_3^{\pm}$. In the following we will assume $b_3 \neq 0$ so that $\lim_{\lambda\to a_3} g(\lambda) = -\infty$ and consequently the pair of c.p.: $\vec{f}_3^{\pm}$ is not present. If $\vec{b} \neq 0$ this can always be achieved by means of a transformation of the type (21), (22). Under this assumption $g(\lambda)$ possesses at least two zeros λ_1, λ_0 , where λ_1 is located on the real axis on the same side of a_3 as a_1 and λ_0 on the opposite side. Correspondingly, there always exist at least two c.p. of the first kind: $\vec{e}(\lambda_0)$, $\vec{e}(\lambda_1)$. In Table I we classify the indices of these two c.p. as well as those of the c.p. of the second kind $\vec{f}_1^{\pm}$ $\vec{f}_2^{\pm}$ according to the signs of $g(a_1)$ and $g(a_2)$. (A star in a column headed by $\vec{f}_k^{\pm}$ $(k = 1,2)$ means that $\vec{f}_k^{\pm}$ is not present.)

The comments following the table deal with the possible existence of additional c.p. of the first kind. Taking these comments into account, the reader easily verifies that, in accordance with a well known topological fact, the sum of all indices of all c.p. of $K^{(2)}$ on S^2 equals two.

TABLE I

sgn $g(a_1)$	sgn $g(a_2)$	Cond. on Vector $\vec{b}$ besides $b_3 \neq 0$	$\vec{e}(\lambda_0)$	$\vec{e}(\lambda_1)$	$\vec{f}_1^{\pm}$	$\vec{f}_2^{\pm}$
-	-	none	+1	+1	*	*
+	+	$b_1 = 0$	+1	$-\varepsilon_1$	ε_1	*
-	+	$b_2 = 0$	ε_3	ε_1	*	ε_2
+	-	$b_1 = b_2 = 0$	ε_3	ε_3	ε_1	ε_2

Comment on row 1: One or two (+ -)-pairs (i.e., one member of the pair has index +1, the other -1) of c.p. of the first kind may be present.

Comment on row 2: If $\varepsilon_1 = -1$ we have $g(b_{23}) > 0$, where

$$b_{23} = \frac{b_2^{2/3}a_3 + b_3^{2/3}a_2}{b_2^{2/3} + b_3^{2/3}} \tag{33}$$

An additional (+ +)-pair of c.p. of the first kind exists.

Comment on row 3: If $\varepsilon_2 = -1$ we have $g(b_{31}) > 0$ (where b_{31} is obtained from b_{23} by a cyclic permutation of the subscripts) and the same conclusion holds as in the previous comment.

Comment on row 4: In this case the explicit formulas

$$\begin{aligned} \lambda_1 &= a_3 + \operatorname{sgn}(a_1 - a_3)\, b_3\,, \qquad & \vec{e}(\lambda_1) &= \operatorname{sgn}[b_3(a_1 - a_3)]\, \vec{k} \\ \lambda_0 &= a_3 - \operatorname{sgn}(a_1 - a_3)\, b_3\,, \qquad & \vec{e}(\lambda_0) &= -\vec{e}(\lambda_1) \end{aligned} \tag{34}$$

hold.

<u>Example</u>. $H = \frac{1}{2}(x_1^2 + y_1^2) + \frac{1}{2}(x_2^2 + y_2^2) + x_1^2 x_2 + \mu x_2^3$. We find $a_1 - a_2 = 2(\mu - 2)$, $a_3 - a_1 = -\frac{15}{2}(\mu + \frac{1}{3})(\mu - 1)$, $a_2 - a_3 = -\frac{15}{2}(\mu - \frac{1}{3})(\mu - \frac{3}{5})$; $b_1 = b_2 = 0$, $b_3 = \frac{15}{2}(\mu^2 - \frac{1}{9})$. It follows that the pair of c.p.

$$\vec{f}_2^{\pm} = \frac{1}{\frac{3}{5} - \mu} \left\{ 0, \pm \left[\frac{28}{15} \left(\frac{2}{15} - \mu \right) \right]^{1/2}, -\frac{1}{3} - \mu \right\}$$

exists for $\mu < 2/15$ with total index 2. The pair

$$\vec{f}_1^{\pm} = \frac{1}{1 - \mu} \left\{ \pm 2 \left[\frac{1}{3} \left(\frac{2}{3} - \mu \right) \right]^{1/2}, 0, \frac{1}{3} - \mu \right\}$$

exists for $\mu < 2/3$, with total index -2 for $\mu < -1/3$ and total index +2 for $-1/3 < \mu < 2/3$. $\vec{k}$ and $-\vec{k}$ are c.p. of the first kind with index

$$\operatorname{ind}(-\vec{k}) = \begin{cases} +1, & \mu < -\frac{1}{3};\ \frac{2}{15} < \mu < \frac{1}{3};\ \mu > \frac{2}{3} \\ -1, & -1 < \mu < \frac{2}{15};\ \frac{1}{3} < \mu < \frac{2}{3} \end{cases}$$

$$\operatorname{ind}(\vec{k}) = \begin{cases} +1, & |\mu| > \frac{1}{3}; \\ -1, & |\mu| < \frac{1}{3}. \end{cases}$$

3. THE HAMILTONIAN H_-

We now turn to the case of a Hamiltonian H_- as it was defined in $(1.1)_-$. We try to imitate the procedure employed in the case of H_+ as closely as possible. Again, the z_k's $(k = 1,2)$ are defined as in (2.1) but now we define

$$Z = \begin{pmatrix} z_1 \\ z_2 \end{pmatrix}, \qquad Z^\dagger = (\bar{z}_1, \bar{z}_2) \tag{1}$$

and

$$M_k = \frac{1}{2} Z^\dagger \tilde{\sigma}_k Z, \qquad k = 0,1,2,3, \tag{2}$$

where the $\tilde{\sigma}_k$'s are the Pauli matrices (2.4) with σ_0, σ_3 interchanged, i.e., we set $\tilde{\sigma}_0 = \sigma_3$, $\tilde{\sigma}_1 = \sigma_1$, $\tilde{\sigma}_2 = \sigma_2$, $\tilde{\sigma}_3 = \sigma_0$. Using this notation the quadratic term in $(1.1)_-$ becomes again $2M_0$ and the fundamental 2-form (1.2) becomes

$$\omega = i d (Z^\dagger \sigma_3 dZ). \tag{3}$$

Hence, the group of linear substitutions leaving M_0 and ω invariant is the group

$$U(1,1) = \left\{ U \in GL(2) \quad U^\dagger \sigma_3 U = \sigma_3 \right\} \tag{4}$$

and the M_k's satisfy the following relations with respect to the bracket (2.7)

$$[M_1,M_2] = -iM_3 , \qquad [M_2,M_3] = iM_1$$
$$[M_3,M_1] = iM_2 , \qquad [M_k,M_0] = 0 \quad (k = 1,2,3) \tag{5}$$

Moreover, the relation

$$M_3^2 - M_1^2 - M_2^2 = M_0^2 \tag{6}$$

holds between the M_k's $(k = 0,1,2,3)$. The flow of the leading term $2M_0$ of H_- decomposes the phase space R^4 into a one parameter family of 3-hyperboloids

$$h_m^3 = \left\{(x_1,y_1,x_2,y_2 \mid M_0 = m\right\} \tag{7}$$

each of which is filled with periodic orbits of period 2π. The map Π which identifies the points of each orbit is defined as above by (2.10). Its domain of definition is the complement in R^4 of the cone $M_0 = 0$ and its image is the two-sheeted hyperboloid in R^3

$$h^2 = \left\{\vec{x} \mid h(\vec{x}) = 1\right\} = h_+^2 \cap h_-^2 , \tag{8}$$

where $h_{\pm}^2 = h^2 \cap \begin{Bmatrix} z > 0 \\ z < 0 \end{Bmatrix}$ and $h(\vec{x}) = -x^2 - y^2 + z^2$. (9)

Π maps h_m^3 onto h_+^2 or h_-^2 depending on whether m is positive or negative.

By means of the formulas

$$\begin{pmatrix} x_1 \\ y_1 \end{pmatrix} = \left(2L + \xi^2 + \eta^2\right)^{1/2} \begin{pmatrix} \cos\alpha_1 \\ -\sin\alpha_1 \end{pmatrix}$$
$$\begin{pmatrix} x_2 \\ y_2 \end{pmatrix} = \begin{pmatrix} \cos\alpha_1 & \sin\alpha_1 \\ \sin\alpha_1 & -\cos\alpha_1 \end{pmatrix} \begin{pmatrix} \xi \\ \eta \end{pmatrix} \tag{10}$$

we introduce a new symplectic chart (α_1,L,η,ξ) in the region $x_1^2 + y_1^2 > 0$ of R^4 in which ω has the expression:

$$\omega = d\left(L\, d\alpha_1 + \xi\, d\eta\right) . \tag{11}$$

The map Π takes the point (L,α_1,ξ,η) into the point

$$\vec{x}(L,\xi,\eta) = \frac{1}{L}\left\{(2L+\xi^2+\eta^2)^{1/2}\ \xi,\ (2L+\xi^2+\eta^2)^{1/2}\ \eta,\ L+\xi^2+\eta^2\right\} . \tag{12}$$

If Π is restricted to the surface

$$h_m^3 \cap \left\{\alpha_1 = \alpha_1^0 = \text{const.}\right\}, \qquad m > 0$$

it becomes a diffeomorphism onto h_+^2 whose inverse is given by the formulas (2.13). It follows that on each energy surface (for small enough absolute energy) the flow of H_- possesses a transversal section that is diffeomorphic to the surface h_+^2 which in turn is diffeomorphic to the plane R^2.

Gustavson averaging will again yield a Hamiltonian of the form (2.14) (where -- because of the new meaning of the M_k's -- the definition of the polynomial $K_n(M_0,\vec{M})$ has to be modified in an obvious way). Again, we define the truncated Hamiltonian $\tilde{H}_-$ by simply dropping the term $\mathcal{O}_{2n+1}$ in (2.14). On h^2 , $\tilde{H}_-$ induces a flow that is governed by the differential equations

$$\frac{d\vec{x}}{d\tau} = \frac{1}{2}\left(\nabla h \times \nabla K^{(n)}\right)(\vec{x}) \tag{13}$$

where $\tau = M_0^{n-1}t$, $K^{(n)}(\vec{x}) = K_n(1,\vec{x})$ and $h(\vec{x})$ was defined in (9). Because the equations (2.13) and (2.17) continue to hold in the present section, a similar argument as in Section 2 shows that the flow on the orbit manifold (in the present case h^2) determines the flow of the turncated Hamiltonian in R^4 completely. A c.p. $\vec{e}$ of $K^{(n)}$ on h^2 is defined by the condition that there exists a real number λ (called the multiplier of $\vec{e}$) such that

$$\nabla K^{(n)}(\vec{e}) = \frac{\lambda}{2}\nabla h(\vec{e}) . \tag{14}$$

Proceeding as in Section 2 we show that each stable/unstable c.p. $\vec{e}$ gives rise to a one parameter family of stable/unstable periodic orbits $\tilde{H}_-$, the family parameter being m. Here the frequency of the orbit lying on h_m^3 is again given by the expression (2.19).

The formula (2.20) for the index of a c.p. is replaced by

$$\Delta(\vec{e}) = -\det\begin{pmatrix}\lambda G - A(\vec{e}) & G\vec{e}\\ (G\vec{e})^T & 0\end{pmatrix} = \left\langle\vec{e},\ \operatorname{adj}\left(\lambda - G\,A(\vec{e})\right)\vec{e}\right\rangle \tag{15}$$

Here, $A(\vec{e})$ has the same meaning as in Section 2, $G = \operatorname{diag}(-1,-1,1)$

and $< \, , \, >$ is the inner product that in the standard basis of R^3 is represented by the matrix G. Notice that the function $h(\vec{x})$ defined in (9) is the associated quadratic form:

$$h(\vec{x}) = <\vec{x},\vec{x}> .$$

In order to obtain a sufficient criterion for the ability to continue a family of stable periodic orbits of the truncated Hamiltonian to a family of orbits of the full Hamiltonian (for sufficiently small absolute energy) we again make the assumption that the stable c.p. which gives rise to the given family of periodic orbits has position vector $\vec{k}$. This assumption in no way represents a loss of generality because if the c.p. first has general position $\vec{e} \in h_+^2$ it can be brought into the position $\vec{k}$ by means of a symplectic transformation

$$Z = U\hat{Z} \ , \qquad U \in SU(1,1) \ , \tag{16}$$

where Z is defined in (1). Indeed, it is well known that such transformations induce transformations O(U) of the restricted Lorentz-group (often denoted by $SO^{\uparrow}(2,1)$) in $\vec{x}$-space via a formula that differs from (2.22) by the replacement of $\vec{\sigma}$ by $\vec{\underset{\sim}{\sigma}}$. Besides bringing $\vec{e}$ into the position $\vec{k}$ we can choose U in (16) in such a way that the Hessian $A(\vec{k})$ has vanishing 12-entry. If finally $\vec{e} \in h_-^2$ then a "time-reversal" i.e., the transformation

$$t \to -t \ , \qquad z_1 \leftrightarrow z_2 \tag{17}$$

which in the differential equations (13) replaces τ by $-\tau$ and $\vec{x}$ by $-\vec{x}$ will replace $\vec{e}$ by $-\vec{e} \in h_+^2$ and we may proceed as before.

Summarizing, we see that it is no loss of generality to assume that our stable c.p. is given by $\vec{k}$ and that its Hessian $A(\vec{k})$ has vanishing 12-entry. Thus,

$$\left(\lambda - G\,A(\vec{k})\right) = \begin{pmatrix} A & 0 & A_{13} \\ 0 & B & A_{23} \\ -A_{13} & -A_{23} & 0 \end{pmatrix} , \quad \Delta(\vec{k}) = A\cdot B > 0 \ ,$$

where $A(\vec{k}) = (A_{ik})_{1\leq i,k\leq 3}$, $A = \lambda + A_{11}$, $B = \lambda + A_{22}$, and $C = \lambda - A_{33}$.

If, in addition, the definitions (2.25) are replaced by

$$B_{30} = \frac{1}{2} A_{13} + \frac{1}{6} A_{111} , \qquad B_{03} = \frac{1}{2} A_{23} + \frac{1}{6} A_{222}$$

$$B_{12} = \frac{1}{2}(A_{13} + A_{122}) , \qquad B_{21} = \frac{1}{2} (A_{23} + A_{112})$$

$$C_{40} = \frac{1}{8} (A - C) + \frac{1}{4} A_{113} + \frac{1}{24} A_{1111}$$

$$C_{04} = \frac{1}{8} (B - C) + \frac{1}{4} A_{223} + \frac{1}{24} A_{2222}$$

$$C_{22} = \frac{1}{8} (A + B - 2C) + \frac{1}{4} (A_{113} + A_{223} + A_{1122})$$

then the first part of Theorem 1 of Section 2 carries over to the case in which H_+ is replaced by H_- and s^2 by h^2.

For n = 2 the expression (1.26) now becomes:

$$12(B^3A_{12}^2 + A^3A_{23}^2) - 4(AB)^2T + 3(A + B)^2 D ,$$

where

$$T = \mathrm{Tr}\left(\lambda - G\,A(\vec{k})\right) \quad \text{and} \quad D = \det\left(\lambda - G\,A(\vec{k})\right) .$$

Theorem 2 has no counterpart in the case of a Hamiltonian H_-.

Also, on account of the indefiniteness of the quadratic part of H_- the stability of the origin of R^4 is no longer guaranteed (as was the case for the Hamiltonian H_+) and a special investigation of its nature becomes necessary. A criterion which allows us in most instances to decide whether this equilibrium point is stable or unstable has been developed by Sokol'skii [2]. In the sequel we present a generalized version of his criterion. To that end we assume that $K_{np-1}(\vec{M})$ $(1 \leq p \leq n)$ is the first polynomial in the expansion (2.15) of K_n in terms of powers of M_0 that is neither zero nor divisible by $M_1^2 + M_2^2 + M_3^3$.

Let $Q(\phi)$ be the trigonometric polynomial defined by

$$Q(\phi) = K_{np-1}\left(\cos\phi, \sin\phi, 1\right) . \qquad (18)$$

The following theorem contains the announced generalized version of Sokol'skii's criterion:

THEOREM 4. (i) If $Q(\phi)$ has a simple zero, then the origin of R^4 is unstable.

(ii) If $Q(\phi)$ does not vanish and if p = 1, then the origin of R^4

is stable.

Finally, we attempt to classify the c.p. of $K^{(n)}$ on the manifold h^2 in the case n = 2. In this case $K^{(2)}(\vec{x})$ can be written in terms of a vector $\vec{b}$ and a symmetric 3 by 3 matrix A as follows

$$K^{(2)}(\vec{x}) = \langle\vec{x},\vec{b}\rangle + \frac{1}{2}(\vec{x}, A\vec{x}) .$$

The question naturally arises whether or not the matrix A can be diagonalized by a transformation (16). Because, as already pointed out above, such transformations induce restricted Lorentz transformations in $\vec{x}$-space we are led to look for conditions that guarantee the existence of a matrix $O \in SO^{\uparrow}(2,1)$ with the property $O^T A\, O = GD$, where $D = \mathrm{diag}(a_1,a_2,a_3)$. It is not difficult to see that such a matrix O exists if and only if GA possesses two independent space-like and one time-like eigenvectors, the a_i's being the corresponding eigenvalues. Here we have used the terminology of relativistic physics according to which a vector $\vec{x}$ is called space-, null- or time-like depending on whether $\langle\vec{x},\vec{x}\rangle$ is smaller, equal or greater than zero.

In particular, if all a_i's are real and mutually distinct -- an assumption that will be made in the sequel(*) -- then the existence of a transformation $O \in SO^{\uparrow}(2,1)$ with the stated property is guaranteed. That is to say, under our assumption it is no loss of generality to write $K^{(2)}(\vec{x})$ in the form

$$K^{(2)}(\vec{x}) = - b_1 x - b_2 y + b_3 z + \frac{1}{2}\left(- a_1 x^2 - a_2 y^2 + a_3 z^2\right), \qquad (19)$$

where the a_i's are pairwise distinct. Again, a c.p. of the first kind is given by the formula (2.31), where λ_0 is now a zero of the function

$$g(\lambda) = 1 + \frac{b_1^2}{(\lambda - a_1)^2} + \frac{b_2^2}{(\lambda - a_2)^2} - \frac{b_3^2}{(\lambda - a_3)^2} \qquad (20)$$

and its index is calculated with the aid of formula (2.32).

If $b_3 = 0$, only the pair of c.p. of the second kind

$$\vec{f}_3^{\pm} = \left(\frac{b_1}{a_3 - a_1}, \frac{b_2}{a_3 - a_2}, \pm [g(a_3)]^{1/2}\right)$$

---

*An important case which is omitted here arises when the matrix GA possesses a pair of complex conjugate eigenvalues (see Ref. 4, p. 108).

exists with multiplier $\lambda = a_3$ and index $\varepsilon_3 = \operatorname{sgn} \Delta_3$, where $\Delta_3 = (a_3 - a_1)(a_3 - a_2)$. Similarly, if $b_3 \neq 0$ and $g(a_i) < 0$ $(i=1,2)$ the pair of c.p. $\vec{f}_i^{\pm}$ $(i = 1,2)$ exists with multiplier a_i and index ε_i, $\varepsilon_i = \operatorname{sgn} \Delta_i$ $(i = 1,2)$. Here, $\Delta_1 = -(a_1 - a_2)(a_1 - a_3)$ and $\Delta_2 = -(a_2 - a_1)(a_2 - a_3)$. Explicitly, we have

$$\vec{f}_1^{\pm} = \left(\pm \left[- g(a_1)\right]^{1/2}, \frac{b_2}{a_1 - a_2}, \frac{b_3}{a_1 - a_3}\right)$$

and an analogous formula holds for $\vec{f}_2^{\pm}$. In the following, we confine ourselves to the case $b_2 \neq 0$. Under this assumption the function g of (20) has at least two zeros λ_0, λ_1, where λ_1 is located on the real axis on the same side of a_3 as a_1, λ_0 on the opposite side. Table II classifies the indices of the corresponding c.p. $\vec{e}(\lambda_0)$, $\vec{e}(\lambda_1)$ as well as those of the c.p. of the second kind $\vec{f}_i^{\pm}$ $(i = 1,2)$ in terms of the signs of $g(a_1)$, $g(a_2)$. In reading this table the same conventions have to be observed as in reading Table I. Again, the comments following the table deal with the question of the possible existence of additional c.p. of the first kind.

TABLE II

sgn $g(a_1)$	sgn $g(a_2)$	Cond. on Vector $\vec{b}$ (besides $b_3 \neq 0$)	$\vec{e}(\lambda_0)$	$\vec{e}(\lambda_1)$	$\vec{f}_1^{\pm}$	$\vec{f}_2^{\pm}$
+	+	none	ε_3	ε_3	*	*
+	-	$b_2 = 0$	1	ε_1	*	*
-	+	$b_1 = 0$	ε_3	$-\varepsilon_1$	ε_1	*
-	-	$b_1 = b_2 = 0$	1	1	ε_1	ε_2

On rows 1 and 4 the same comments are applicable as on the corresponding rows of Table I.

Comment on row (i+1) $(i = 1,2)$: If $\varepsilon_i = -1$ one more (+ -)-pair of c.p. of the first kind *may* be present. If $\varepsilon_i = +1$ one additional (+ +)-pair of c.p. of the first kind *must* be present.

If we take into account the following relations between the ε_i's: $\varepsilon_1\varepsilon_2 = -\varepsilon_3$, $(\varepsilon_1 + 1)(\varepsilon_2 + 1) = 0$, $1 + \varepsilon_1 + \varepsilon_2 = \varepsilon_3$, where the last relation is implied by the first two, then an inspection of Table II reveals the following fact: Under our assumptions ($b_3 \neq 0$, all a_i's pairwise distinct) the total index of all c.p. that the function $K^{(2)}$ (as defined in (19)) possesses on h^2 is always $2\varepsilon_3$.

Moreover, there exists an interesting connection (expressed in Theorem 5) between this total index and the stability of the origin of R^4. Indeed, applying our recipe (18) for extracting the Sokol'skii function $Q(\phi)$ from the polynomial K_n to the present case in which $n = 2$ and in which

$$K_2(M_0,M) = M_0\langle\vec{b},\vec{M}\rangle + \frac{1}{2}(-a_1M_1^2 - a_2M_2^2 + a_3M_3^2) ,$$

where all a_i's are pairwise distinct (implying $n = 2$, $p = 1$ in (18)) we obtain:

$$Q(\phi) = [a_3 - \frac{1}{2}(a_1 + a_2)] + \frac{1}{2}(a_1 - a_2)\cos 2\phi .$$

The following theorem is now an immediate consequence of Theorem 4:

THEOREM 5. The origin of R^4 is stable (unstable) precisely if the total index of all c.p. is 2 (-2) (i.e. if $\varepsilon_3 = 1$ (-1)).

Example. $H = \frac{1}{2}(x_1^2 + y_1^2) - \frac{1}{2}(x_2^2 + y_2^2) + x_1^2x_2 + \mu x_2^3$.

We find

$$a_1 - a_2 = 2(2+\mu) \qquad , \; a_3 - a_1 = \frac{15}{2}(\mu+1)(\mu - \frac{1}{3})$$

$$a_2 - a_3 = -\frac{15}{2}(\mu + \frac{1}{3})(\mu + \frac{3}{5}),\; b_1 = b_2 = 0,\; b_3 = \frac{15}{2}(\frac{1}{9} - \mu^2) .$$

It follows that the pair of c.p.

$$\vec{f}_2^{\pm} = \frac{1}{\frac{3}{5}+\mu}\left\{0,\ \pm\,\mathrm{sgn}\,(\tfrac{3}{5}+\mu)\left[-\tfrac{28}{15}(\tfrac{2}{15}+\mu)\right]^{1/2},\ \mu - \tfrac{1}{3}\right\}$$

exists for $\mu < -2/15$, $\mu \neq 3/5$ with index

$$\mathrm{ind}\ \vec{f}_2^{\pm} = \begin{cases} +1 & \text{for} \quad \mu < -2 \ , \ -\frac{3}{5} < \mu < -\frac{1}{3} \\ -1 & \text{for} \quad -2 < \mu < -\frac{3}{5} \ , \ -\frac{1}{3} < \mu < -\frac{2}{15} \end{cases}$$

The pair of c.p. of the second kind

$$\vec{f}_1^{\pm} = \frac{1}{1+\mu}\left\{\pm 2\,\mathrm{sgn}\,(1+\mu)\left[-\tfrac{1}{3}(\mu+\tfrac{2}{3})\right]^{1/2},\ 0,\ \mu + \tfrac{1}{3}\right\}$$

exists for $\mu < -2/3$, $\mu \neq -1$ with index,

$$\text{ind } \vec{f}_1^{\pm} = \begin{cases} +1 & \text{for} \quad -2 < \mu < -1 \\ -1 & \text{for} \quad \mu < -2, \quad -1 < \mu < -\frac{2}{3} \end{cases}$$

The only c.p. of the first kind are $\pm \vec{k}$. They are present for all (real) μ and we have

$$\text{ind } (-\vec{k}) = \begin{cases} +1 & \text{for} \quad \mu < -\frac{2}{3}, \; -\frac{1}{3} < \mu < -\frac{2}{15}, \quad \mu > \frac{1}{3} \\ -1 & \text{for} \quad -\frac{2}{3} \leq \mu < -\frac{1}{3}, \; -\frac{2}{15} \leq \mu < \frac{1}{3} \end{cases}$$

$$\text{ind } \vec{k} = \begin{cases} +1 & \text{for} \quad |\mu| > \frac{1}{3} \\ -1 & \text{for} \quad |\mu| < \frac{1}{3} \end{cases}$$

Theorem 5 implies: the origin of R^4 is unstable for $-1 < \mu < -3/5$ and $|\mu| < 1/3$ and stable for μ lying in the interior of the complementary intervals.

REFERENCES

[1] Moser, J., Regularization of Kepler's problem and the averaging method on a manifold, Comm. Pure Appl. Math. 23, (1970) 609-636.

[2] Sokol'skii, A. G., On stability of an autonomous Hamiltonian system with two degrees of freedom in the case of equal frequencies, Appl. Math. and Mech. 38 (1974) 791-799.

[3] Kummer, M., On resonant nonlinearly coupled oscillators with two equal frequencies, Comm. Math. Phys. 48 (1976) 53-79. (The expression (35), p. 63 of Ref. [3] is only correct for n = 2, whereas the expression (2.26) of the present paper is correct for general n.)

[4] Kummer, M., On resonant classical Hamiltonians with two equal frequencies, Comm. Math. Phys. 58 (1978) 85-112.

For a more complete bibliography see References [3], [4].

ACKNOWLEDGMENT

The author would like to express his sincerest thanks to Dr. George Kertz for a critical reading of this work and to Mrs. Kay Locke for her expeditious and superb typing of the original manuscript, and to Ms. Connie Engle for the final version.

A SURVEY OF THE HÉNON-HEILES HAMILTONIAN WITH APPLICATIONS TO RELATED EXAMPLES

by

R. C. Churchill
Department of Mathematics, Hunter College, C.U.N.Y.
New York, N.Y.

G. Pecelli**
Department of Mathematics, S.U.N.Y. at Albany
Albany, N.Y.

D. L. Rod*
Department of Mathematics and Statistics
University of Calgary
Calgary, Alberta

Abstract

This paper surveys some recent techniques and results for Hamiltonian systems in the context of the Hénon-Heiles Hamiltonian. The paper attempts to give some reasonable interpretation of various numerical (computer) results alongside recent mathematical techniques and results presented as they apply to this Hamiltonian. In particular, various computed periodic orbits are identified with geometrically constructed periodic orbits and rigorous conclusions concerning the stability status of the orbit as the energy of the system changes are indicated whenever possible. The paper concludes with a discussion of some related Hamiltonians and suggested computer experiments.

**On leave from Hunter College, C.U.N.Y.

*Author's research supported in part by National Research Council of Canada, Grant A8507.

0. Introduction

This paper surveys some new techniques and results which can be applied to many Hamiltonian systems of the form

$$H(x,y) = (1/2)|y|^2 + V(x), \quad x,y \in R^2, \tag{0.1}$$

where the potential $V: R^2 \to R$ is assumed sufficiently differentiable. Our objective will always be to gain geometrical insight into the nature of solutions of the corresponding differential equation

$$\ddot{x} = -V_x, \tag{0.2}$$

where V_x = grad V. Thus, explicit representation of solutions of (0.2) by convergent series or otherwise will play a secondary role. For this reason much current research along such lines will not be surveyed in this paper. For another approach to these problems, we refer to the work of R. Helleman, E. Montroll, and students [7,30,31,38,39,52] and the references therein (see also [56]). The reader is also referred to the surveys [32,33,34] of J. Ford which give the physical problems whose model equations display stochasticity. We also mention the surveys by M. Berry [4] and R. Devaney [72].

The differential equation (0.2) is equivalent to the first order Hamiltonian system

$$\begin{cases} \dot{x} = y = H_y, \\ \dot{y} = -V_x = -H_x. \end{cases} \tag{0.3}$$

Any solution x(t) of (0.2) is thus seen to satisfy $(d/dt)H(x(t),\dot{x}(t)) = 0$. This "conservation of energy" implies that the energy manifolds $\{(x,y) \in R^4: H(x,y) = h\}$, denoted H=h, are invariant under the corresponding flow. Thus equation (0.2), or equivalently (0.3), can be studied by examining the flows on these energy manifolds. We will exploit the geometrical properties and symmetries of the potential V for this.

We will use the Hénon-Heiles Hamiltonian [40] as a model for (0.1), rather than simply cite theorems; in Section 8 further examples will be discussed. Interest in the Hénon-Heiles system stemmed initially from the search for further "conservation laws" in certain galactic potentials admitting an axis of symmetry, and Hénon and Heiles presented their potential as a "model problem" which had the advantage of computational simplicity. Later Lunsford and Ford [48] observed that by truncating after the cubic terms, a wide class of three particle systems could be reduced to the Hénon-Heiles example simply by diagonalizing the quadratic terms of the potential and adjusting the time scale.

At energies $h<0$ the Hénon-Heiles Hamiltonian is relatively uninteresting, since all solutions become unbounded in both time directions. When $h=0$ a critical point appears at the origin, and as h increases past zero this critical point bifurcates into (at least) eight periodic orbits. All these periodic orbits lie on a compact component of $H=h$, $0<h<(1/6)$, and it is the flow on this component which has received most of the attention in the literature. In particular, at low energies $h>0$ numerical studies indicate that this flow appears to be "integrable," while at higher energies $h<(1/6)$ the flow displays "stochastic" behavior in large regions of the phase space. It was precisely this contrasting behavior of solutions at different energies which interested those who first examined the potential, as it suggested that additional conservation laws did exist at certain energies, but did not exist at other energies. At $h=(1/6)$ three new symmetrically placed critical points appear, each bifurcating into a hyperbolic periodic orbit at energies $h>(1/6)$.

We will report on progress in explaining the "integrable" and "stochastic" phenomena that were first observed numerically, but we mention that much work remains to be done. In particular, we will see that for small $h>0$ the study of the flow can be reduced to that of a mapping which admits invariant curves [8,43], while at high energies we will see that the Smale horseshoe mapping can be embedded into the flow [14], thus proving the nonexistence of a second global analytic integral for the equations.

The authors wish to thank the typist, Ms. Terri Casey, for her valuable assistance in preparing this manuscript.

1. Preliminary Remarks.

We first present several general techniques for investigating an arbitrary Hamiltonian system

$$\dot{x} = H_y, \quad \dot{y} = -H_x, \tag{1.1}$$

where H is a sufficiently differentiable real-valued function defined on some open subset $U \subset R^{2n}$.

For any solution (x(t), y(t)) of (1.1), one has $(d/dt)H(x(t),y(t)) \equiv 0$. Thus the energy manifolds H=h given by $\{(x,y) \varepsilon U: \ H(x,y) = h\}$ are invariant under the corresponding Hamiltonian flow. A corollary of the implicit function theorem together with Sard's Theorem implies that for almost all energies h the set H=h is in fact a differentiable manifold. Given a compact component of H=h, the "Recurrence Theorem" of Poincaré implies that "almost all" orbits within the component are Poisson stable [58, p. 448]. Recall that an orbit $\gamma(t)$ is Poisson stable provided that for every $\varepsilon > 0$ there are sequences of times $t_n \uparrow \infty$ and $t_n' \downarrow -\infty$ such that the distances $d(\gamma(t_n), \gamma(0)) < \varepsilon$ and $d(\gamma(t_n'), \gamma(0)) < \varepsilon$. The phrase "almost all" in this result refers to the induced invariant measure on H=h (see [2, p. 4]). This recurrence result was generalized to the non-compact case by E. Hopf, who showed that in this case for almost all points in the space the corresponding orbit through the point was either Poisson stable or had no α and ω-limit set [58, pp. 454-459].

For Hamiltonian systems of the form (0.1), which will be our principal concern, the energy manifolds are quite easy to describe. For a solution (x(t),y(t)) of (0.3) with energy h we have from (0.1) that

$$V(x(t)) \leq h. \tag{1.2}$$

Conversely, if x, $y \varepsilon R^2$ satisfy

$$|y|^2 = 2(h - V(x)), \tag{1.3}$$

then the solution of (0.3) with initial value (x,y) will have energy h. For fixed x the set of points (x,y) where y satisfies (1.3) is a circle in the energy manifold H=h with radius $r = \sqrt{2(h-V(x))}$ that goes to zero on the boundary V=h. Thus H=h is topologically the product of the set

$$\{x \in R^2: \quad V(x) \leq h\} \tag{1.4}$$

with a circle, where the circle is identified to a point for those x for which V(x) = h. In particular, H=h is compact, or has a compact component, if and only if the set (1.4) has this property.

We refer to the set (1.4) as the <u>Hill's region</u> corresponding to energy h. It is simply the x-plane projection of the manifold H=h. When there are no singularities the level curve V=h is called the <u>zero velocity</u> curve since y=0 there by (1.3).

The topological type of the energy manifolds H=h can change only at critical points of H [51, Theorem 3.1, p. 12], and thus the obvious place to begin studying the flow generated by (1.1) is at such points. Notice from (0.3) that, in the special case of (0.1), these occur only at pairs (a,0), where $V_x(a) = 0$, and thus we are reduced to studying the critical points of V. These can be located quickly after sketching the level curves of this potential.

One of the earliest stability theorems in the qualitative theory of differential equations, generally attributed to Lagrange, concerns the behavior of the flow generated by a Hamiltonian system in a neighborhood of a critical point which we may take to be the origin. The result is simply that if the quadratic terms H^2 of the Hamiltonian H are definite, then the origin is "stable", in the sense that solutions with initial value near the origin will remain near the origin for all time [54, p. 35]. In the case that the Hamiltonian is of the form (0.1), then the definiteness of H^2 is necessary and sufficient. In the case that H^2 is not definite, a sufficient condition for stability is formulated in a result of Arnold [54, p. 36].

Many results about flows in a neighborhood of a critical point P are formulated in terms of the eigenvalues of P which we now define. Let H_{**} denote the Hessian matrix of second partials of H, and let $J = \begin{pmatrix} 0 & I \\ -I & 0 \end{pmatrix}$ where I is the identity matrix. The eigenvalues of P are then defined to be the eigenvalues of the matrix $J\, H_{**}(P)$, and can be shown to be independent of coordinates.

Recall that in the case of (0.1) the critical points are at the points P=(a,0) where $V_x(a) = 0$. Here we have

$$\det(JH_{**}(P) - \lambda I) = \lambda^4 + \lambda^2 \text{trace } V_{xx}(a) + \det V_{xx}(a) = 0, \tag{1.5}$$

where $V_{xx}(a)$ denotes the matrix $(\partial^2 V/\partial x_i \partial x_j)(a)$, and thus these eigenvalues can be computed directly from V. In fact one sees that if μ_1, μ_2 are the eigenvalues of $V_{xx}(a)$, then the eigenvalues of $P=(a,0)$ are

$$\pm\sqrt{-\mu_1}\ , \qquad \pm\sqrt{-\mu_2}\ . \tag{1.6}$$

A knowledge of the eigenvalues of a critical point can sometimes give the existence of nearby periodic orbits. A classical theorem of Liapunov, which places restrictions on the purely imaginary eigenvalues, was for a long time the most quoted [54, p. 6]. A recent result of A. Weinstein, however, drops these restrictions (see [69, 70] or [57]).

Theorem: (Weinstein) If the Hamiltonian $H(x,y)$ is of class C^2 near $(x,y) = (0,0)$, where $x,y\varepsilon R^n$, and the Hessian matrix $H_{**}(0,0)$ is positive definite, then for ε sufficiently small any energy surface $H(x,y) = H(0,0) + \varepsilon^2$ contains at least n periodic orbits of the corresponding Hamiltonian equations $\dot{x} = H_y$, $\dot{y} = -H_x$ whose periods are close to those of the linear system $\dot{z} = JH_{**}(0,0)z$.

We refer the reader to [57] for a discussion of and references to related and more global results. We do not pursue the matter here, as the symmetries in the Hénon-Heiles potential allow a direct construction of many more periodic orbits than the two guaranteed by Weinstein's results (see Section 3). When the quadratic terms H^2 are indefinite, we will often be able to apply [53] to give detailed analytical information about the periodic orbit generated from the critical point (see Section 3).

We mention a standard trick, known as stretching variables, by which information about a flow near a critical point can often be obtained. Here for simplicity we assume H in (1.1) is analytic, we choose $0<\varepsilon<1$, and make the non-canonical substitution

$$x = \varepsilon\tilde{x}\ , \qquad y = \varepsilon\tilde{y}\ , \tag{1.7}$$

into the equations (1.1). The result is a new Hamiltonian system, with Hamiltonian

$$\tilde{H}(\tilde{x},\tilde{y}) = \varepsilon^{-2}H(\varepsilon\tilde{x},\varepsilon\tilde{y}). \tag{1.8}$$

Moreover, the flows generated by H and $\tilde{H}$ on

$$H = \varepsilon^2 h , \qquad \tilde{H} = h \tag{1.9}$$

respectively, are conjugate via (1.7). If we assume that n=2 and H has the form

$$H(x,y) = (\alpha_1/2)(x_1{}^2+y_1{}^2)+(\alpha_2/2)(x_2{}^2+y_2{}^2)+\mathcal{O}_3 \tag{1.10}$$

where $\mathcal{O}_3$ denotes terms of order ≥ 3, then

$$\tilde{H}(\tilde{x},\tilde{y}) = (\alpha_1/2)(\tilde{x}_1{}^2+\tilde{y}_1{}^2)+(\alpha_2/2)(\tilde{x}_2{}^2+\tilde{y}_2{}^2)+\mathcal{O}(\varepsilon). \tag{1.11}$$

Thus at energy h>0, the flow governed by (1.11) can be considered a perturbation of that governed by

$$\tilde{H}_0(\tilde{x},\tilde{y}) = (\alpha_1/2)(\tilde{x}_1{}^2+\tilde{y}_1{}^2)+(\alpha_2/2)(\tilde{x}_2{}^2+\tilde{y}_2{}^2). \tag{1.12}$$

Using this fact, information about (1.11) can often be obtained. In view of (1.9), this gives information about (1.1) at energies $\varepsilon^2 h$, i.e. small positive energies. This trick is an integral part of many existence results for periodic orbits at low energies, including Weinstein's theorem (see [57]). We mention that the ratio (α_1/α_2) being rational or irrational often plays a crucial role (though not in Weinstein's paper); in the first case one speaks of resonance, and in the second of non-resonance.

After critical points, the most important orbits are the periodic ones, and the nature of these is generally determined by the associated Poincaré mapping which, on energy manifolds, can be described as follows for the case n=2 in (1.1) (for a general treatment see [37, pp.

251-253]). If $(x(t),y(t))$ is a periodic solution of energy h of (1.1), then, within H=h, place a disk D transversal to $(x(t),y(t))$ at t=0, and place coordinates on D so that $(x(0),y(0))$ lies on D at the origin. Next, map a neighborhood of the origin in D back into D by following a point of D along the solution through it until it next intersects D. Notice that the origin is fixed under this mapping, which takes the following form:

$$\zeta^{(1)} = A\zeta + O_2(\zeta), \quad \zeta = (\zeta_1,\zeta_2) \in \mathbb{R}^2. \tag{1.13}$$

Because of the Hamiltonian character of the equations, the 2×2 matrix A must be symplectic. Thus the two eigenvalues λ_1 and λ_2 of A, called the <u>characteristic values</u> of $(x(t),y(t))$, satisfy $\lambda_1 \cdot \lambda_2 = +1$ and hence are either [2, pp. 215-218] (a) unequal real reciprocals; (b) both +1 or both -1; or (c) unequal complex conjugates (of unit modulus). If (a) holds, the periodic orbit is called <u>hyperbolic</u>, and two manifolds will intersect along $(x(t),y(t))$, one containing all orbits positively asymptotic to this orbit (the "stable" manifold), and the other containing all orbits negatively asymptotic (the "unstable" manifold). If (b) holds, the periodic orbit is called <u>parabolic</u>, and without further information nothing can be said about its stability or instability. If (c) holds, the periodic orbit is <u>elliptic</u>. Under certain additional facts one can guarantee that such an orbit is encased in invariant tori within H=h that shrink down upon it, thus guaranteeing orbital stability. First one needs to assume that the eigenvalues $\bar{\lambda}_2 = \lambda_1 = e^{2\pi i\Theta}$ of the matrix A in (1.13) have Θ "badly approximable by rationals" (such λ_1 form a set of full measure 2π on the unit circle). Then (1.13) can be transformed by a <u>formal</u> power series $C = C(\zeta_1,\zeta_2)$ into Birkhoff normal form [68, Section 23]. If this normal form is <u>linear</u>, then C is a convergent power series by a result of Rüssmann [67]. In this case (1.13) is conjugate to a rotation. Hence the origin is a <u>stable</u> elliptic fixed point for the mapping (1.13), and the original periodic orbit is encased in invariant tori that nest to it. If, however, the Birkhoff normal form is <u>not</u> linear, then using the first non-vanishing term in [55, Theorem 2.13, p. 56], we obtain the same result on stability. In either case the set of invariant tori about the periodic orbit will have nonzero measure.

Following periodic orbits, the next most interesting class is that of quasi-periodic orbits. These comprise the flow on the invariant tori that encase an elliptic stable periodic orbit. Their existence is shown by the techniques of Kolmogorov-Arnold and Moser [68, Section 36].

Any C^1 function G: $U \to R$ is called an <u>integral</u> of (1.1) if for any solution (x(t),y(t)) of (1.1) we have $\frac{d}{dt}G(x(t),y(t)) \equiv 0$. (1.1) is called "integrable" if n such "independent" G exist [2, pp. 210-214]. Accordingly, a Hamiltonian $H(x,y,\varepsilon) = H_0(x,y) + \varepsilon H_1(x,y) + \mathcal{O}(\varepsilon^2)$, where ε is a small parameter, is referred to as "near-integrable" if the equations (1.1) corresponding to $H_0(x,y)$ are integrable. Often the technique of stretching variables can put one into a near integrable context, at least at low energies. It is here that many current applications of the Kolmogorov-Arnold-Moser results occur. We will give in this paper some cases where these results can be applied in situations that are <u>far</u> from integrable (for example see [15,61]) by choosing energy levels at which one has the right eigenvalue conditions to apply the above discussion on the Birkhoff normal form associated to (1.13).

Following the work of Robinson [62,63] and Markus and Meyer [50], what was suspected by Poincaré is now confirmed: generically Hamiltonian systems are not integrable. However, since many Hamiltonian systems are of practical importance, it is of interest to show that a specific Hamiltonian cannot be integrable (or is integrable). Here the method for non-integrability consists in showing that the Smale Horseshoe mapping can be imbedded, usually by analyzing a "nondegenerate" homoclinic orbit that is doubly asymptotic to a hyperbolic periodic orbit. By a theorem in [55, pp. 188-189], when this can be done, at least in the case n=2, the system will have no second analytic integral. We will indicate a geometric method for proving the shift can be imbedded into many specific systems of the form (0.1), and hence show the non-existence of second analytic integrals for these examples. In this regard we would like to also point out the work of R. Devaney on imbedding the shift into Hamiltonian systems [21-27, 72].

2. The Hénon-Heiles Hamiltonian.

For our purposes this will be the Hamiltonian

$$H(x,y) = (1/2)(y_1^2+y_2^2)+(1/2)(x_1^2+x_2^2)+(1/3)x_1^3-x_1x_2^2 \tag{2.1}$$

where $x = (x_1,x_2)$, $y = (y_1,y_2)$. In their paper [40], Hénon and Heiles considered

$$H(q,p) = (1/2)(p_1^2+p_2^2)+(1/2)(q_1^2+q_2^2)+q_1^2q_2-(1/3)q_2^3 \tag{2.2}$$

but (2.1) and (2.2) are equivalent via the canonical transformation

$$(q_1,q_2,p_1,p_2) \leftrightarrow -(x_2,x_1,y_2,y_1). \tag{2.3}$$

Notice that (2.1) is in the form of (0.1) with the potential given by

$$V(x) = (1/2)(x_1^2+x_2^2)+(1/3)x_1^3-x_1x_2^2. \tag{2.4}$$

In the literature, (2.1) and (2.2) are sometimes encountered with the (1/3) replaced by a parameter λ, in which case the name "Hénon-Heiles Hamiltonian" may or may not be used [8,43,61]. Also, as in [13,65], one may encounter (2.1) with a parameter $\varepsilon>0$ multiplying both cubic terms. By the technique of stretching variables, this is equivalent to (2.1) at energies $\varepsilon^2 h$ (see (1.9)).

The equations associated with (2.1) are (see (0.3))

$$\begin{cases} \dot{x}_1 = y_1, & \dot{y}_1 = -x_1 - x_1^2 + x_2^2, \\ \dot{x}_2 = y_2, & \dot{y}_2 = -x_2 + 2x_1x_2, \end{cases} \tag{2.5}$$

or, equivalently (see (0.2)),

$$\begin{cases} \ddot{x}_1 = -x_1 - x_1^2 + x_2^2 , \\ \ddot{x}_2 = -x_2 + 2x_1x_2 . \end{cases} \tag{2.6}$$

If we set $z = x_1 + ix_2$, then the potential becomes

$$V(z,\bar{z}) = (1/2)|z|^2 + (1/3)\mathrm{Re}(z^3), \tag{2.7}$$

and (2.6) is replaced by

$$\ddot{z} = -z - (\bar{z})^2. \tag{2.8}$$

The level curve $V(x) = 1/6$ consists of the three straight lines $x_1 = 1/2$ and $x_1 + 1 \pm 3^{\frac{1}{2}}x_2 = 0$. The level curves of V at other energies are sketched in Figure 1, from which certain symmetries become evident.

Indeed, from (2.7) we see that V is invariant under a rotation by $\omega = e^{(2\pi i/3)}$.

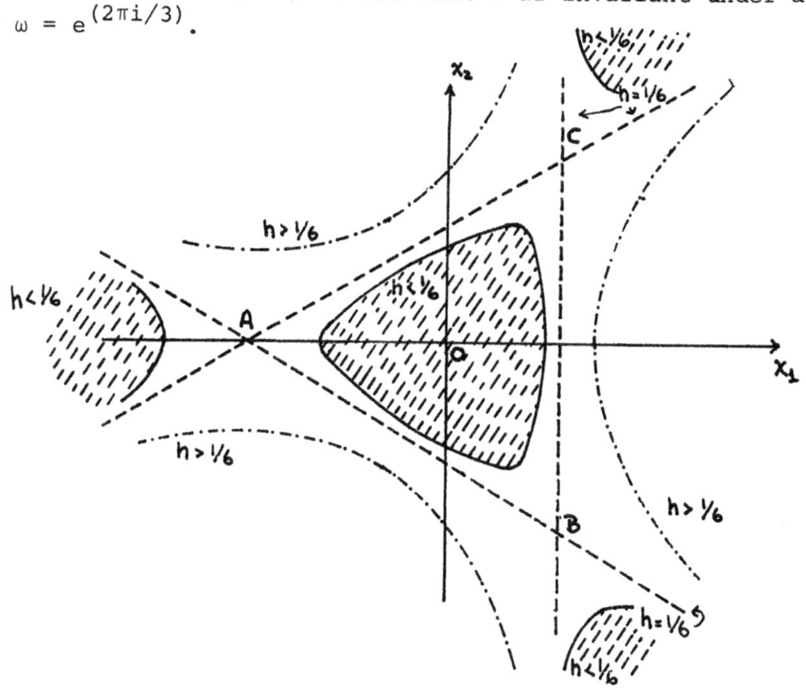

Figure 1.

Thus if $z(t)$ is a solution of (2.8), then so are $\omega z(t)$ and $\omega^2 z(t)$. Also V is symmetric with respect to the x_1-axis as $V(x_1,x_2) = V(x_1,-x_2)$. The differential equations in the form (0.2) show that if $x(t)$ is a solution, then so is $x(-t)$. This implies that if $(x(t),y(t))$ is a solution of (2.5), then so is $(x(-t),-y(-t))$, hence the system is "reversible".

The four critical points of V are easily found to be the origin, $A = (-1,0)$, $B = \omega A$, and $C = \omega^2 A$ (where we let the complex number ω act on real vectors by rotating them counter-clockwise through an angle $(2\pi/3)$, see Figure 1). The origin lies on H=0 and has repeated eigenvalues $\pm i$. The quadratic terms of the Hamiltonian at the origin are positive definite, and hence the origin is stable in the sense of Lagrange as outlined in Section 1. The other three critical points lie on $H=(1/6)$ and have eigenvalues ± 1 and $\pm i(3)^{\frac{1}{2}}$. The quadratic terms of the Hamiltonian expanded about these three critical points give an indefinite quadratic form, but a theorem of Moser [53] applies at these points. We will discuss this further in the next section.

The topological type of the energy manifolds H=h changes only at these two values of h. In fact, from our discussion of Hill's regions and their relation to energy manifolds in Section 1, we see from Figure 1 that for $h<0$ the manifold H=h has three unbounded components projecting onto the outer shaded regions of Figure 1. As h increases above zero, a new compact component is added, corresponding to the bounded shaded region of Figure 1; it was within this region that the work of Hénon and Heiles [40] was carried out. Finally, as h increases above (1/6), these four components coalesce, resulting in an unbounded, connected energy manifold with three symmetrically placed "exit legs" from the central region about the origin to infinity.

The authors' study of (2.1) emerged from the fact that at energies $h>(1/6)$ many of the phenomena in the Hénon-Heiles Hamiltonian are modeled by (0.1) with the simpler potential (see [64])

$$W(x) = (1/3)x_1^3 - x_1 x_2^2 . \qquad (2.9)$$

3. Periodic Orbits in the Hénon-Heiles Hamiltonian.

This section presents a geometric construction of a number of periodic orbits that arise out of the critical points in the Hénon-Heiles Hamiltonian.

(a) Periodic orbits arising out of the origin $(0<h<(1/6))$:

At the origin the Hessian matrix $H_{**} = I$; thus Weinstein's Theorem, discussed in Section 1, shows that this system has at least two periodic orbits at each positive energy level h close enough to zero. However, for each energy h in the range $0<h<(1/6)$, we can actually construct eight natural periodic solutions of (2.5) that arise out of the origin. We remark that to construct such a periodic orbit $(x(t),y(t))$, where $\dot{x}(t) = y(t)$ and $\ddot{x}(t) = \dot{y}(t) = -V_x(x(t))$, it suffices to construct $x(t)$. This allows us to work solely with the x-plane projection of H=h and exploit the symmetries of the potential V.

The x_1-axis is a gradient line of the potential. Hence there is a periodic solution $(\Pi_1,\dot{\Pi}_1)$ of (2.5) with energy h whose x-plane projection Π_1 lies on the x_1-axis with endpoints on the level curve $V(x) = h$. By symmetry, there are two other gradient lines of the potential that pass through the origin. They can be obtained by rotating the x_1-axis through the angles $\pm(2\pi/3)$. There are corresponding periodic orbits $(\Pi_2,\dot{\Pi}_2)$ and $(\Pi_3,\dot{\Pi}_3)$ whose x-plane projections lie along these gradient lines.

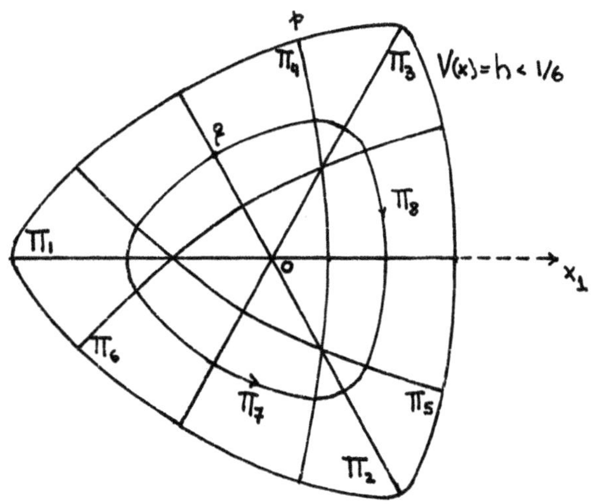

Figure 2.

Recall that orbits with energy h that start on the level curve $V(x)=h$ have velocity $y=0$ there. Such orbits that start on that portion of $V(x)=h$ with $x_2>0$ must fall to the x_1-axis due to the values of the acceleration field $-V_x$ *above* the x_1-axis. For these orbits we can define the intersection angle with the x_1-axis to be the angle measured counterclockwise from the x_1-axis to the tangent line to the orbit at the point of *first* intersection of the orbit with the x_1-axis after leaving that portion of $V(x)=h$ with $x_2>0$.

The gradient line orbits Π_2 and Π_3 have obtuse and acute intersection angles respectively with the x_1-axis (see Figure 2). By continuity in initial conditions, there is a point p on the arc of $V(x)=h$ between the intersections of Π_2 and Π_3 with $V(x)=h$, $x_2>0$, such that the orbit Π_4 with energy h starting from p intersects the x_1-axis perpendicularly. By symmetry about the x_1-axis, Π_4 is a periodic orbit with two endpoints on $V(x)=h$ (see Figure 2). Rotating Π_4 through angles $\pm(2\pi/3)$ gives two more periodic orbits Π_5 and Π_6 (see Figure 2). The fact that Π_4 when followed with $\dot{x}_2>0$ (rising orbit) has positive curvature (except at its endpoints on $V=h$) in the x-plane, and intersects the x_1-axis to the right of the origin as depicted in Figure 2, will be shown in Section 10. The techniques of such a proof involve concepts that have had a wide application to Hamiltonians of the form (0.1) (see [11] and [65]). The corresponding curvature properties and placement of Π_5 and Π_6 in Figure 2 then follow by symmetry.

We next construct two periodic orbits whose x-plane projection bounds a disc containing the origin. Consider the class of orbits $x(t)$ with energy h that intersect the gradient line labelled by Π_2 in Figure 2 perpendicularly with $x_2>0$, $\dot{x}_1 = y_1<0$. Such orbits fall to the negative x_1-axis by the values of the acceleration field $-V_x$ in the region to the left of this gradient line and above the x_1-axis. If the initial point on Π_2 with $x_2>0$ is close to the origin, such an orbit has an acute intersection angle with the x_1-axis. If this initial point with $x_2>0$ is close to $V(x)=h$, such an orbit has an obtuse intersection angle with the x_1-axis as does Π_2 itself. Again, by continuity in initial conditions there is a point q on Π_2 in Figure 2 between the origin and $V(x)=h$ $(x_2(q)>0)$, such that the orbit Π_7 with energy h starting at q, with velocity vector perpendicular to this gradient line, intersects the negative x_1-axis perpendicularly. By symmetry in the x_1-axis, Π_7 can be continued to intersect perpendicularly the gradient line on which Π_3 lies. Now, using the invariance of the potential under rotation of points x through an angle $(2\pi/3)$ or $(4\pi/3)$, we see that Π_7 can be continued as a periodic orbit back to q. Note that the x-plane projection of $(\Pi_7,\dot{\Pi}_7)$ goes counterclockwise about the origin as time increases. By the reversibility of the differential equations $\ddot{x} = -V_x$, there is a periodic orbit $(\Pi_8(t),\dot{\Pi}_8(t)) = (\Pi_7(-t),-\dot{\Pi}_7(-t))$, also with energy h, with $\Pi_8(t)$ goint clockwise around the origin in Figure 2 as time increases. Moreover, the orbits $(\Pi_8,\dot{\Pi}_8)$ and $(\Pi_7,\dot{\Pi}_7)$ link not only each other in phase space but also the three orbits $(\Pi_i,\dot{\Pi}_i)$, i=1,2,3. One merely notes that in C^1-continuously deforming the x-plane projections of these orbits so as to separate these projections into disjoint discs in the plane, one always meets a point of intersection x where the corresponding velocity vectors $\dot{x}=y$ coincide. The same argument shows that all the periodic orbits $(\Pi_1,\dot{\Pi}_1)$ through $(\Pi_6,\dot{\Pi}_6)$ link each other in phase space. Note that we have drawn Π_7 and Π_8 in Figure 2 so as to suggest that they also link each of Π_4 through Π_6 in phase space. However, this has not yet been proven, and it is conceivable that Π_7 and Π_8 lie interior to the disc containing the origin that is bounded by arcs of Π_4 through Π_6 in Figure 2. The placement of Π_7 and Π_8 in Figure 2 is based on some computer evidence that we will discuss in Section 4. The positive curvature of $\Pi_7(t)$ in Figure 2 as time increases follows readily from results in Section 10.

The periodic orbits Π_1 through Π_8 exist at each energy level h in the range $0<h<(1/6)$. As $h\downarrow 0$ all these periodic orbits collapse to the origin. As $h\uparrow(1/6)$ the orbit Π_1 limits to a homoclinic orbit asymptotic to the critical point (equilibrium point) A = (-1,0) (see Figure 1). By symmetry Π_2 and Π_3 have similar properties as $h\uparrow(1/6)$. By some additional argumentation one can show that Π_4 through Π_6 exist at $h=(1/6)$ and can even be extended through a small energy range above (1/6). Whether they exist at all energies $h>0$ is an open question. By contrast, Π_7 and Π_8 do exist at all energies $h>0$, and the proof for energies $h\geq(1/6)$ is the same as for the case $0<h<(1/6)$.

(b) <u>Periodic orbits arising out of the three critical points A, B, and C $((1/6)<h<\infty)$:</u>

We label the three gradient lines of the potential V as G_1, G_2, and G_3 in Figure 3. At $h=(1/6)$ we have three symmetrically placed critical points A=(-1,0), B, and C on these gradient lines. The eigenvalues of A are ± 1 and $\pm i\sqrt{3}$, and hence [53] applies to give a 1-parameter family of <u>hyperbolic</u> periodic orbits, parametrized by energy h, in some energy range $(1/6) < h < (1/6)+\delta$ for some $\delta > 0$. By symmetry similar results hold at B and C. We will directly construct these periodic orbits for all $h > (1/6)$. Their hyperbolicity in this extended energy range was shown in [65,66].

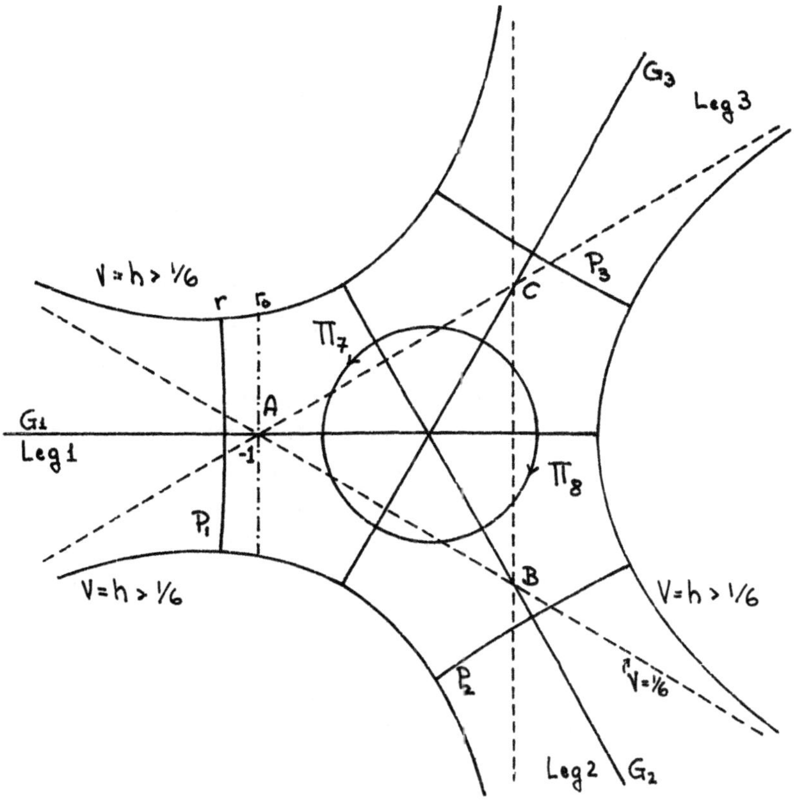

Figure 3.

At any energy $h > (1/6)$ let the line $x_1 = -1$ intersect the branch of the level curve $V(x)=h$ that lies entirely above the x_1-axis at $r_0 = r_0(h)$ (see Figure 3). By the values of the acceleration field above the x_1-axis and between the lines $x_1 = -1$ and G_2, the solution of $\ddot{x} = -V_x$ with energy h that starts with velocity 0 at r_0 falls to intersect the negative x_1-axis in an obtuse angle. Similarly, one can show that orbits with energy h that start at points on this upper branch of $V(x)=h$ sufficiently far to the left of r_0 fall to intersect the negative x_1-axis in acute angles. By continuity in initial conditions, there is an orbit P_1 with energy h starting from a point r on $V(x)=h$, with $x_1(r) < x_1(r_0)$, that falls to intersect the negative x_1-axis perpendicularly (see Figure 3). By symmetry in the x_1-axis, P_1 can be continued to a periodic orbit that has endpoints on those two branches of V=h that do not intersect the x_1-axis. By the types of arguments given in Section 10, it can be shown that P_1 has positive curvature (except at its endpoints on V=h) in the x-plane when traversed with $\dot{x}_2>0$ (rising orbit); for a proof see [11, Lemma 2.3]. Moreover, one can show by a consideration of the values of $-V_x$ in the region to the left of the line $x_1 = -1$, that P_1 lies strictly to the left of this line as depicted in Figure 3. At any given energy $h>(1/6)$, P_1 is also the only bounded orbit with energy h that remains for all time in the region to the left of G_2 and G_3 in Figure 3; for a proof see [11, Theorem 1.1]. By symmetry the periodic orbits P_2 and P_3 given in legs 2 and 3 of Figure 3 have corresponding properties.

The proof of the existence of Π_7 and Π_8 at energies $h\geq(1/6)$ is the same as that given for $0<h<(1/6)$. The placement of these periodic orbits in Figure 3 is sheer convention; no evidence is available as to whether or not they lie interior to the region containing the origin bounded by the three straight lines composing the level curve $V=(1/6)$. Thus their precise curvature properties are also in doubt in this energy range. However, one can easily show that Π_7 and Π_8 lie interior to the disc containing the origin bounded by the P_i, i=1,2,3, and the connecting arcs along the three branches of the level curve $V=h>(1/6)$ as depicted in Figure 3. This follows from the values of $-V_x$ to the left of P_1; any orbit that crosses P_1 can be shown to go to infinity in leg 1.

We conclude this section with some preliminary remarks on the stability status of these periodic orbits. Most existence theorems for periodic orbits say little about the stability status of the orbit. The theorem of Moser [53] does, however, state that for some $\delta>0$ the orbits P_i, i=1,2,3, will by hyperbolic in the energy range $(1/6)<h<(1/6)+\delta$. For the periodic orbits arising out of the origin the work of Braun [8] and Kummer [43] applies, and we will discuss this further in Section 5. For a detailed analysis of the Hénon-Heiles Hamiltonian one needs to know the stability status of such periodic orbits at energies far from the critical point energy level, and thus in general those results depending on perturbation theory will have limited value. In [65,66] it has been shown by techniques independent of perturbation theory that the P_i, i=1,2,3, are hyperbolic at all energies $h>(1/6)$. In [61] it is shown that the Π_i, i=1,2,3, go through an infinite sequence of transitions in stability type between ellipticity and hyperbolicity as $h\uparrow(1/6)$. The stability status of the other periodic orbits is unknown, but certain numerical evidence to be discussed in the next section indicates that the Π_i,

$i=4,5,6$, may by hyperbolic at "most" energies in the range $0<h<(1/6)+\alpha$ for some small $\alpha>0$, and that Π_7 and Π_8 are elliptic for positive h close to 0 and possibly hyperbolic at $h=(1/6)$.

4. Numerical Explorations in the Hénon-Heiles Hamiltonian, $0<h<(1/6)$.

Due to the values of the acceleration field $-V_x$ on both sides of the x_1-axis in Figure 2, every solution of (2.6) with energy h in the range $0<h<(1/6)$, other than $\Pi_1(t)$, must repeatedly intersect the x_1-axis with $\dot{x}_2>0$ (rising orbit) in future time. Similar remarks hold in backwards time by the reversibility of (2.6). Thus to get a global picture of the flow in phase space in this energy range, it suffices to look at the induced mapping on the section of phase space that projects to the x_1-axis in Figure 2. On this section we can use x_1 and y_1 as coordinates, and by (2.1) x_1 and y_1 are then restricted by the inequality $(1/2)(x_1^2+y_1^2)+(1/3)x_1^3 \le h$, since on the x_1-axis we have $x_2=0$ and we are considering orbits with $\dot{x}_2=y_2 \ge 0$. The boundary of this region, given by $x_2=y_2=0$, is the periodic orbit $(\Pi_1,\dot{\Pi}_1)$ which has an oval shape on this section somewhat like the level curve V=h of Figure 2. In their paper [40] Hénon and Heiles numerically integrated the differential equations (2.5) in forward time, with initial values in the interior of this oval section, and plotted the successive points of intersection of the orbit with the section. Their numerical results indicated that for positive energies h close to zero these successive points were organized on curves as if the system were "integrable." At higher energy levels the curve pattern disintegrated rather suddenly and "stochastic motion" ensued in large parts of the phase space.

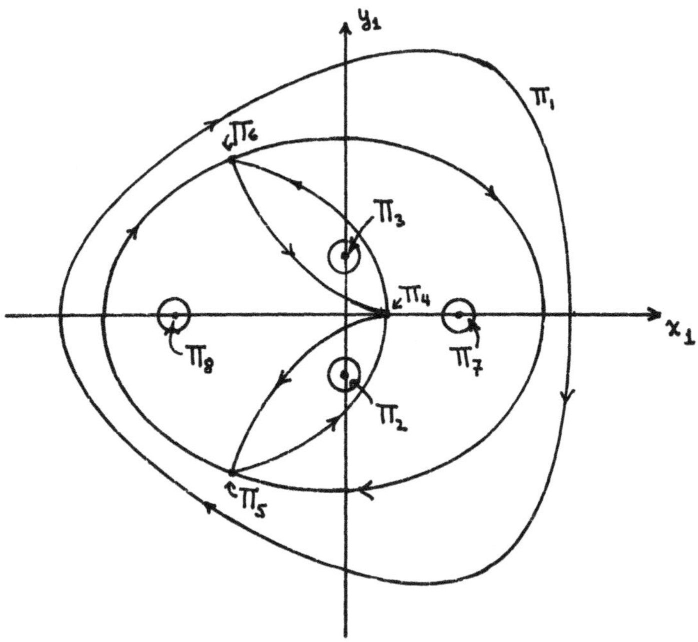

Figure 4.

In Figure 4 we have given a sketch of the results which Hénon and Heiles observed at energy h=(1/24); a similar picture was obtained at h=(1/12). Notice that in this sketch we are using the x_1-y_1 coordinates of (2.1); hence by (2.3) we have a "mirror image" through the vertical axis of the corresponding diagrams of [36,40,48] and [54, pp. 16-19].

Recall that the potential (2.4) satisfies $V(x_1,x_2) = V(x_1,-x_2)$. We thus see that if $(x_1(t),x_2(t),y_1(t),y_2(t))$ is a solution with energy h of (2.5), then so is $(x_1(-t),-x_2(-t),-y_1(-t),y_2(-t))$. The symmetry about the x_1-axis in Figure 4 is easily explained by comparing the forward intersections of this first solution, and the backward intersections of the second, with the section represented in Figure 4.

We have indicated by arrows in Figure 4 the direction of motion of points on invariant curves at low energies under iterates of the mapping as suggested by the motion of the boundary orbit $(\Pi_1,\dot{\Pi}_1)$. Also, to save notation in that figure we have labelled the periodic orbits as Π_i rather than $(\Pi_i,\dot{\Pi}_i)$. This labelling is largely conjectural, but the angles that these periodic orbits make with the x-axis in Figure 2, thereby specifying the values of y_1, together with the naturality of of the construction of these orbits in Section 3, give some justification for the picture. Accepting this labelling as a natural hypothesis, one can then immediately conjecture that at low positive energies the periodic orbits Π_i, i=1,2,3, are elliptic stable; that is, are surrounded by invariant tori that nest to the respective periodic orbits. However, this has not yet been mathematically proven, although there is some other unpublished numerical evidence for this (see below).

It also appears that Π_7 and Π_8 are elliptic stable at low positive energies. If one can identify these orbits with the ones considered by Braun [8] and Kummer [43], this would have been shown in their work for sufficiently small positive energies. Some computational evidence for this tentative identification will be given in Section 5.

One sees that the conjectural identification of the periodic orbits of Figure 2 with those obtained by Hénon and Heiles in Figure 4 is the reason for the placement of Π_7 and Π_8 relative to the Π_i, i=4,5,6, in Figure 2. From Figure 4 we also conjecture that the Π_i, i=4,5,6, are hyperbolic. This has been shown numerically at energy level samples of $(\Delta h) = 10^{-2}$ starting from 0 to slightly above (1/6) by numerically integrating the Poincaré map along the orbit after a search routine has located the orbit. However, one must be wary of sampling at such discrete energy levels even ignoring the attendant error analysis required for a proof. For example, numerical work originally "indicated" that the Π_i, i=1,2,3, were elliptic for $0<h<.13$, changed to being hyperbolic for $.13<h<.16$, and then were again elliptic for $.16<h<(1/6)$. It is now known, however, that the stability status of these periodic orbits oscillates infinitely often between ellipticity and hyperbolicity as $h\uparrow(1/6)$ (see Section 6). However, the numerical evidence presented by Hénon and Heiles and other researchers has generally proved to be a good source of interesting conjectures.

As an aside we remark that any attempt to convert numerical computations on a computer into hard mathematical proofs must use a numerical integration scheme that allows truncation error estimates. For this reason the Runge-Kutta method was not used in [13,15].

Moreover, the Taylor series method used in those papers was particularly well adapted to the polynomial potentials considered there, and the order of the series could be adjusted at will by suitable programming. This allowed for a machine calculation of error in those cases where fine estimates were required as in [15]. We have generally found that the Taylor series method of numerical integration also preserved the total energy as constant to more decimal places than the Runge-Kutta method for polynomial potentials (see the remarks in [13, Section 3]).

The onset of large scale observable stochasticity is reported by Hénon and Heiles in [40] to begin at $h\approx.11$. In their published diagram at $h=.125$ this random motion appears to be in a neighborhood of the "leaves" of the stable and unstable manifolds of the periodic orbits Π_i, i=4,5,6; here we are assuming that those orbits are hyperbolic and that the orbits and their asymptotic manifolds have the placement given in Figure 4. This suggests that these asymptotic manifolds are transverse to one another in this energy range. If such were the case, this would give the existence of non-degenerate heteroclinic orbits connecting any pair of the Π_i, i=4,5,6. In such circumstances one can embed the Smale horseshoe map and locally obtain the random motion that Hénon and Heiles observed numerically (see [55]). One possibility is that these leaves dovetail, as depicted in Figure 4 for low energies, and then become transverse at higher energies. This might seem perplexing in view of the analyticity of the differential equations. One explanation is to postulate a transition energy range in which the Π_i, i=4,5,6, become elliptic, and then resume being hyperbolic at higher energies with asymptotic manifolds then transverse. Alternately, it could be that the Π_i, i=4,5,6, are always hyperbolic with asymptotic manifolds always transverse throughout the energy range $0<h<(1/6)$, there being a fine structure of stochasticity not observable in the published diagrams at low energies. Between these extremes are many other possibilities, and numerical exploration of the stability status of these orbits as the energy changes would be useful. We note that the construction of the orbits corresponding to the Π_i, i=4,5,6, in another potential,

$$W(x) = (1/2)(x_1^2+x_2^2) - (1/2)x_1^2x_2^2 , \qquad (4.1)$$

does not lead to any new periodic orbits, but gives back some gradient line orbits that were shown in [15] to be elliptic for positive energies up to and above the escape energy. Moreover, Hénon and Heiles [40] reported this latter potential as having no "observable" stochastic properties. We will comment more on this in Section 8.

Equally interesting are the possible changes in stability status of Π_7 and Π_8. Assuming the identifications above, they appear to be elliptic at low energies (recall the remarks on Braun's and Kummer's work above, and see Section 5 below). Now, the Poincaré map along a periodic orbit is defined on any sufficiently small disc D transverse to the orbit. The intersection of the periodic orbit with D is then a fixed point of the map and thus has an index which is preserved under changes in the stability status of the orbit as the energy changes. There is some problem with the domains of the Poincaré maps, so we can accept this argument only heuristically. However, if an elliptic orbit with index +1 for the fixed point of the induced map on D became at a higher energy hyperbolic with index -1 (if hyperbolic with reflection, then the square of the Poincaré map will give index -1), one might expect to see two 1-parameter families of elliptic orbits (each with index +1) appearing near the hyperbolic orbit in order to preserve the total index. Conversely, if two additional families of periodic orbits suddenly appear near another periodic family as the energy is raised, one could conjecture that such a bifurcation had taken place and then proceed to intensively investigate the given orbits. Such heuristic reasoning can be applied in the Hénon-Heiles Hamiltonian at h=(1/6) to provide a number of conjectures. In their remarks on p. 76 of [40], Hénon and Heiles note that the periodic orbits we have denoted as Π_7 and Π_8 in Figure 4 appear to have become unstable with the appearance of two new sets of closed invariant curves on the section in phase space above the x_1-axis. One could account for this by hypothesizing the appearance of two elliptic stable orbits $(Q_i, \dot{Q}_i)$, i=1,2, that both intersect this section twice before closing. Such an orbit has the pretzel-like appearance in the x-plane given in Figure 5; Birkhoff [6, p. 23; p. 682 in reprint] had already come across such orbits. Here we have conjectured that Π_7 has "given birth" to Q_1 and

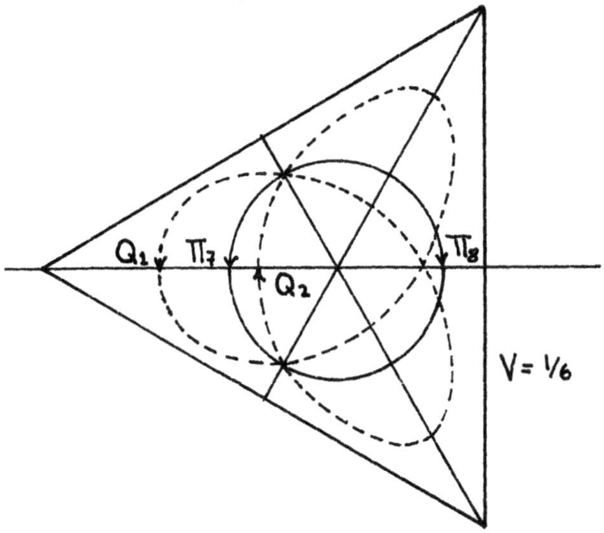

Figure 5.

Π_8 to Q_2. The reader should check the published diagram at h=(1/6) in [40] to see that the placement of the invariant curves on the section in phase space above the x_1-axis is around periodic orbits whose x-plane projection would intersect the x_1-axis in angles as shown in Figure 5. In particular each of Q_1 and Q_2 have two perpendicular intersections with the negative x_1-axis (recall, however, that one must flip Hénon and Heiles's pictures about their vertical axis to correspond with ours). Now, if Π_7 and Π_8 have changed stability status from elliptic to hyperbolic by h=(1/6), one is immediately directed to inquire about their stability status at h>(1/6).

At energies .11<h<(1/6) the large-scale observable "stochastic" behavior of the flow generated by (2.5) is suggestive of the C-systems discussed in [2], the prime example of which is the geodesic flow on a compact manifold of negative sectional curvature. But it is a classical result that the solutions of (0.1) with energy h may be regarded as reparametrized geodesics of the "Jacobi metric"

$$(ds)^2 = 2(h-V(x))((dx_1)^2 + (dx_2)^2) \tag{4.2}$$

on the interior of the corresponding Hill's region [46, p. 172]. Thus the stochastic behavior could be explained, at least on invariant sets avoiding the zero-velocity curve $V(x)=h$, if the corresponding curvature were negative. However, the curvature $K(x)$ in the metric (4.2) when $V(x)<h$ is given by

$$K(x) = (1/4)(h-V(x))^{-2}[\nabla^2 V(x)+(h-V(x))^{-1}|\nabla V(x)|^2] \tag{4.3}$$

[59, p. 279] from which it is easy to check that for V as in (2.4) we have $K(x)>0$ everywhere; this is pointed out in [1]. Of course, it is conceivable that the metric could somehow be adjusted so as to render the curvature negative, and yet still admit solutions of (2.5) as reparametrized geodesics. However, as a corollary of the Gauss-Bonnet Theorem [59, p. 388], a closed geodesic on a 2-manifold of negative curvature cannot bound a disc in the manifold. Since the central periodic orbits Π_7 and Π_8 do bound a disc interior to the Hill's region of Figure 2, one could never adjust the metric so that the curvature was everywhere negative. This argument was applied in [64] to the harmonic potential of (2.9), in which case the curvature K of (4.3) is everywhere positive, except at $x=0$ where it vanishes.

In connection with these curvature observations, we should mention results of Klingenberg [42] which "seem to indicate" that any compact manifold with geodesic flow of Anosov type "must admit a metric of negative sectional curvature." Of course, in our case the manifold $\{x: V(x)<h\}$ is not compact; hence Klingenberg's results are not directly applicable. Nevertheless, in combination with the remarks of the previous paragraph they do suggest that the stochastic behavior observed by Hénon and Heiles cannot occur in the entire energy manifold, and that "islands of stability" must exist. Indeed, this has already been pointed out by Lunsford and Ford [48], and in Section 6 we will report on a proof that these islands do exist at many energies between 0 and (1/6).

Geodesic flows on non-compact manifolds of negative sectional curvature have been studied by P. Eberlein [28,29], and since Hill's regions are in general not compact, his results may be of interest. However, as yet the authors are unaware of any utilizations of his theorems in this direction.

5. Braun's and Kummer's Results.

So far as the authors are aware, Braun in [8] was the first to attempt a rigorous mathematical explanation of the results observed numerically by Hénon and Heiles in [40]. His method was to transform the Hamiltonian into Gustavson normal form (defined below), stretch variables, (thereby considering only small positive energies), and reduce the study of the induced flows on the energy surfaces to the study of certain mappings. He showed his mappings could be regarded as perturbations of some simple maps, which admitted obvious invariant curves. By a clever introduction of action-angle variables, he was then able to use the Moser twist theorem to prove that "many" of these invariant curves must actually persist after the perturbations are made.

Although Braun established the existence of invariant curves in mappings associated with the Hénon-Heiles potential, it is unfortunately not clear that his mappings are the same as those studied by Hénon and Heiles. We shall elaborate on this point later in the section, offering suggestions as to how the two relate.

Braun's methods were reformulated by Kummer in [43], and we will give a brief sketch of Kummer's techniques in this section. First we must discuss the Gustavson normal form, a form which can be achieved for the Hénon-Heiles Hamiltonian, at least through low order terms, after a rather complicated transformation of variables [43].

Let α_1, α_2 be real numbers, and consider a Hamiltonian of two degrees of freedom

$$H(x,y) = (\alpha_1/2)(x_1^2+y_1^2)+(\alpha_2/2)(x_2^2+y_2^2)+H^{(3)}+H^{(4)}+\ldots, \tag{5.1}$$

where $H^{(j)}$ is a homogeneous polynomial of degree j in x_1, x_2, y_1, y_2. Now set

$$D = \sum_{j=1}^{2} \alpha_j \left(x_j \frac{\partial}{\partial y_j} - y_j \frac{\partial}{\partial x_j}\right). \tag{5.2}$$

We say that (5.1) is in Gustavson normal form through terms of order $k \geq 3$ if

$$DH^{(j)} = 0, \quad j=3,\ldots,k. \tag{5.3}$$

Now set

$$\begin{cases} J = M_0 = \frac{1}{4}(x_1^2+x_2^2+y_1^2+y_2^2) \\ M_1 = \frac{1}{2}(x_1y_2-x_2y_1) \\ M_2 = \frac{1}{2}(x_1x_2+y_1y_2) \\ M_3 = \frac{1}{4}(x_2^2+y_2^2-x_1^2-y_1^2). \end{cases} \tag{5.4}$$

The following is implicit in Kummer [43]; it can be proven by induction on the degree of the monomials involved.

Theorem 5.1. Suppose $\alpha_1 = \alpha_2 = 1$ in (5.1). Then H is in Gustavson normal form through terms of order $k \geq 3$ if and only if for each odd $3 \leq j \leq k$ we have $H^{(j)} \equiv 0$, and for each even $3 \leq j \leq k$ there is a homogeneous polynomial $P_{j/2}$ of degree $(j/2)$ such that

$$H^{(j)} = P_{j/2}(J,M_1,M_2,M_3). \tag{5.6}$$

Kummer considers a Hamiltonian

$$H = 2J+K_n(J,M_1,M_2,M_3)+\mathcal{O}_{2n+1}(x_1,x_2,y_1,y_2), \tag{5.7}$$

which without the $\mathcal{O}_{2n+1}$ terms is in Gustavson normal form by Theorem 5.1. First he studies the "truncated" system

$$H_0 = 2J+K_n(J,M_1,M_2,M_3) \tag{5.8}$$

in the following manner. So long as

$$J > 0 , \tag{5.9}$$

we can use the homogeneity of K_n to write (5.8) as

$$H_0 = 2J+J^n K^{(n)}(M_1/J, M_2/J, M_3/J), \tag{5.10}$$

where $K^{(n)}(x,y,z) = K_n(1,x,y,z)$, $(x^2+y^2+z^2)>0$. Now make the canonical substitution

$$\begin{cases} x_1 = (2L-(\xi^2+\eta^2))^{\frac{1}{2}} \sin\alpha \\ x_2 = \xi\cos\alpha + \eta\sin\alpha \\ y_1 = (2L-(\xi^2+\eta^2))^{\frac{1}{2}} \cos\alpha \\ y_2 = -\xi\sin\alpha + \eta\cos\alpha \end{cases} \tag{5.11}$$

$L>0$, $\xi^2+\eta^2<2L$, $\alpha\varepsilon R$, and write the resulting Hamiltonian as

$$\tilde{H}_0 = L + (\tfrac{L}{2})^n G, \tag{5.12}$$

where

$$G = K^{(n)}(M_1/J, M_2/J, M_3/J) \tag{5.13}$$

$$\begin{cases} (M_1/J) = -(\xi/L)(2L-[\xi^2+\eta^2])^{\frac{1}{2}} \\ (M_2/J) = (\eta/L)(2L-[\xi^2+\eta^2])^{\frac{1}{2}} \\ (M_3/J) = (1/L)(-L+\xi^2+\eta^2). \end{cases} \tag{5.14}$$

Notice the equations for (5.12) are

$$
\begin{cases}
\dot{\alpha} = 1+(\frac{n}{2^n})L^{n-1}G+(\frac{L}{2})^n \frac{\partial G}{\partial L} \\
\dot{L} = 0 \\
(a) \begin{cases} \dot{\eta} = (\frac{L}{2})^n \frac{\partial G}{\partial \xi} \\ \dot{\xi} = -(\frac{L}{2})^n \frac{\partial G}{\partial \eta} \end{cases}
\end{cases}
\qquad (5.15)
$$

from which we see that L is an integral. In fact L = (constant)>0 is obviously a 3-sphere, and from (5.13) and (5.14) we see the final two equations are independent of α.

We can geometrically view equations (5.15) as follows. First consider a configuration, as in Figure 6, formed by two solid cones having base radii 2L>0, placed together over the disc $\eta^2+\xi^2 \leq 2L$ of the η-ξ plane. Next, in this configuration identify (i.e. consider as the same) two points on the boundary having the same η-ξ plane projection (e.g. u and v in Figure 6). The result is topologically a 3-sphere, as is not difficult to see, and we can "coordinatize" most of this sphere as in Figure 6 by letting $(\alpha_p, \eta_p, \xi_p)$ be the coordinates of the point P shown. More specifically, given P we first construct the plane containing P and the vertical axis; then we construct Q, then we construct the line through P and Q so as to obtain α_p.

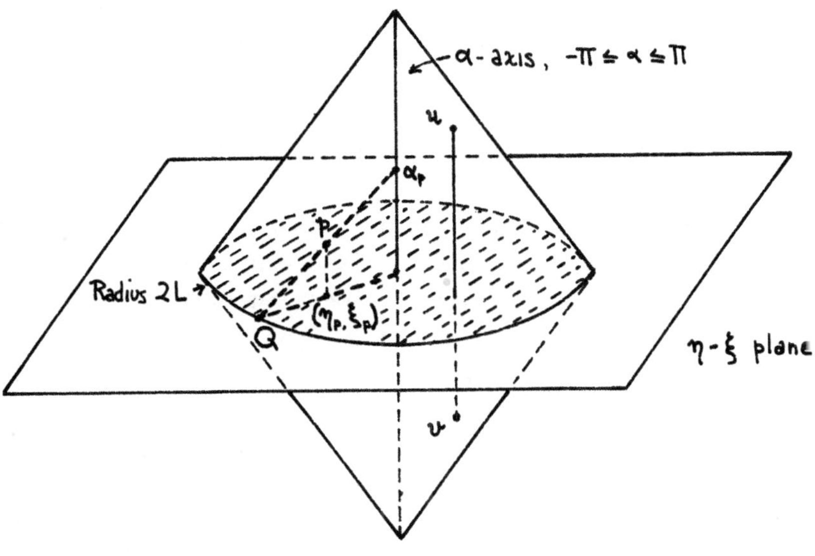

Figure 6.

For fixed $L>0$, (5.15a) defines a flow on the shaded disc of Figure 6. Moreover, if $\dot{\alpha}>0$, which Kummer eventually guarantees by stretching variables, then any critical point of the flow generated by (5.15a) must correspond to a "vertical" periodic orbit of (5.15). Intuitively, this orbit is hyperbolic or elliptic if and only if the corresponding critical point on the disc has this property.

To analyze the critical points of the disc flow, Kummer maps the disc onto S^2-{north pole} by

$$\begin{cases} x = (M_1/J) \\ y = (M_2/J) \\ z = (M_3/J) \end{cases} \tag{5.16}$$

(note $M_1^2+M_2^2+M_3^2 = J^2$, which is why the mapping is into S^2), where it can be easily verified that the equations governing the resulting flow are

$$\vec{x}' = \nabla K^{(n)} \times \vec{x}, \tag{5.17}$$

$\vec{x} = (x,y,z)$. Here the ' is with respect to a rescaling of t. But note that (5.17) will have critical points (on S^2) exactly at unit vectors e where $\nabla K^{(n)}$ is proportional to e. By rotating such e to the south pole, Kummer then shows how to determine if e is hyperbolic or elliptic.

Finally, Kummer returns to (5.7), stretches variables, and shows that at low energies the periodic orbits he located in the truncated system actually correspond to periodic orbits in the original equations, provided a certain index does not vanish. Moreover, he offers computable algorithms to determine if these orbits are hyperbolic or elliptic stable, the latter terminology meaning that the orbit is encased in invariant tori which shrink down upon it.

In the case of the Hénon-Heiles Hamiltonian, Kummer's (as well as Braun's) methods give two elliptic critical points and a line of fixed points in the associated disc flow, and thus in the full system we are guaranteed at least two elliptic stable orbits for small energies. Since the invariant tori surrounding these orbits will intersect a transversal section in invariant curves, this gives a rigorous proof at sufficiently low positive energies of the existence of invariant curves analogous to those observed numerically by Hénon and Heiles in Figure 4.

Because of the complicated changes of variables involved in going into Gustavson normal form, it is difficult to associate the elliptic stable periodic orbits found by Braun and Kummer with any that we have constructed in Figure 2. Nevertheless, we now indicate some evidence for associating the two stable equilibrium points of the disc flow in Figure 6, obtained by truncating the Gustavson normal form of the original Hamiltonian, with the points labeled Π_7 and Π_8 in Figure 4. This comes from an examination of [8, Figure 4] coupled with Braun's remarks on p. 305 of [8] which offer an explanation of why two families of closed curves in the analogues of our Figure 4 begin to degenerate into a string of fixed points as the energy $h\downarrow 0$. This phenomenon is not reported in [40], and apparently was a private communication to Braun. However, in Gustavson's paper [36, Figure 7] one can see, in the only published diagram at really low positive energies, that this appears to be happening in the section in phase space above the x_1-axis of $H=h$ at $h=10^{-4}$. This diagram is, admittedly, labelled as a plot of the level lines of a truncated <u>formal</u> second integral for (2.5), but at low positive energies the agreement between such plots and the actual orbit plots in the analogues of our Figure 4 has been asserted to be quite good, at least at those energies sampled. It appears, then, that as $h\downarrow 0$ the asymptotic manifolds that seem to connect Π_5 and Π_6 to Π_4 in Figure 4 move towards the vertical axis. This is brought out by a rescaling of Figure 4 which Gustavson does in his Figure 7. Indeed, this would make the vertical axis look like a string of fixed points since the induced flows on these asymptotic manifolds are in opposite directions and hence would cancel in the limit as $h\downarrow 0$ under this continual rescaling of Figure 4. Such a picture with two stable fixed points surrounded by invariant curves and an axis of fixed points is exactly like that obtained by Braun and Kummer for the associated disc flow of Figure 6, suggesting that their stable fixed points correspond to the points labelled Π_7 and Π_8 in Figure 4. The above is highly conjectural, but suggests which orbits one should concentrate on obtaining when tracking through the canonical transformations involved in converting the original Hamiltonian into Gustavson normal form up through a given order. Therefore, we conjecture that the periodic orbits generated by applying [8,43] are Π_7 and Π_8. Thus Braun's and Kummer's results have direct relevance in piecing together the global picture for explaining Hénon and Heiles' results displayed schematically in Figure 4.

Kummer also applied his techniques to other examples in [44,45].

6. Stability transitions of the periodic orbits Π_i, i=1,2,3, for $0<h<(1/6)$.

In this section we will sketch the general arguments of [61], which, when applied to the periodic orbits Π_i, i=1,2,3, of the Hénon-Heiles Hamiltonian (see Figure 2), show that each Π_i goes through an infinite sequence of transitions in stability type between ellipticity and hyperbolicity as $h\uparrow(1/6)$. When the orbits are elliptic over an energy range $h_1<h<h_2$, one can further show that they are elliptic stable (i.e. surrounded by invariant tori in H=h that nest down to the respective orbit) for a set of energies of full measure in this interval.

We consider Hamiltonians of the form (0.1) where the potential V is real analytic. We assume that the corresponding differential equations (0.3) admit a 1-parameter family of periodic orbits $\Pi_h^*=(\Pi_h,\dot{\Pi}_h)$, parametrized by energy h, satisfying the following two conditions:

(6.1a) Π_h^* limits as $h\uparrow h_e$ = (escape energy) to a homoclinic orbit, asymptotic as the time $t\to\pm\infty$ to a critical point of the Hamiltonian, with eigenvalues $\pm\alpha$, $\pm i\beta$, α and β positive.

(6.1b) All the Π_h lie along a fixed gradient line of the potential V.

In [61] there are extensions of the theory to such cases as: (1) Π_h^* limits to a pair of heteroclinic orbits connecting two critical points of the Hamiltonian, and: (2) $h\downarrow h_e$ and Π_h^* limits to a pair of homoclinic orbits both asymptotic to the same critical point (double bowl potentials). In all such cases, however, condition (6.1b) is retained.

Under suitable relabeling by a canonical transformation, we can assume that the gradient line of (6.1b) is the x_1-axis. Letting $V_i = (\partial V/\partial x_i)$ and $V_{ij} = (\partial^2 V/\partial x_i \partial x_j)$, this implies that $V_2(x_1,0) \equiv 0$, and hence that $V_{12}(x_1,0) \equiv 0$. The equation for the linearized Poincaré map Z(t) along $\Pi_h^*(t)$ is $\dot{Z} = JH_{**}(\Pi_h^*(t))Z$, with $Z(0) = I$. Since Π_h lies along the x_1-axis, the differential equations of $Z(t) = (z_{ij}(t))$ reduce to the uncoupled Hill's equations (j=1,2,3,4):

$$\ddot{z}_{1j} + V_{11}(\Pi_h(t))z_{1j} = 0, \tag{6.2}$$

$$\ddot{z}_{2j} + V_{22}(\Pi_h(t))z_{2j} = 0, \tag{6.3}$$

with $z_{3j} = \dot{z}_{1j}$ and $z_{4j} = \dot{z}_{2j}$. The initial conditions $Z(0) = I$ become $z_{11}(0) = z_{22}(0) = \dot{z}_{13}(0) = \dot{z}_{24}(0) = 1$ and $\dot{z}_{11}(0) = \dot{z}_{22}(0) = z_{13}(0) = z_{24}(0) = 0$ (these give the normalized solutions of (6.2) and (6.3)). Let Π_h^* have period T_h, where $T_h \to \infty$ as $h \uparrow h_e$. Since we have a Hamiltonian system of differential equations, $\det(Z(T_h) - \lambda I) = 0$ has a double root +1 and $\det(Z(T_h)) = 1$. The remaining two eigenvalues of $Z(T_h)$, denoted by μ_1 and μ_2, then satisfy $\mu_1 \cdot \mu_2 = +1$ with $\mu_{1,2} = (1/2)[(D-2) \pm (D(D-4))^{\frac{1}{2}}]$ and $D = \text{Trace } Z(T_h)$. We thus see that Π_h^* is elliptic if $0<D<4$ and hyperbolic if $D<0$ or $D>4$.

Let $E_h = \dot{\Pi}_h^*(0)$ be the tangent to the orbit at $t=0$, and let N_h be the normal to $H=h$ at this point. Then it is known that $Z(T_h)E_h = E_h$ and $Z^*(T_h)N_h = N_h$, where the * denotes the transpose of the matrix. These two relations, when written out in coordinates, show that $[z_{11}(T_h) + \dot{z}_{13}(T_h)] = 2$, where z_{11} and $\dot{z}_{13} = z_{33}$ are on the diagonal of Z. Thus the conditions on D above can be rephrased in terms of the quantity $d = (D-2) = [z_{22}(T_h) + \dot{z}_{24}(T_h)]$, where $\dot{z}_{24} = z_{44}$. We conclude that Π_h^* is elliptic when $|d|<2$ and hyperbolic when $|d|>2$. The stability status of Π_h^* is now controlled solely by the single Hill's equation (6.3) and its normalized solutions which determine d. Henceforth (6.2) will be irrelevant to our discussion.

Let Δ_H = the Hill discriminant of (6.3) (see [49, Chapter 2] for a definition). If $V_{22}(\Pi_h(t))$ has the same minimal period as $\Pi_h^*(t)$, then $d = \Delta_H$. Further reductions can be made if we assume $V_{22}(\Pi_h(t))$ is a symmetric function of time. By using certain identities in [49, Chapter 1], one can obtain:

(c1) If $V_{22}(\Pi_h(t))$ has minimal period T_h, then $d = \Delta_H = 2z_{22}(T_h)$ and Π_h^* is elliptic when $|z_{22}(T_h)|<1$, and hyperbolic when $|z_{22}(T_h)|>1$.

(c2) If $V_{22}(\Pi_h(t))$ has minimal period $(T_h/2)$, one can show that $d = (\Delta_H^2 - 2)$, where $\Delta_H = 2z_{22}(T_h/2)$. Then Π_h^* is elliptic if and only if $0<|\Delta_H|<2$. Note that we always have $d \geq -2$ for this case, hence Π_h^* can never pass from being elliptic to hyperbolic through the eigenvalues $\mu_1 = -1 = \mu_2$ as h changes.

The periodic orbits $\Pi_i^* = (\Pi_i, \dot{\Pi}_i)$, i=1,2,3, in the Hénon-Heiles Hamiltonian satisfy (c1), and the transitions between ellipticity and hyperbolicity will take place at both the eigenvalues $\mu_1 = -1 = \mu_2$ and $\mu_1 = 1 = \mu_2$ as $h \uparrow (1/6)$. The other potential (4.1) considered by Hénon and Heiles will be discussed in Section 8. It has a pair of periodic orbits satisfying (c2) that in the x-plane lie along two diagonal gradient lines of the potential (one has to rotate the diagonals to the axes to apply the discussion of the beginning of this section). These periodic orbits also go through an infinite sequence of transitions in stability status as $h \uparrow h_e$, but the transitions take place only at $\mu_1 = 1 = \mu_2$. Another "degenerate" example is given in [61]: the eigenvalues μ_1 and μ_2 of the Poincaré map along Π_h^* wind around the unit circle in the complex plane, repeatedly taking the values ± 1 as $h \uparrow h_e$, while Π_h^* never becomes hyperbolic. To avoid such degenerate cases, one must show that an infinite number of the instability intervals of the Hill's equation (6.3) do not collapse (this is called the "coexistence problem"; see [49, Chapter 7] for details). We now sketch this part of the problem.

For the examples in [61], the Hill's equation (6.3) is, on letting $z = z_{2j}$,

$$\ddot{z} + (a + bk^2 \mathrm{sn}^2(t;k^2))z = 0, \tag{6.4}$$

where the <u>positive</u> parameters a, b, and k are functionally dependent on the energy h with $k(h) \uparrow 1$ as $h \uparrow h_e$. $\mathrm{sn}(t;k^2)$ is the Jacobi elliptic sine with modulus k. In the examples, the energy h ranges over an interval $(0,h_e)$ where, as $h \downarrow 0$, the periodic orbit Π_h^* collapses to a critical point, usually the origin, of the Hamiltonian.

Transforming (6.4) to Ince's equation (see [49, Chapter 7] for details), one can check that if the parameter $b=b(h)$ is positive, then (6.4) has for each fixed h no collapsing interval of instability. This is in contrast with the classical Lamé equation where $b = -n(n+1)$ (with n a positive integer) and all but $(n+1)$ intervals of instability collapse.

Recall that the 2-parameter Mathieu equation has <u>stability curves</u>, on which the Hill discriminant has value ± 2, that divide the plane into stable and unstable regions. Similarly, for (6.4) we have <u>stability surfaces</u> in (a,b,k)-space on which the Hill discriminant $\Delta_H = \pm 2$. These surfaces are the boundaries between the stable ($|\Delta_H| < 2$) and unstable ($|\Delta_H| > 2$) regions for (6.4). To conclude that (6.4) experiences infinitely many transitions between stability and instability we need only show that the arc $\gamma(h) = (a(h), b(h), k(h))$, $0 < h < h_e$, cuts through infinitely many of these stability surfaces as $h \uparrow h_e$. These stability transitions for (6.4) then induce stability transitions in the original periodic orbits Π_h^* as $h \uparrow h_e$. For example, if $V_{22}(\Pi_h(t))$ has minimal period T_h, then $d = \Delta_H$ and Π_h^* is elliptic when (6.4) is in a stable region, and hyperbolic when (6.4) is in an unstable region. Alternately, if $V_{22}(\Pi_h(t))$ is as in (c2), then Π_h^* is elliptic when (6.4) is in a stable region except when $\Delta_H = 0$, and hyperbolic otherwise. The reader is referred to [15] for a related argument dealing with the 2-parameter Mathieu equation in the context of the potential (4.1). Other applications are given in [60].

We now discuss the problem of showing that the parameter arc $\gamma(h)$ traverses, as $h \uparrow h_e$, an infinite number of the stability surfaces. By the "Oscillation Theorem" [49, Theorem 2.1], the arc $\gamma(h)$ is on such a stability surface if and only if, at these values of the parameters, (6.4) has a real periodic solution of the same period as $sn^2(t;k^2)$ or twice the period (recall that we have $k(h) \uparrow 1$, and hence the period goes to infinity as $h \uparrow h_e$). But the number of zeros such periodic solutions can have in one period is constant along any given stability surface by a result of Haupt [49, Theorem 2.14], with the "higher order" stability surfaces allowing more zeros. Thus we can gain our result if we can show that (6.4) has, as $h \uparrow h_e$, a sequence of periodic solutions with periods as above, whose number of zeros in one period tends to infinity. It is here that we finally make use of hypothesis (6.1a). The eigenvalue condition on the critical point in (6.1a) allows us to use the Moser coordinates of [53] in a phase space neighborhood of the critical point. The particular form that we require has been detailed in [16]. In these coordinates one can show that there is a twisting which increases to infinity as $h \uparrow h_e$. This forces the number of zeros of the appropriate periodic solutions of (6.4) to go to infinity as $h \uparrow h_e$. Thus the arc $\gamma(h)$ crosses an infinity of stability and non-trivial instability regions as $h \uparrow h_e$. This induces an infinite sequence of transitions between ellipticity and hyperbolicity for the Π_i^*, i=1,2,3, as $h \uparrow (1/6)$ in the Hénon-Heiles Hamiltonian.

During those energy intervals where the Π_i^*, i=1,2,3, are elliptic, the analyticity of the system implies that the eigenvalues μ_1 and μ_2 of the Poincaré map along Π_i^* move around the unit circle as the energy changes so as never to be constant over any non-trivial energy interval. Thus, in any such energy interval there is a set of energy values of full measure at which the eigenvalue conditions of the Moser-Rüssmann criterion for elliptic stability detailed in Section 1 are met. This gives the set of invariant tori in H=h which nest to the Π_i^*, i=1,2,3, at such energies. It would seem reasonable that the invariant tori exist over an entire such energy interval, but unfortunately this cannot be proven at present. The existence of these invariant tori at a number of energies in the interval $0 < h < (1/6)$

confirms in part some conjectures of Lunsford and Ford in [48] on the Hénon-Heiles Hamiltonian.

The above stability oscillation results parallel the results of Devaney [23,72] and Henrard [41]. They consider the Hamiltonian flow in a neighborhood of a critical point P at energy h_e with eigenvalues $(\pm\alpha\pm i\beta)$, with α and β positive. Again one can use Moser coordinates [53] in a neighborhood of P. If the flow admits a non-degenerate homoclinic orbit asymptotic as time $t\to\pm\infty$ to P, then one can prove the existence of a 1-parameter family of periodic orbits Π_h^* satisfying the analogue of (6.1a) such that Π_h^* goes through an infinite sequence of oscillations in stability type between ellipticity and hyperbolicity as $h\uparrow h_e$. In our case the critical point P satisfies a different eigenvalue condition, and our homoclinic orbit is not non-degenerate. In place of these we have to assume the existence of the periodic orbits Π_h^*. Also, in the present situation the condition (6.1b) enters in a technical way when computing the Poincaré map in the Moser coordinates about P. Moreover, the actual transition to hyperbolicity assumes that the Hill's equation (6.3) one obtains from the Poincaré map does not have collapsing of intervals of instability. That this hypothesis is not trivial is shown by a counterexample in [61].

When the critical point P has eigenvalues $(\pm\alpha\pm i\beta)$, $\alpha,\beta>0$, and possesses a non-degenerate homoclinic orbit, then Devaney shows that the Poincaré map eigenvalues have lim sup = $+\infty$ and lim inf = $-\infty$ (see [23, Theorem B] or [72, Theorem E]). However, this result does not necessarily hold in the case where Π_h^* limits to a pair of heteroclinic orbits connecting two distinct critical points with eigenvalues $\pm\alpha$, $\pm i\beta$, α and $\beta>0$ (see [61]). With either eigenvalue condition on P, at those values of the energy for which the Poincaré map eigenvalues are n-th roots of unity, we can apply results of Meyer [75,76]. These state that generically one or two additional families of periodic orbits branch off from Π_h^* at these values of h. It would be interesting to use a computer to find such possible additional families of periodic orbits. Such a search could be extended to look for the families of periodic orbits that arise when the Π_i^*, i=1,2,3, change from elliptic to hyperbolic (recall the discussion on the preservation of the Poincaré index in Section 4).

The problem of the existence or non-existence of second integrals in the energy range $0<h<(1/6)$ in the Hénon-Heiles Hamiltonian has so far eluded proof. If one could rigorously show the existence of non-degenerate homoclinic or heteroclinic orbits between, for example, the

Π_i, i=4,5,6, in a certain energy range, then there would be no second independent analytic integral at such energies [55, pp. 188-189]. There is a total lack of theory to explain why the formal integrals which Gustavson [36] plots in truncated form appear to represent the corresponding pictures of Figure 4 so well at low positive energies. Preliminary numerical results indicate a transition from ellipticity to hyperbolicity for the Π_i^*, i=1,2,3, occurring at the Poincaré map eigenvalues $\mu_1 = -1 = \mu_2$ at energy $h \approx .13$. Note that this is <u>after</u> $h \approx .11$ at which Hénon and Heiles first observed stochasticity starting on a large scale. Thus, these stability transitions for the Π_i^* would not appear to be interfering with all such <u>lower</u> energy results showing "integrability" for the flow. We will comment more on this in the discussion of Example (b) in Section 8.

The appearance and disappearance of invariant tori as the energy changes has been noted in numerical explorations of the restricted three-body problem by both M. Hénon and J.M.A. Danby (private communications).

7. Stochasticity at Energies $h>(1/6)$.

At energies $h>(1/6)$ the energy manifold $H=h$ is connected and unbounded, and projects to the Hill's region in the x-plane given in Figures 3 and 7. Let L_i be the minimum distance line segment in leg i that connects the two branches of $V=h$ that bound leg i, and note that L_i depends on h (see Figure 7).

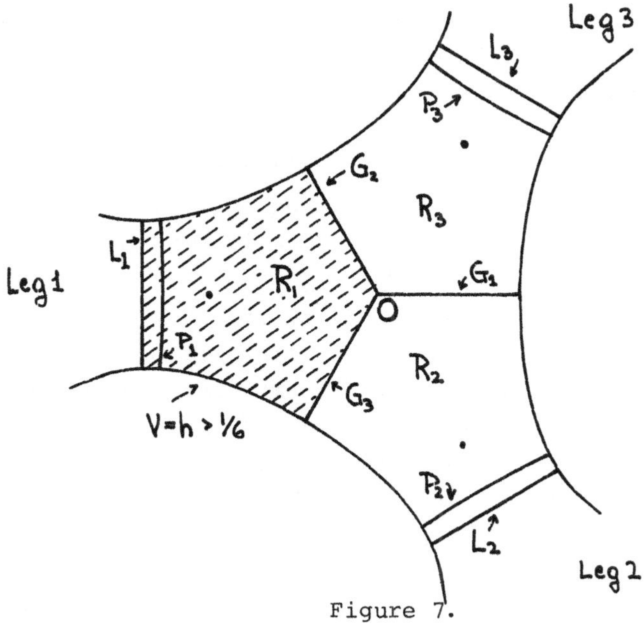

Figure 7.

One can easily show that the values of $-V_x$ to the left of and on the line segment L_1 in leg 1 force any solution of (2.6) with energy $h>(1/6)$ that enters the region to the left of L_1 with $\dot{x}_1 < 0$ to go to infinity in this leg (possibly in finite time). Thus the flow in the legs beyond the L_i is relatively uninteresting. It can actually be shown that any orbit that crosses the periodic orbit P_1 with $\dot{x}_1<0$ must go to infinity in leg 1, but this is more difficult to prove (see [11, Theorem 1.1] or [64, Corollary A.7]). Another consequence of [11, Theorem 1.1] is that P_1 is the only bounded solution of (2.6) with energy h that remains for all time in leg 1 to the left of the segments of the gradient lines G_2 and G_3 that run from $V=h$ to the origin as in Figure 7. We define the compact region R_1 to be that part of the Hill's region $V(x)\le h$ in leg 1 containing P_1 that is bounded on the left by L_1 and on the right by the above segments of G_2 and G_3 (see Figure 7). By rotation through the angles $\pm(2\pi/3)$ we define the regions R_2 and R_3, and set $R=(R_1 \cup R_2 \cup R_3)$. By what we have stated above, all bounded solutions with energy $h>(1/6)$ of (2.6) must lie in R.

Using the ideas outlined in Section 1, we can give a topological model of the energy manifold H=h that lies above R. Recall that above each point x there is a circle in H=h that projects to x (see (1.3)), and that the radius of this circle goes to zero on the boundary V=h. Thus above each arc in the region $V(x) \leq h$ that intersects V=h in only its two distinct endpoints, there is a 2-sphere in H=h that projects to this arc. Denote the 2-sphere above L_i by Σ_i^*. Hence the compact region R^* in H=h that projects to R can be modelled as a closed 3-ball with two open 3-balls removed, the three boundary components being the Σ_i^*, i=1,2,3. The bounded solutions with energy h of (2.5) then lie in R^*. The mathematical techniques of [13] involve relating information derived from the solutions of (2.6) in the x-plane to the corresponding flow of (2.5) on H=h. However, the results of this interplay of x-plane and phase space information can be given in the x-plane in terms of the notation introduced in Figure 7.

In [65] it is shown that the periodic orbits P_i, i=1,2,3, are hyperbolic at each $h > (1/6)$ when considered within the manifold of all solutions having the same energy. (Recall that such a result is immediate from [53] for $(1/6) < h < (1/6) + \delta$, where $\delta > 0$ is small.) We now wish to generate some non-degenerate homoclinic and heteroclinic orbits relating the P_i to one another. This will give Theorem 7.1 below. Denote by Σ_i the 2-spheres in H=h that project to P_i. Since P_i has its endpoints on two distinct branches of V=h in Figure 7, each of the spheres Σ_i must separate H=h. The periodic orbit $(P_i, \dot{P}_i)$ is an equator that separates Σ_i into the two open hemispheres $\Sigma_i^{\pm}$ consisting of points on orbits with energy h of (2.5) that respectively go to infinity in leg i in $\pm$ time (see Figure 8).

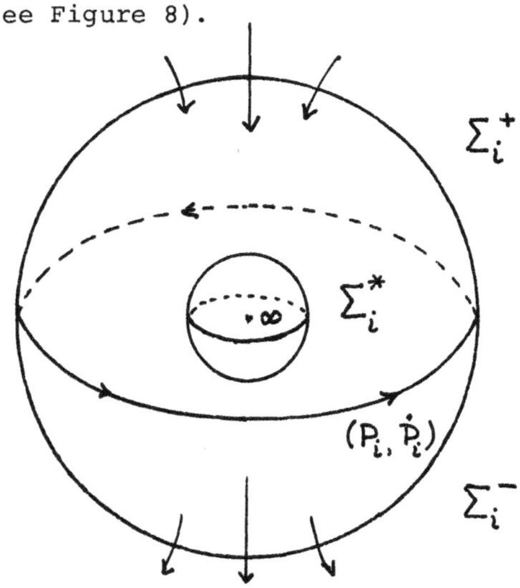

Figure 8.

In Figure 8 the "infinity" of leg i is at the "center" of the sphere Σ_i^* inside Σ_i. It is shown in [13] that the open hemispheres $\Sigma_i^{\pm}$, i=1,2, when carried forward in $\mp$ time by the flow intersect the cylinder (pinched at one end over V=h) G_3^* in H=h, that projects to G_3, in open discs whose circular boundaries are the intersection of the asymptotic manifolds of the $(P_i,\dot{P}_i)$, i=1,2, with G_3^*. Symmetry gives the construction for any other pair of the P_i, i=1,2,3. We denote the stable (unstable) asymptotic manifold of $(P_i,\dot{P}_i)$ by $P_i^+ (P_i^-)$. Thus, there is a heteroclinic orbit connecting P_2 to P_1 if $(P_2^- \cap G_3^*) \cap (P_1^+ \cap G_3^*)$ is not empty (see Figure 9). In the case where there are Jordan

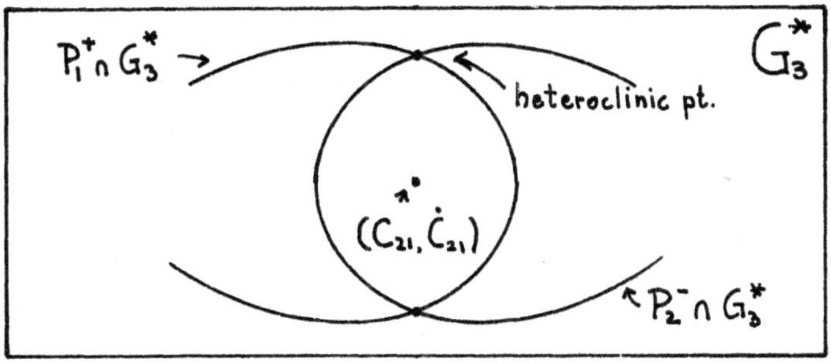

Figure 9.

arcs of $(P_1^+ \cap G_3^*)$ and $(P_2^- \cap G_3^*)$ that cross as in Figure 9, but not <u>transversely</u>, one can construct transversal intersections of P_1^+ and P_2^- in any neighborhood to obtain <u>non-degenerate</u> heteroclinic orbits if the Hamiltonian is <u>analytic</u> (see [17] and [14]).

To show that there is such an intersection in Figure 9 giving heteroclinic orbits, it suffices to construct a <u>crossing orbit</u> $(C_{21}, \dot{C}_{21})$ with energy h that intersects Σ_2^- and Σ_1^+; hence C_{21} comes from infinity in leg 2 and crosses over to go to infinity in leg 1 in Figure 7. Such a crossing orbit intersects G_3^* in the interior of the area bounded by arcs of $P_1^+ \cap G_3^*$ and $P_2^- \cap G_3^*$ as shown in Figure 9. Thus, the whole problem of finding non-degenerate heteroclinic orbits connecting the P_i <u>reduces</u> to the problem of finding such a crossing orbit C_{21} from leg 2 to leg 1; all other required crossing orbits can be obtained by symmetry and reversibility of the differential equations. By a variant of the stretching of variables argument given in Section 1, one can show that such crossing orbits exist for all sufficiently high energies [14]. This is done by obtaining the Hénon-Heiles Hamiltonian as a perturbation of the potential (2.9) which is known to have crossing orbits (see [64, Lemma 3.7]). Alternatively, one can work in the x-plane and simply numerically integrate

(2.6) to construct such a crossing orbit; one need only show that the constructed C_{21} comes from below L_2 and goes to the left of L_1 in Figure 7. Using the symmetry of Figure 7 with respect to G_3, it suffices to consider orbits starting perpendicular to G_3 and numerically integrate them only in forward time until they cross L_1. This was done starting at the origin at various energies in [13] with a corresponding error analysis to show that the computed orbit represented an actual solution of (2.6). Because of this error analysis we used the Taylor series method for numerical integration in which one has truncation error estimates. The problem of numerically constructing crossing orbits was refined in [15], where the computer program was set up to have the machine itself provide the error estimates. We will return at the end of this section to the whole area of crossing orbit construction. Before giving the consequences of the existence of such a crossing orbit, and hence heteroclinic orbits connecting the P_i, we first summarize. We have reduced the problem of finding non-degenerate heteroclinic orbits connecting distinct hyperbolic periodic orbits to a problem that can be tackled by a computer at any given energy level. Corresponding homoclinic orbits connecting each P_i to itself can be similarly constructed. We will comment in the next section on some other Hamiltonians in which this can be done.

At any energy $h>(1/6)$ at which there exists a crossing orbit from any leg to another (different) leg in the Hénon-Heiles Hamiltonian, the analogue of Figure 9 will hold. The heteroclinic orbits, if not already non-degenerate, can be used to construct non-degenerate ones. Around these latter, one can embed the Smale horseshoe map and construct a Cantor set on which the flow in H=h will be conjugate to a shift operator on bi-infinite sequences (see [55] as a general reference). The consequences of this can be simply stated in terms of the x-plane regions R_i, i=1,2,3, of Figure 7:

Theorem 7.1. At any energy $h>(1/6)$ for which there is a crossing orbit C_{21} from leg 2 to leg 1 we have the following results describing the solutions of (2.6) that intersect the interior of the region R:

(a) Given any bi-infinite sequence of the regions R_1, R_2, and R_3, in which no region repeats itself, there are uncountably many solutions of (2.6) with energy h which pass through the interior of these regions in the prescribed order as time increases from $-\infty$ to $+\infty$. If the bi-infinite sequence is periodic with period $K\geq 2$, then there is a non-empty subset of these solutions that can be analyzed by using the shift operator on a space of bi-infinite sequences of finitely many letters. Moreover, for each $\ell>0$ there are periodic orbits that go through the sequence of K regions (constituting the period of the bi-infinite sequence) ℓ times and then close.

(b) Given any finite sequence of the regions R_i, $i=1,2,3$, in which no region repeats itself, there are uncountably many solutions of (2.6) with energy h which come from infinity in the leg corresponding to the first region, pass from region to region in the specified order, and then return to infinity in the leg indicated by the last region (possibly in finite time). By reversibility, a similar result holds in backwards time.

(c) Given any infinite sequence of the regions R_i, $i=1,2,3$, in which no region repeats itself, with starting region R_{i_0}, there are uncountably many solutions of (2.6) with energy h which come from infinity in leg i_0 and then pass from region to region in the specified order as time increases from $-\infty$ to $+\infty$. By reversibility, a similar result holds in backwards time.

(d) At this energy h the flow generated by (2.5) has no second analytic integral.

The proof of Theorem 7.1 can be found in [13], except the proof of part (d) given there is incorrect (see [14] for a correct proof). [14] also shows how to obtain such crossing orbits at sufficiently high energies, and how to obtain non-degenerate homoclinic and heteroclinic orbits in any neighborhood of a corresponding homoclinic or heteroclinic orbit.

We see that at such energies $h>(1/6)$ at which we have a crossing orbit the description of the flow in R given by Theorem 7.1 is somewhat "stochastic". The first rigorous proof of the existence of a non-degenerate homoclinic orbit in a Hamiltonian system with two degrees of freedom was given by Cherry [10], though the consequences for embedding the shift were known to Birkhoff [5, Chapter 4]. We also mention the interesting survey by Cherry [9].

We conclude this section with some general remarks about the construction of crossing orbits. Some of the problems are brought out in a counterexample in a potential where crossing orbits might at first be thought to exist but in fact do not (see [13, Section 8]).

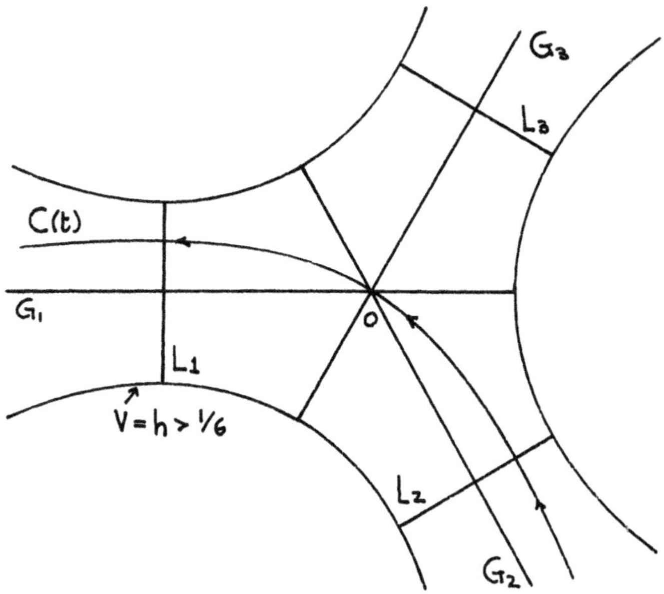

Figure 10.

Using the symmetry of Figure 10 with respect to the gradient line G_3, a number of crossing orbits $(C,\dot{C})$ were constructed in [13] by numerical integration in forwards time starting from the origin perpendicular to G_3. As h changes we have a 1-parameter family of orbits with these initial conditions. At h=1, (1/2), and (1/3), they had the general appearance of the orbit C(t) sketched in Figure 10. At h=(1/4), however, the orbit turned around and exited the region R via leg 2. In phase space each such orbit $(C,\dot{C})$ links the orbit that projects to G_3. Both the number of such links and the exit leg were observed to change as $h\downarrow(1/6)$. A detailed listing is given in [13, Table II, p. 107], and some additional results are reported in [15, Addendum to Section 4]. The conjecture is that for the above family $(C,\dot{C})$, this linking number goes to ∞ as $h\downarrow(1/6)$. As this linking number changes, the flow on H=h for (2.5) is apparently not conjugate at all energies above (1/6). Since the linking number becomes one at all sufficiently high energies, perhaps the flows on H=h are the "same" at these energies. That is, things may eventually "settle down" in a manner reminiscent of the phenomena at small positive h.

For another possible approach to the construction of crossing orbits, we refer to a paper of Gordon [35].

8. Further Examples.

The techniques of [11,13,61,65] apply to a number of potentials. In this section we contrast the Hénon-Heiles potential with two further examples.

(a) Let

$$W_1(x) = (1/2)(x_1^2 + x_2^2) - x_1x_2^2. \tag{8.1}$$

This potential has the origin as a critical point at energy 0, and two additional critical points $p^{\pm} = (1/2)(1, \pm\sqrt{2})$ at energy (1/8). The level curve $W_1 = (1/8)$ consists of the straight line $x_1 = (1/2)$ and the parabola $(4x_1 - 8x_2^2 + 2) = 0$ (see Figure 11). The potential $U = (1/2)(Ax_1^2 + Bx_2^2) - \varepsilon x_1 x_2^2$ was studied numerically for various values of A, B, and ε by Contopoulos [18,19,20] and for A=B by Barbanis [71]. One can put an ε with the cubic terms of (8.1) by stretching variables, so the only real restriction is our taking A = 1 = B. Both critical points $p^{\pm}$ at $h = (1/8)$ have eigenvalues $\lambda = \pm 1, \pm i\sqrt{2}$. Thus the result of Moser in [53] implies that each of $p^{\pm}$ gives birth to a hyperbolic periodic orbit at energies $(1/8) < h < (1/8) + \delta$

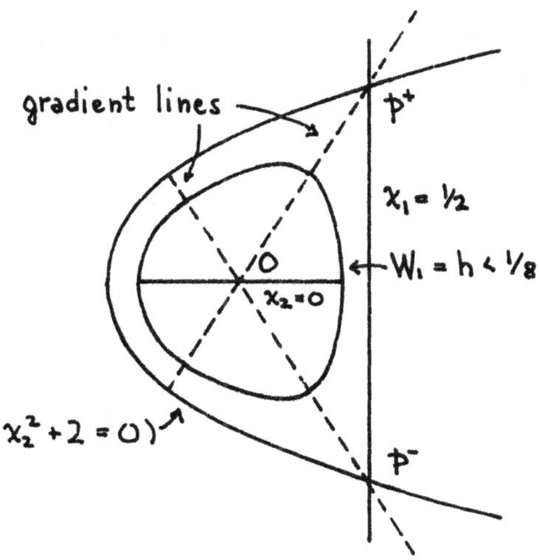

Figure 11.

for positive δ sufficiently small. However, due to certain technicalities in [11], we cannot extend these results as before to obtain corresponding hyperbolic periodic orbits at all energies $h > (1/8)$. In particular, Hypothesis 3 of [11, p. 333] does not apply since the level curve $W_1 = (1/8)$ does not consist solely of straight lines. Therefore, though it is probable that there are analogues of the results in Section 7 at energies $h>(1/8)$ for this potential, nothing has yet been shown. Thus, in the following discussion we restrict ourselves to $0<h<(1/8)$. In this energy range the central region about the origin is enclosed in an oval-shaped level curve $W_1=h$ similar to Figure 2. Moreover, the potential W_1 has three gradient lines, one of which is the x_1-axis, the other two passing through the origin and $p^{\pm}$ respectively as in Figure 2. Thus there are three periodic orbits Π_i, $i=1,2,3$, along these gradient lines as in Figure 2. However, due to a lack of rotational symmetry, we cannot geometrically construct the analogues of the circular orbits Π_7 and Π_8 and also Π_5 and Π_6. We remark on Π_4 below.

By direct calculation one can show that the periodic orbit Π_1 of (8.1) along the x_1-axis is hyperbolic for $0<h<h_o\approx 8$ and thereafter alternates between ellipticity and hyperbolicity infinitely often as $h\to\infty$. This comes from well-known results on the stability chart for a 2-parameter Mathieu equation (see [15] for the setup, applies there to the potential we will consider below in (b)).

The eigenvalues of the two critical points at $h=(1/8)$ satisfy the requirements of Section 6, and one obtains similar results for the diagonal periodic orbits Π_2 and Π_3. Both alternate infintely often in stability status between ellipticity and hyperbolicity as $h\uparrow(1/8)$. The transition to hyperbolicity takes place alternately through the Poincaré map eigenvalues $\mu_1 = 1 = \mu_a$ and $\mu_1 = -1 = \mu_2$. In each energy range $h_1<h<h_2$ in which the orbits are elliptic, one can apply the Moser-Rüssmann criterion sketched in Section 1 to obtain invariant tori in $H=h$ that nest to the respective periodic orbits.

The construction of the analogue of Π_4 of Figure 2 goes through for (8.1) with placement and curvature properties as illustrated there. The proofs of these latter statements follow from the same type of considerations used in Section 10 to prove these results in the Hénon-Heiles potential. As in the Hénon-Heiles potential, all Π_i, $1\le i\le 4$, link one another in phase space.

(b) Let

$$W_2(x) = (1/2)(x_1^2 + x_2^2) - (1/2)x_1^2x_2^2. \tag{8.2}$$

This potential has the origin as a critical point at energy 0, and four symmetrically placed critical points $(\pm1,\pm1)$ at energy $(1/2)$. The level curve $W_2=(1/2)$ consists of the four straight lines $x_1 = \pm1$ and $x_2 = \pm1$. The eigenvalues of the critical points at $h=(1/2)$ are

$\pm(2)^{\frac{1}{2}}$ and $\pm i(2)^{\frac{1}{2}}$. Thus, again [53] shows that from each of these four critical points one obtains a hyperbolic periodic orbit in some small energy range above (1/2). Unlike Example (a) above, the techniques of [11] and [65] do apply here, at least in the energy range $(1/2)<h\leq(9/2)$, to prove the existence at each such h of hyperbolic periodic orbits P_i, i=1,2,3,4, which are the only bounded orbits remaining in their respective legs for all time (see Figure 12).

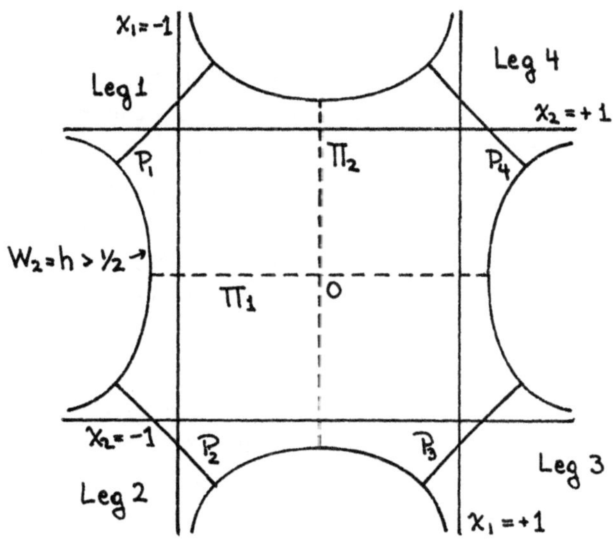

Figure 12.

There are technical reasons in [11] as to why this result has not yet been proven for $h>(9/2)$. The curvature properties of the P_i follow from the techniques in Section 10. The periodic orbits Π_1 and Π_2 that respectively run along the x_1- and x_2-axes are elliptic for $0<h<h_o$, where $h_o\approx 1.15$, and hyperbolic thereafter for all $h>h_o$ (see [15] for a proof). Again one can use the Moser-Rüssmann criterion to obtain invariant tori in H=h that nest to $(\Pi_1,\dot{\Pi}_1)$ and $(\Pi_2,\dot{\Pi}_2)$ respectively for a set of energies of full measure in $0<h<h_o$.

At any energy $(1/2)<h\leq(9/2)$ at which there is a crossing orbit coming from infinity in leg 2, say, and going to infinity in the non-opposite legs 1 or 3, [13, Theorem 7.2] and [14] apply. We then have non-degenerate heteroclinic orbits connecting all the P_i and can embed the Smale horseshoe map in multiple ways. A description of the orbit structure then parallels Theorem 7.1 (a)-(d). Thus, the flow will be somewhat "stochastic," and will have no second independent analytic integral at such energies. [13] lists the results of several computer runs for constructing crossing orbits at various energies with their

attendant error analyses. [15] constructs such crossing orbits near $h=.99$, thus obtaining the analogue of the "pathology" of Theorem 7.1 alongside the invariant tori about $(\Pi_1,\dot{\Pi}_1)$ and $(\Pi_2,\dot{\Pi}_2)$ in phase space (one can choose h near .99 to satisfy the Moser-Rüssmann criterion as the eigenvalues of the linearized Poincaré map are dependent on h). We now consider the energy range $0<h<(1/2)$.

We have already stated above that Π_1 and Π_2 are elliptic for $0<h<h_o$ where $1.15\approx h_o>(1/2)$. If one applies the construction given in Section 3 for the Π_i, $i=4,5,6$, of Figure 2, one obtains only Π_1 and Π_2 back again. Thus, these additional orbits that are presumably hyperbolic in the Hénon-Heiles case are absent from the potential W_2. Since the stochasticity for the Hénon-Heiles Hamiltonian appears in numerical experiments to be at first located in an area near the presumed stable and unstable manifolds in phase space of the Π_i, $i=4,5,6$, of Figure 2, the report [40, p. 77] by Hénon and Heiles that the potential W_2 had no observable large scale stochasticity becomes potentially significant (compare also the numerical results for the potential W_1 in (a) above).

The analogues of the central periodic orbits Π_7 and Π_8 of Figure 2 can also be constructed at all energies $h>0$, since W_2 is invariant under rotations through an angle $(\pi/2)$. However, nothing is known about their stability status. For $0<h<(1/2)$ there are two gradient line periodic orbits G_1 and G_2 that run along $x_1=x_2$ and $x_1=-x_2$ respectively. Both orbits limit in phase space as $h\uparrow(1/2)$ to a pair of heteroclinic orbits connecting the two opposite critical points. A modification of the theory discussed in Section 6 (see [61]) applies to show that the G_i, $i=1,2$, alternate infinitely often in stability status between ellipticity and hyperbolicity as $h\uparrow(1/2)$. The change to hyperbolicity, however, can only take place through the Poincaré map eigenvalues $\mu_1 = +1 = \mu_2$. The alternation in stability status is surprising in light of the numerical result in [40] reporting no observable stochasticity in the energy range $0<h<(1/2)$. Perhaps there is some small scale stochasticity, and the absence of this on a large scale may be connected with the absence in the potential W_2 of periodic orbits corresponding to the Π_i, $i=4,5,6$, of Figure 2 as discussed above. Moreover, recall that in the Hénon-Heiles Hamiltonian preliminary numerical results indicate that the first transition to hyperbolicity of the gradient line orbits Π_i, $i=1,2,3$, of Figure 2 was at $h\approx.13$, and that this was <u>after</u> the introduction of large scale observable stochasticity at $h\approx.11$ as reported by Hénon and Heiles [40, p. 77]. If these numerical results are accurate, then the contribution to stochasticity of these transitions to hyperbolicity would, at least temporarily, appear to be "secondary" in importance. Note that as in the Hénon-Heiles potential, all six periodic orbits constructed for the potential W_2 in the energy range $0<h<(1/2)$ link one another in phase space. The Moser-Rüssmann criterion also applies to the G_i, $i=1,2$, when they are elliptic.

For other examples the reader is referred to [11,13,14,15,61,65] and [60].

9. Remarks on Problems for Computer Exploration.

Many papers work with Hamiltonians (0.1) solely in the x-plane, where one can investigate the properties of the potential V(x), or solely with a Poincaré map defined on a section of a compact component of H=h in phase space. We have seen in this survey that a combination of both approaches yields many results. In particular, if one has located a 1-parameter family of periodic orbits Π_h, parametrized by energy h, it is natural to seek the possible changes in stability status of the Π_h as h changes. Mathematical techniques exist for analyzing critical points and periodic orbits, but not much is known that can be explicitly applied to these examples apart from this. Since critical points are relatively easy to locate and analyze, we concentrate on the problems surrounding periodic orbits. Here the numerical explorations can be invaluable. For example, in Figure 3 the P_i, i=1,2,3, were shown numerically to be hyperbolic at particular energies $h>(1/6)$ long before this was rigorously shown in [65]. Similarly the Π_i, i=4,5,6, in Figure 2 have been shown numerically at a number of energies in the range $0<h<(1/6)$ to be hyperbolic, but this has not yet been rigorously proven. Also, some of the changes in the stability status of the Π_i, i=1,2,3, of Figure 2 were initially picked up by a computer run, and this led eventually to the results reported in Section 6.

As discussed in Section 4, when working with polynomial potentials the Taylor series method of numerical integration is recommended. In particular, if one wishes to obtain rigorous proofs from numerical calculations, this is the only method in which we can rigorously obtain bounds for the truncation errors (asymptotic estimates for these errors are useless for proving theorems). In the following we list some problems in the Hénon-Heiles potential that we feel would be worth exploring numerically. Such problems, of course, also appear in each of the two examples (a) and (b) of Section 8 and in many other potentials.

(1) Investigate the stability status and any changes therein of the central periodic orbits Π_7 and Π_8 of Figure 2. This can be done by numerically integrating the linearized Poincaré map along the orbit through one full period and then finding the eigenvalues. An efficient search routine for the periodic orbits should use the fact that they are perpendicular to the three gradient lines G_i, i=1,2,3. The values of the energy at the first and last changes of stability status (if any) would be of interest.

(2) Determine the energy range (possibly infinite) in which the Π_i, i=4,5,6, of Figure 2 exist and check their stability status in this energy range. Check for homoclinic and heteroclinic orbits connecting them when they are hyperbolic.

(3) Find the energy of the first few transitions in stability type of the gradient line periodic orbits Π_i, i=1,2,3, of Figure 2. Locate possible 1-parameter families of elliptic periodic orbits that would appear in the transition from ellipticity to hyperbolicity (recall the discussion on the preservation of the index of the fixed point of the Poincaré map in Section 4). Follow these elliptic families and see if they continue into the elliptic orbits one excpects just as one goes into the next transition from hyperbolicity back to ellipticity as the energy increases. Distinguish the cases of "hyperbolic" versus "hyperbolic with reflection."

(4) Detailed studies are needed of the linking and unlinking phenomena reported for crossing orbits in Section 7 when $h>(1/6)$. Here one is looking for evidence that the flow "settles down" after some energy $h'>(1/6)$ so that at distinct energies above h' the flows are essentially "conjugate." What happens as $h\downarrow(1/6)$? Also needed is a check on the placement of Π_7 and Π_8 in Figure 2; namely, do these orbits link the Π_i, i=4,5,6, in phase space or not? This question is equivalent to whether Π_7 and Π_8 should appear as in Figure 2 or should lie interior to the connected open disc containing the origin that is bounded by closed arcs of the Π_i, i=4,5,6. The result of such an investigation has ramifications for Figure 4 and the interpretations offered in Section 4 (this assumes that Π_7 and Π_8 are in some sense unique).

To summarize, in a problem involving a Hamiltonian of the type (0.1), numerical explorations concerned with the location of families of periodic orbits, their stability status and changes therein as the energy changes, and their respective linking properties (placement in the x-plane and phase space) are most valuable to the mathematician trying to unravel the phenomenon of stochasticity (or its lack as reported for example (b) of Section 8). When applicable, the determination of homoclinic and heteroclinic orbits, or numerical evidence for their existence, is also valuable. To obtain information about higher dimensional objects like invariant tori presents problems in pictorial representation. For a start in this direction we refer to the paper of Baxter, Eiserike, and Stokes [3]. With the presently available mathematical tools, however, the investigation of periodic orbits and possible homoclinic and heteroclinic phenomena is of utmost importance. The problems surrounding the "sizes" of islands (estimates for how "large" a torus one can find around a given elliptic stable periodic orbit) lie in the mathematical future.

10. Appendix: Curvature and Placement of the Periodic Orbits Π_i, $i=4,5,6$, for $0<h<(1/6)$.

In this Appendix we justify our placement of the periodic orbit Π_4 in Figure 2, and show that this planar curve has positive curvature (except for its endpoints on $V=h$ where the curvature is undefined) when followed in the direction having $\dot{x}_2>0$ (rising orbit). Corresponding results hold for Π_5 and Π_6 in Figure 2 by symmetry. Thus we have to show that Figure 14(b) is the case, as opposed to Figure 14(a).

The reader will note that this statement of positive curvature is the analogue of the central Lemma 2.3 of [11] which was applied to prove the existence and isolatedness (as an invariant set) of the periodic orbits P_i, $i=1,2,3$, of Figure 3 [11, Section 5], and thence their hyperbolicity for all $h>(1/6)$ [65, Section 4, Example C]. The reasoning below provides an alternate proof of Lemma 2.3 of [11], and applies to many other potentials (see [11], [65], and the examples in Section 8 above).

Let V_x be the gradient, and V_{xx} the Hessian, of the potential V. Set $T = -JV_{xx}JV_x$, where the 2×2 matrix $J=\begin{pmatrix}0 & 1\\ -1 & 0\end{pmatrix}$. The gradient V_x restricted to an integral curve of the T-vectorfield has constant direction along this curve [11, Lemma 2.1]. Let $\langle\,,\rangle$ be the standard inner product for planar vectors. A direct calculation then shows that for the Hénon-Heiles potential V of (2.4) we have,

$$\langle T,JV_x\rangle=2x_2[x_2+(3)^{\frac{1}{2}}x_1]\cdot[x_2-(3)^{\frac{1}{2}}x_1]\cdot[1-|x|^2]. \qquad (10.1)$$

The gradient lines G_2 and G_3 (see Figures 2 and 3) respectively have equations $x_2 + (3)^{\frac{1}{2}}x_1 = 0$ and $x_2 - (3)^{\frac{1}{2}}x_1 = 0$. The three gradient lines G_i, $i=1,2,3$, subdivide the Hill's region $V(x)\le h<(1/6)$ into six subregions in which we specify the signum of (10.1) in Figure 13 (recall $|x|^2<1$ in this central region when $h<(1/6)$).

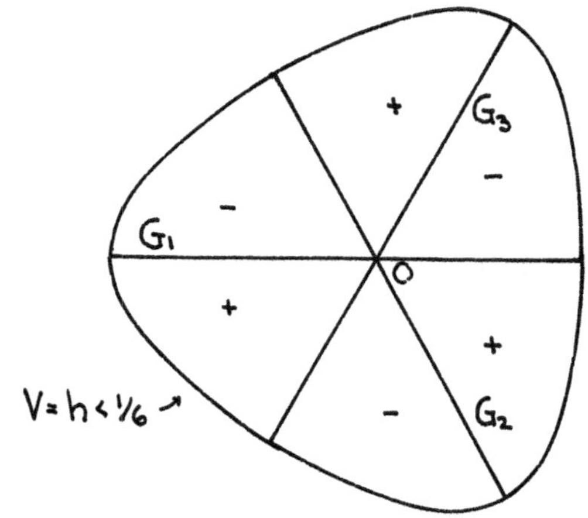

Figure 13: Signum of $<T, JV_x>$.

The curvature of a solution x(t) with energy h of $\ddot{x} = -V_x$ is given by $k = \frac{<V_x, J\dot{x}>}{|\dot{x}|^3}$ where $|\dot{x}|^2 = 2(h-V(x))$. We assume x(0) lies on the level curve V=h (hence $\dot{x}(0) = 0$), and consider x(t) as <u>rising</u> to V=h as $t\uparrow 0$ (thus $\dot{x}_2(t)>0$ for t close to but <0). We wish to calculate $\lim_{t\uparrow 0} k(x(t))$. For a plane vector $v \neq 0$ we denote the unit vector $(v/|v|) = (v)^*$. Recall that $\lim_{t\uparrow 0} \dot{x}(t) = 0$ and $\lim_{t\uparrow 0} (\dot{x})^* = (V_x(x(0)))^*$. Using L'Hospital's Rule and $\ddot{x} = -V_x$ we have:

$$\lim_{t\uparrow 0} k(x(t)) = \lim_{t\uparrow 0} \frac{<V_{xx}\dot{x}, J\dot{x}> + <V_x, J\ddot{x}>}{3|\dot{x}| <\ddot{x}, \dot{x}>}$$

$$= (1/3)|V_x|^{-3}<T, JV_x> = (1/3)k_G. \tag{10.2}$$

Here $k_G = \text{div}[(JV_x)^*] = |V_x|^{-3}<T, JV_x>$ evaluated at x(0) on V=h is the curvature of the gradient curve with tangent V_x. Thus, the periodic orbit $\Pi_4(t)$ approaches V=h with $\dot{x}_2>0$, and has, in the limit as $t\uparrow 0$, curvature equal to $(1/3)k_G$, which is positive in this region by

Figure 13. Now that we have a portion of $\Pi_4(t)$ having positive curvature, we need only choose between Figures 14(a) and (b) for the placement of $\Pi_4(t)$ in the x-plane. Note that the curvature of $\Pi_4(t)$ goes through a discontinuity on V=h (from $(1/3)k_G$ to $-(1/3)k_G$).

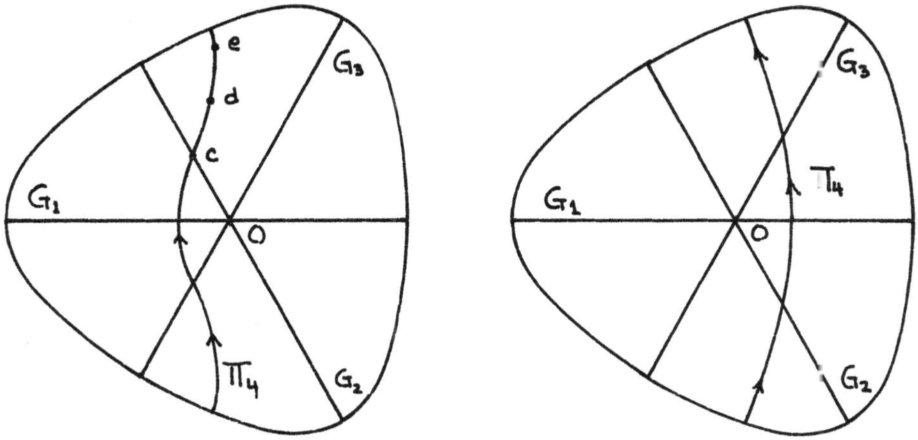

(a) (b)

Figure 14.

Consider $\Pi_4(t)$ as rising ($\dot{x}_2>0$) and possibly crossing the gradient line G_2 at c with $x_2(c)>0$ as in Figure 14(a). At $\Pi_4(t_0)$ = c the curvature k<0 since $<V_x, J\dot{x}><0$ there. By the calculations for (10.2) $\Pi_4(t)$ approaches V=h with positive curvature. Hence there must be a point d along the orbit length at which k=0 and thus $\dot{x}=\alpha V_x$ for some scalar $\alpha\neq 0$ (see Figure 14(a)). Then at d we have

$$(d/dt)<V_x, J\dot{x}> = \alpha^2<V_{xx}\cdot V_x, JV_x> = -\alpha^2<T, JV_x> < 0. \tag{10.3}$$

Thus, the curvature of $\Pi_4(t)$ becomes negative again in a region along the orbit above the point d, and can never become positive as required at points e sufficiently close to V=h in Figure 13(a). A remaining possibility is that $\Pi_4(t)$ passes through the origin. But then there are nearby orbits x(t) that cross G_2 at points c with $x_2(c)>0$. Such orbits will have negative curvature at c, but positive curvature at some point e as in Figure 14(a) provided $x_2(c)$ is sufficiently small and e is taken near V=h. But, again, this change of curvature is impossible by the calculations above. Thus $\Pi_4(t)$ must cross the positive x_1-axis as in Figure 14(b). Moreover, between G_1 and G_3 the orbit $\Pi_4(t)$ with $\dot{x}_2>0$ also has positive curvature due to the values of V_x in this region. Above G_3 this orbit has positive curvature by the above arguments. Thus, by symmetry $\Pi_4(t)$ has positive curvature throughout its orbit length except for its endpoints on V=h where the curvature goes through a discontinuity. This completes the proof of the curvature properties and placement of $\Pi_4(t)$.

References

1. Y. Aizawa, Instability of a classical dynamical system with the negative curvature region. J. Physical Soc. Japan 33, No. 6, (1972), 1693-1696.

2. V. I. Arnold and A. Avez, "Ergodic Problems of Classical Mechanics," Benjamin, New York, 1968.

3. R. Baxter, H. Eiserike, and A. Stokes, A pictorial study of an invariant torus in phase space of four dimensions, in "Ordinary Differential Equations," (Leonard Weiss, Ed.), 331-349, Academic Press, New York, 1972.

4. M. V. Berry, Regular and irregular motion, in "A Primer in Nonlinear Mechanics," (R. Helleman and J. Ford, Eds.), to appear.

5. G. D. Birkhoff, Nouvelles recherches sur les systemes dynamiques, Memoriae Pont. Acad. Sci. Novi Lyncaei, S. 3, Vol. 1, (1935), 85-216. Reprinted in "Collected Mathematical Papers of George David Birkhoff," Vol. 2, 530-661, Dover, New York, 1968.

6. G. D. Birkhoff, Sur le problème restreint des trois corps, Annali Scuola Normale Superiore di Pisa, S. 2, Vol. 5, (1936), 1-42. Reprinted in "Collected Mathematical Papers of George David Birkhoff," Vol. 2, 668-709, Dover, New York, 1968.

7. T. Bountis, "Nonlinear Models in Dynamics and Statistical Mechanics," Thesis, Univ. of Rochester, to appear 1978.

8. M. Braun, On the applicability of the third integral of motion, J. Differential Equations 13 (1973), 300-318.

9. T. M. Cherry, The pathology of differential equations, J. Austral. Math. Soc. 1 (1959), 1-16.

10. T. M. Cherry, Asymptotic solutions of analytic Hamiltonian systems, J. Differential Equations 14 (1968), 142-159.

11. R. C. Churchill, G. Pecelli, and D. L. Rod, Isolated unstable periodic orbits, J. Differential Equations 17 (1975), 329-348.

12. R. C. Churchill and D. L. Rod, Pathology in dynamical systems I: General theory, J. Differential Equations 21 (1976), 39-65.

13. R. C. Churchill and D. L. Rod, Pathology in dynamical systems II: Applications, J. Differential Equations 21 (1976), 66-112.

14. R. C. Churchill and D. L. Rod, Pathology in dynamical systems III: Analytic Hamiltonians, to appear.

15. R. C. Churchill, G. Pecelli, S. Sacolick, and D. L. Rod, Coexistence of stable and random motion, Rocky Mt. J. of Math. 7 (1977), 445-456.

16. C. Conley, On the ultimate behavior of orbits with respect to an unstable critical point (I), J. Differential Equations 5 (1969), 136-158.

17. C. Conley, Twist mappings, linking, analyticity and periodic solutions which pass close to an unstable periodic solution, in "Topological Dynamics," (Joseph Auslander, Ed.), W. A. Benjamin, New York, 1968.

18. G. Contopoulos, Orbits in highly perturbed dynamical systems I. Periodic orbits, Astronom. J. 75 (1970), 96-107.

19. G. Contopoulos, Orbits in highly perturbed dynamical systems II. Stability of periodic orbits, Astronom. J. 75 (1970), 108-130.

20. G. Contopoulos, Orbits in highly perturbed dynamical systems III. Nonperiodic orbits, Astronom. J. 76 (1971), 147-156.

21. R. L. Devaney, Homoclinic orbits in Hamiltonian systems, J. Differential Equations 21 (1976), 431-438.

22. R. L. Devaney, Reversible diffeomorphisms and flows, Trans. Amer. Math. Soc. 218 (1976), 89-113.

23. R. L. Devaney, Blue sky catastrophes in reversible and Hamiltonian systems, Indiana Univ. Math. Jour. 26, No. 2, (1977), 247-263.

24. R. L. Devaney, Transversal homoclinic orbits in an integrable system, to appear in Amer. J. Math.

25. R. L. Devaney, Non-regularizability of the anisotropic Kepler problem, to appear in J. Differential Equations.

26. R. L. Devaney, Collision orbits in the anisotropic Kepler problem, Inv. Math. 45 (1978), 221-251.

27. R. L. Devaney, Subshifts of finite type in linked twist mappings, to appear in Proc. A.M.S.

28. P. Eberlein, Geodesic flows on negatively curved manifolds I, Ann. of Math. 95 (1972), 492-510.

29. P. Eberlein, Geodesic flows in negatively curved manifolds II, Trans. Amer. Math. Soc. 178 (1973), 57-82.

30. C. R. Eminhizer, "Convergent Perturbative and Iterative Methods Applied to the Duffing Equation and Anharmonic Oscillator

Systems," Thesis, Univ. of Rochester, 1975.

31. C. R. Eminhizer, R.H.G. Helleman, and E. W. Montroll, On a convergent nonlinear perturbation theory without small denominators or secular terms, J. Math. Physics 17, No. 1, (1976), 121-140.

32. J. Ford, The transition from analytic dynamics to statistical mechanics, Advances in Chemical Physics 24 (1973), 155.

33. J. Ford, Stochastic behavior in nonlinear oscillator systems, in "Lecture Notes in Physics," Vol. 28, 204-245, Springer-Verlag, New York, 1974.

34. J. Ford, The statistical mechanics of classical analytic dynamics, in "Fundamental Problems in Statistical Mechanics," Vol. 3, (E.D.G. Cohen, Ed.), North-Holland, Amsterdam, 1975.

35. W. B. Gordon, The existence of geodesics joining two given points, J. Differential Geometry 9 (1974), 443-450.

36. F. G. Gustavson, On constructing formal integrals of a Hamiltonian system near an equilibrium point, Astronomical J. 71, No. 8, (1966), 670-686.

37. P. Hartman, "Ordinary Differential Equations," Wiley, New York, 1964.

38. R.H.G. Helleman, On the iterative solution of a stochastic mapping, in "Statistical Mechanics and Statistical Methods, Theory and Applications," (U. Landman, Ed.), 343-370, Plenum Pub. Co., New York, 1977.

39. R.H.G. Helleman and E. W. Montroll, On a nonlinear perturbation theory without secular terms I: Classical coupled anharmonic oscillators, Physica 74 (1974), 22-74.

40. M. Hénon and C. Heiles, The applicability of the third integral of motion; some numerical experiments, Astronom. J. 69 (1964), 73-79.

41. J. Henrard, Proof of a conjecture of E. Strömgren, Celestial Mechanics 7 (1973), 449-457.

42. W. Klingenberg, Riemannian manifolds with geodesic flow of Anosov type, Ann. of Math. 99 (1974), 1-13.

43. M. Kummer, On resonant nonlinearly coupled oscillators with two equal frequencies, Commun. Math. Phys. 48 (1976), 53-79.

44. M. Kummer, An interaction of three resonant modes in a nonlinear lattice, J. Math. Analysis and Applications 52 (1975), 64-104.

45. M. Kummer, On resonant classical Hamiltonians with two equal frequencies, Comm. Math. Phys. 58 (1978), 85-112.

46. D. Laugwitz, "Differential and Riemannian Geometry," Academic Press, New York, 1965.

47. W. S. Loud, Stability regions for Hill's equations, J. Differential Equations 19 (1975), 226-241.

48. G. H. Lunsford and J. Ford, On the stability of periodic orbits for nonlinear oscillator systems in regions exhibiting stochastic behavior, J. Math. Phys. 13 (1972), 700-705.

49. W. Magnus and S. Winkler, "Hill's Equation," Wiley-Interscience, New York, 1966.

50. L. Markus and K. R. Meyer, "Generic Hamiltonian Dynamical Systems are neither Integrable nor Ergodic," Memoirs Amer. Math. Soc., No. 144, 1974.

51. J. Milnor, "Morse Theory," Annals of Mathematics Studies, No. 51, Princeton University Press, N.J., 1963.

52. E. Montroll and R.H.G. Helleman, On a nonlinear perturbation theory without secular terms, in "Topics in Statistical Mechanics and Biophysics: A Memorial to Julius L. Jackson," (R. A. Piccirelli, Ed.), 75-110, American Institute of Physics, New York, 1976.

53. J. Moser, On the generalization of a theorem of A. Liapounoff, Commun. Pure Appl. Math. 11 (1958), 257-271.

54. J. Moser, "Lectures on Hamiltonian Systems," Memoirs Amer. Math. Soc., No. 81, (1968), 1-60.

55. J. Moser, "Stable and Random Motions in Dynamical Systems," Annals of Mathematics Studies, No. 77, Princeton University Press, Princeton, N.J., 1973.

56. J. Moser, Stability theory in celestial mechanics, in "The Stability of the Solar System and of Small Stellar Systems," (Y. Kozai, Ed.), 1-9, D. Riedel, Boston, 1974.

57. J. Moser, Periodic orbits near an equilibrium and a theorem of A. Weinstein, Commun. Pure Applied Math. 29 (1976), 727-747.

58. V. V. Nemytskii and V. V. Stepanov, "Qualitative Theory of Differential Equations," Princeton University Press, Princeton, N.J., 1960.

59. B. O'Neill, "Elementary Differential Geometry," Academic Press,

New York, 1966.

60. G. Pecelli and E. S. Thomas, An example of elliptic stability with large parameters: Lamé's equation and the Arnold-Moser-Rüssmann criterion, to appear in Quart. of Applied Math.

61. G. Pecelli, D. L. Rod, and R. C. Churchill, Stability transitions in Hamiltonian systems, to appear.

62. R. C. Robinson, Generic properties of conservative systems, Amer. J. Math. 92 (1970), 562-603.

63. R. C. Robinson, Generic properties of conservative systems II, Amer. J. Math. 92 (1970), 897-906.

64. D. L. Rod, Pathology of invariant sets in the Monkey saddle, J. Differential Equations 14 (1973), 129-170.

65. D. L. Rod, G. Pecelli, and R. C. Churchill, Hyperbolic periodic orbits, J. Differential Equations 24 (1977), 329-348.

66. D. L. Rod, G. Pecelli, and R. C. Churchill, Addendum to "Hyperbolic Periodic Orbits," J. Differential Equations 28 (1978), 163-165.

67. H. Rüssmann, Über die Normalform analytischer Hamiltonscher Differentialgleichungen in der Nähe einer Gleichgewichtslösung, Math. Annalen 169 (1967), 55-72.

68. C. L. Siegel and J. Moser, "Lectures on Celestial Mechanics," Springer-Verlag, New York, 1971.

69. A. Weinstein, Lagrangian submanifolds and Hamiltonian systems, Ann. of Math. 98 (1973), 377-410.

70. A. Weinstein, Normal modes for non-linear Hamiltonian systems, Inv. Math. 20 (1973), 47-57.

71. B. Barbanis, On the isolating character of the "third" integral in a resonance case, Astronom. J. 71 (1966), 415-424.

72. R. L. Devaney, Homoclinic orbits to hyperbolic equilibria, Annals of the New York Academy of Sciences, to appear.

73. R. W. Easton, Homoclinic phenomena in Hamiltonian systems with several degrees of freedom, to appear in J. Differential Equations.

74. R. McGehee and K. Meyer, Homoclinic points of area preserving diffeomorphisms, Amer. J. Math. 96 (1974), 409-421.

75. K. Meyer, Generic bifurcation of periodic points, Trans. Amer. Math. Soc. 149 (1970), 95-107.

76. K. Meyer, Generic bifurcation in Hamiltonian systems, in

"Dynamical Systems-Warwick 1974," (Anthony Manning, Ed.), 62-70, Lecture Notes in Mathematics, Vol. 468, Springer-Verlag, New York, 1975.

ERGODIC COMPONENTS IN THE STOCHASTIC REGION IN A HAMILTONIAN SYSTEM

Nobuhiko SAITÔ and Atsushi ICHIMURA
Department of Applied Physics
Waseda University
Tokyo 160, Japan

Synopsis

Computer calculations on a classical dynamical system with hamiltonian

$$H_\varepsilon = \frac{1}{2}(p_1^2 + p_2^2) + \frac{1}{2}(q_1^2 + q_2^2) + q_1^2 q_2 + \frac{1}{3}q_2^3 - \frac{2}{3}\varepsilon q_2^3 .$$

are presented. The Poincaré mapping on (q_2 , p_2) plane revealed that there exist several ergodic components in the stochastic region, and furthermore they have a hierarchy structure. The Lyapunov characteristic numbers are also shown to have different values for different ergodic components.

I. Introduction

A hamiltonian system is either integrable or non-integrable. An integrable system has periodic solutions and furthermore has invariant tori around these periodic solutions. On the other hand a non-integrable system is quasi-integrable in the sense that there exist invariant tori near the stable periodic solutions, in accordance with the celebrated KAM theory.[1] This theory claims that almost of the invariant tori of the unperturbed system remain to exist in the presence of small perturbation. Consequently there exist invariant tori in the neighborhood of the stable periodic solution of non-integrable hamiltonian system. This is clearly demonstrated by computer experiments done by Hénon and Heiles[2] in a system with hamiltonian

$$H = \frac{1}{2}(p_1^2 + p_2^2) + \frac{1}{2}(q_1^2 + q_2^2) + q_1^2 q_2 - \frac{1}{3}q_2^3 . \qquad (1)$$

They confirmed the existence of the invariant tori, as far as the energy is not large and the system is close enough to its periodic solution. But they also showed that there appear stochastic regions where the invariant tori cease to exist when the above conditions are not fulfilled. The KAM theory, however, cannot predict how large the energy has to be, and how far the initial conditions deviate from the periodic solution, in order that these tori are destroyed and the system behaves irregular. It only says that there are irregular regions around the unstable periodic solutions. The irregular stochastic region thus attained is sometimes supposed to be ergodic, but no proof of it is given yet.

We now consider the systems with hamiltonian

$$H_\varepsilon = \frac{1}{2}(p_1^{\,2}+ p_2^{\,2}) + \frac{1}{2}(q_1^{\,2}+ q_2^{\,2}) + q_1^{\,2}q_2 + \frac{1}{3}(1 - 2\varepsilon)q_2^{\,3}, \tag{2}$$

where H_1 ($\varepsilon=1$) is the Hénon-Heiles system, and H_0($\varepsilon=0$) is shown integrable.[3] Thus it will be interesting to see the nature of the stochastic region by changing ε from $\varepsilon=0$ to $\varepsilon=1$. By means of Poincaré mapping of the system $\varepsilon=0.5$, it is found that the stochastic region consists of several ergodic components, and it has hierarchy structure. The same is also found in the case $\varepsilon \simeq 0.5$ as well as in a Lotka-Volterra system.

II. Decomposability of stochastic region

The stochastic behaviors in the Hénon-Heiles system ($\varepsilon=1$) strongly suggest that it is ergodic in this region. Is this property expected to hold, when ε is close to zero which makes the system integrable?[2] When ε is close to zero, the irregular region becomes small. Therefore it is rather better to select the value of ε close neither to 0 nor to 1 (say $\varepsilon=0.5$) in order to study the behaviors of the system in the irregular stochastic region.

Figure 1 is the Poincaré mapping of the system of $\varepsilon=0.5$ on the (q_2, p_2) plane. If the system lies initially at a point A($q_2=0.2$, $p_2=0$) with $p_1>0$, then the successive intersections of the orbit at the (q_2 , p_2) plane with $p_1>0$, $q_1=0$, or the image points lie on islands (or tori). If the orbit starts at a point B($q_2=0$, $p_2=0$) the image points are rather dispersed, and seem to be located on a band with finite width, but it is not clear whether the band consists of islands or not. If the orbit starts at a point C($q_2=-0.2$, $p_2=0$), the image points are more irregularly dispersed, but they are confined in a region bordered by two dotted lines (1) and (2), which are drawn just for indicating roughly the boundaries. It is almost impossible to find the exact closed curves for the boundaries, but if we take a point inside or outside of the line (2) but very close to it , the mapping starting from this point never escapes from the inside or outside of the line (2). These facts suggest the existence of the metrically decomposable regions or the ergodic components.

If the system starts at a point D(q_2 =0, p_2= −0.25) inside of the line(2), the mapping is confined in this region and still leaves three small regions unmapped as shown in Fig.1. If the mapping starts at a point E inside the middle region, we have a similar irregular mapping which is just like a miniature of the mapping inside the line (2), as shown in Fig.2. We thus have a hierarchy of the structure of mapping.

III. Lyapunov characteristic number

As we have seen above there exist several regions which are mutually unconnected in the sense that the mapping from one region to another cannot be achieved. This property is reflected into the Lyapunov characteristic number.

We consider an orbit given by $\mathbf{X} = \mathbf{X}(t)$ of our hamiltonian system $H_{0.5}$(Eq.(2)) in (q_1 , p_1 , q_2 , p_2) space, and put $\mathbf{X}_i = \mathbf{X}(i\tau)$, $i=1,2,\cdots,n$. Take a point $\mathbf{y}'$ near $\mathbf{X}(0)$ with distance $d=|\mathbf{d}|$, $\mathbf{d} = \mathbf{y}' - \mathbf{X}(0)$, and let $\mathbf{y}_1$ be the position at $t=\tau$ of the orbit passing through $\mathbf{y}'$ at $t=0$, and put $\mathbf{d}_1 = \mathbf{y}_1 - \mathbf{X}_1$. Next we take a point $\mathbf{y}_1'$ on the the vector from $\mathbf{X}_1$ to $\mathbf{y}_1$ such that

$$\mathbf{y}_1' - \mathbf{X}_1 = \frac{d}{|\mathbf{d}_1|}(\mathbf{y}_1 - \mathbf{X}_1) ,$$

and let $\mathbf{y}_2$ be the position at $t=2\tau$ of the orbit passing through $\mathbf{y}_1'$ at $t=\tau$, and put $\mathbf{d}_2 = \mathbf{y}_2 - \mathbf{X}_2$. In the same way we can define successively $\mathbf{y}_2', \mathbf{y}_3, \mathbf{d}_3, \mathbf{y}_3', \mathbf{y}_4, \mathbf{d}_4, \cdots$ (see Fig.3). Then the Lyapunov characteristic number can be identified with the limit of $n \to \infty$ and $d \to 0$ of the quantity

$$k_n(\tau, x, d) = \frac{1}{n\tau} \sum_{i=1}^{n} \ln \frac{|\mathbf{d}_i|}{d} .$$

In the numerical study, we can take d as sufficiently small, and then we will find $k_n(\tau, x, d)$ becomes independent of the magnitude and the direction of the vector $\mathbf{d}$.

Benettin, Galgani and Strelcyn[4] showed in computer calculation of Hénon-Heiles system (Eq.(1)) that the quantities $k_n(\tau, x, d)$'s tend to zero in the regular region, but they remain finite (not zero) if X lies in the stochastic region. We found that the same is true in the hamiltonian Eq.(2) as shown in Fig.4.

The Lyapunov characteristic number has different value in different stochastic region as shown in Fig. 5 . The plots of 1,2 and 3 which belong to the same stochastic region are likely to tend to the same characteristic number, but the plots 4 and 5 whose initial points lie in the regions different to each other and to that of the group 1, 2 and 3 tend to different characteristic numbers. In other words every stochastic region has its own Lyapunov characteristic number.

IV. Discussion

We have shown by computer calculations that in a special hamiltonian system the irregular stochastic region is metrically decomposable and thus has several ergodic components. Furthermore the metrically separable regions have some kind of hierarchy structure, just like the hierarchy structure in the regular region where invariant tori exist. The existence of the hierarchy structure in the stochastic region has also been suggested by Hénon as is refered to in ref. 4 .

The stochastic behavior and the ergodic components are found in Lotka-Volterra system with four interacting species. The details will be discussed elsewhere.

References

(1) An excellent review of the KAM theory was given by M.V. Berry in this Como conference, see also this Proceedings.
See also, for example, V.I. Arnold and A. Avez: *Ergodic Problems of Classical Mechanics* (Benjamin Inc. 1968).

(2) M. Hénon and C. Heiles: Astron. J. 69, 73(1964).

(3) Y. Aizawa and N. Saitô: J. Phys. Soc. Japan 32, 1636(1972).

(4) G. Benettin, L. Galgani and J. M. Strelcyn: Phys. Rev. A14, 2338(1976).

(5) M. Casartelli, E. Diana, L. Galgani and A. Scotti: Phys. Rev. A13, 1921(1976).

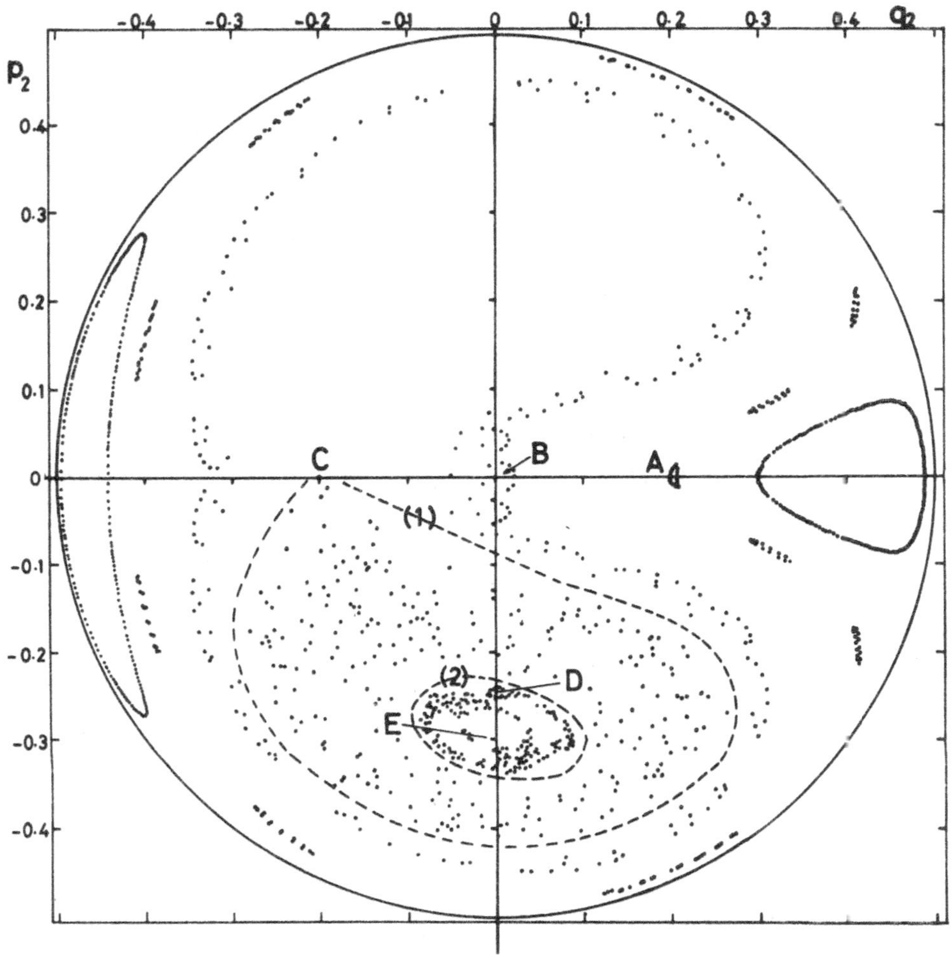

Fig. 1 Poincaré mapping for $\varepsilon=0.5$ and energy $h=0.125$. Mapped points are those for initial points A(0.2,0), B(0,0), C(-0.2, 0), D(0, -0.25).

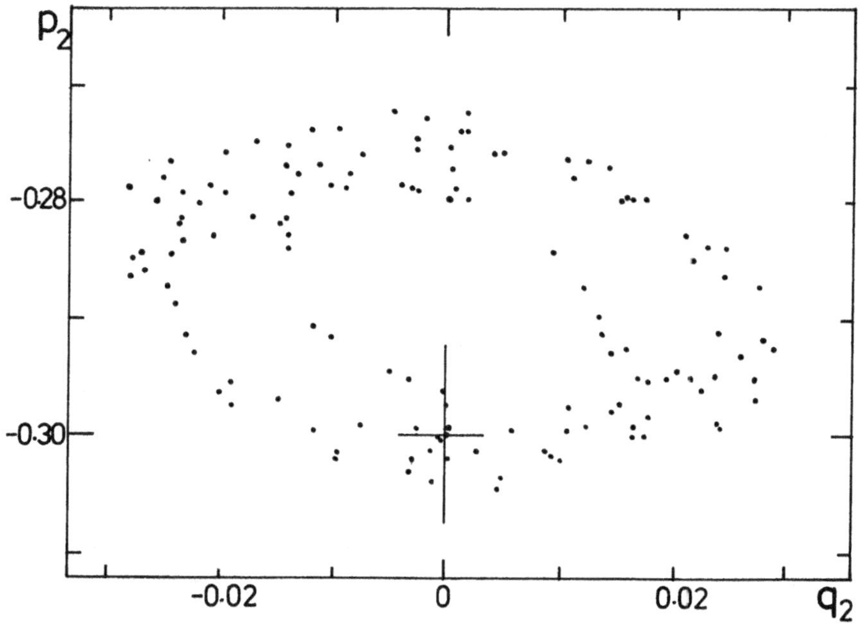

Fig. 2 Poincaré mapping in the region denoted by E in Fig.1.

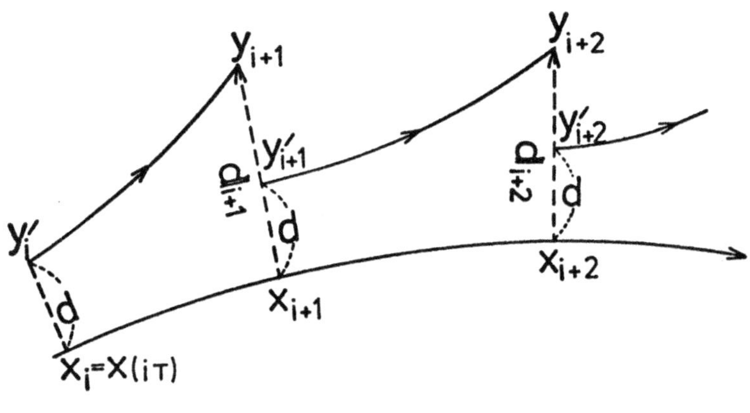

Fig. 3 Schematic diagram for calculating Lyapunov characteristic number.

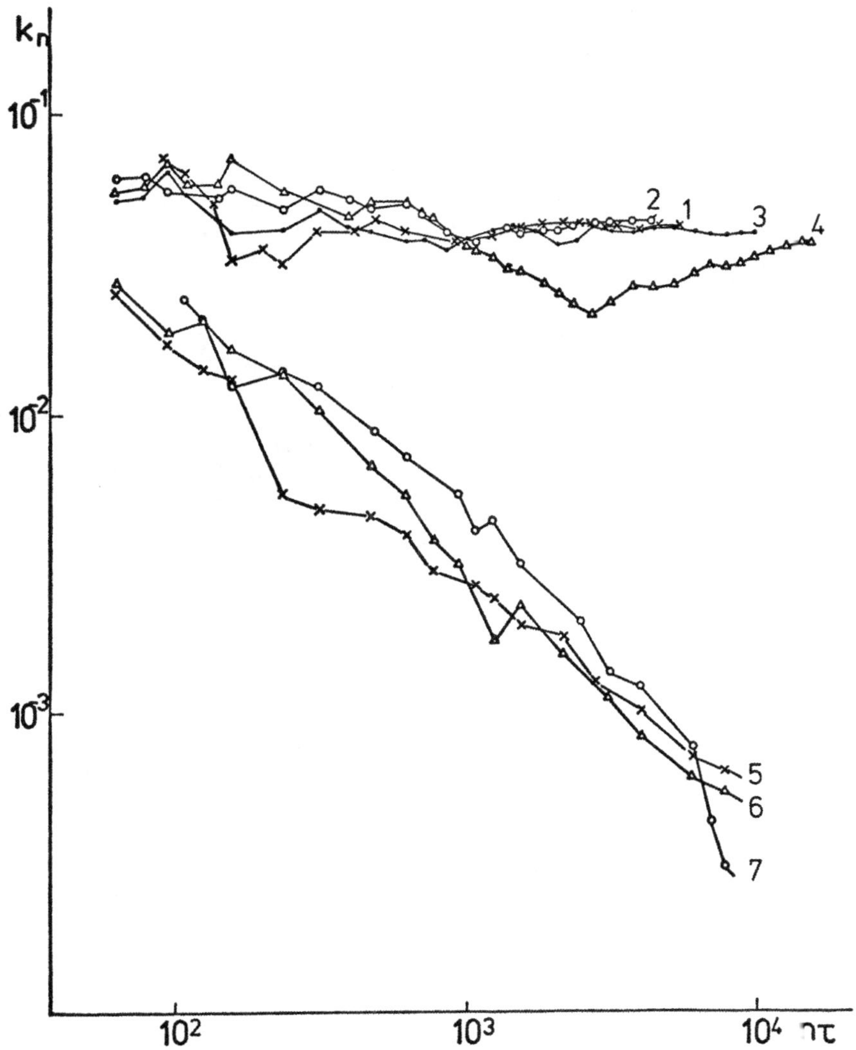

Fig. 4 Behavior of k_n's in the stochastic region (curves 1-4) and in the regular region (curves 5-7).

initial points $x_0 = (q_2, p_2)$

1	(0. , 0.2)
2	(-0.2 , 0.025)
3	(-0.2 ,-0.025)
4	(0. ,-0.2)
5	(0.2 , 0.)
6	(0.25 , 0.)
7	(0.3 , 0.)

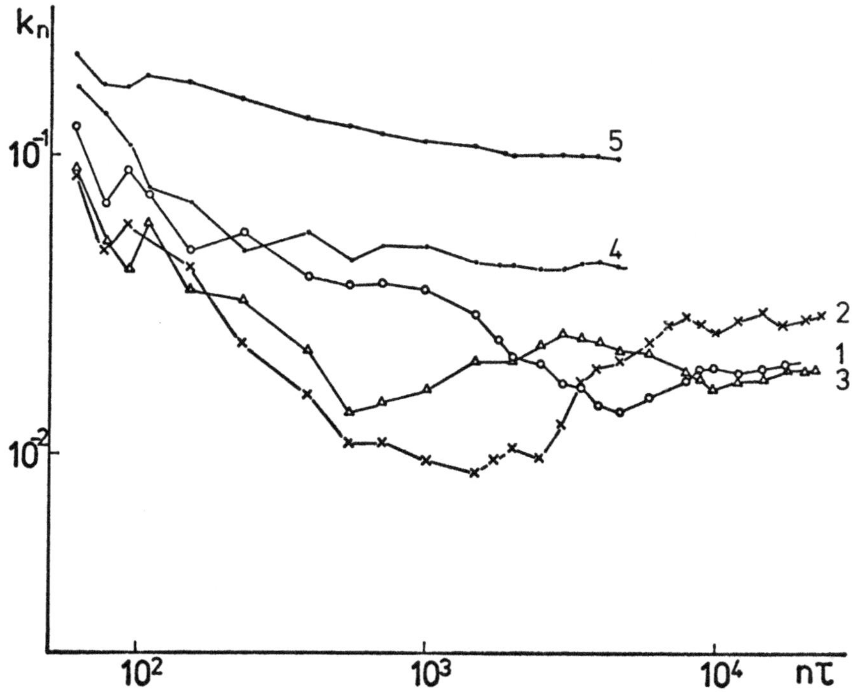

Fig. 5 Behavior of k_n's in the stochastic region.

A QUESTION ABOUT THE LOCALIZED MODE DUE TO A LIGHT IMPURITY

M.TODA
Faculty of Engineering,
Yokohama National University
Yokohama 232, Japan

1. The aim of may present talk is related to the effect of an impurity in a nonlinear lattice. Recently it has been revealed that a nonlinear lattice with a light impurity behaves more or less stochastically. In this connection it will be a basic problem to examine the effect of an impurity in an infinite nonlinear lattice, especially the effect on the so-called localized mode due to a light impurity.

2. More than two decades have passed since Prof. Montroll and others studied in detail the effect of an impurity and impurities in harmonic lattices. One of the outstanding discoveries was the localized mode. Suppose we have a one-dimentional harmonic lattice with an isotopic impurity of the mass m_0. Then we have the equation of motion

$$m_0\ddot{y}_0 = \kappa(y_1-2y_0+y_{-1})$$

and for the host particles of the mass m, we have

$$m\ddot{y}_n = \kappa(y_{n+1}-2y_n+y_{n-1}) \qquad (n\neq 0).$$

The localized mode is of the form

$$y_n = Ae^{-\alpha|n|}e^{i\omega_{loc}t}$$

with the frequency ω_{loc}, which can be written as

$$\omega_{loc} = 2\sqrt{\frac{\kappa}{m}}\cosh\frac{\alpha}{2}, \quad e^{\alpha} = 2\frac{m}{m_0} - 1.$$

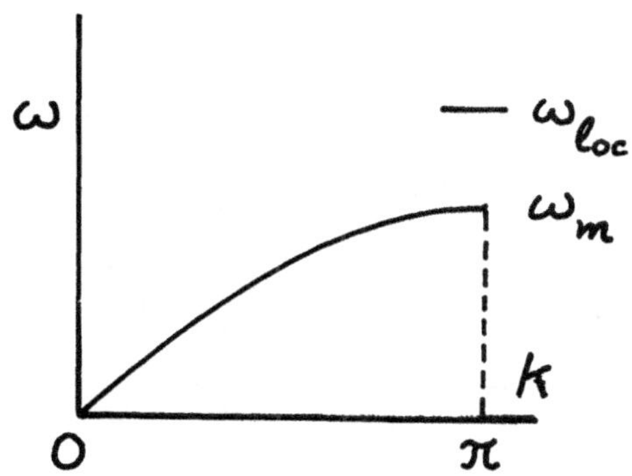

Fig.1 Frequency band and the localized mode frequency of a harmonic lattice.

Fig.2 The localized mode around a light impurity.

The localized mode is above the frequency band

$$\omega = 2\sqrt{\frac{K}{m}}\sin\frac{k}{2} \qquad 0 \le k \le \pi$$

where k is the wave number of the normal mode (extended mode)

$$y_n = Ae^{i(kn-\omega t)}.$$

It is interesting to note the fact that the localized mode frequency is derived from the extended mode frequency by replacing k by a complex number

$$k = \pi + i\alpha \ .$$

But this kind of manipulation does not seem to apply to nonlinear lattices.

3. While a harmonic lattice is the limit of small nonlinearity, the other limit is the system of <u>hard spheres</u>. In this case we have also an exact expression of the mode which corresponds to the localized mode of a harmonic lattice. This mode can be visualized by drawing $t \sim x$ diagram, where x_n denotes the position of the n-th particle

$$x_n = \bar{x}_n + y_n$$

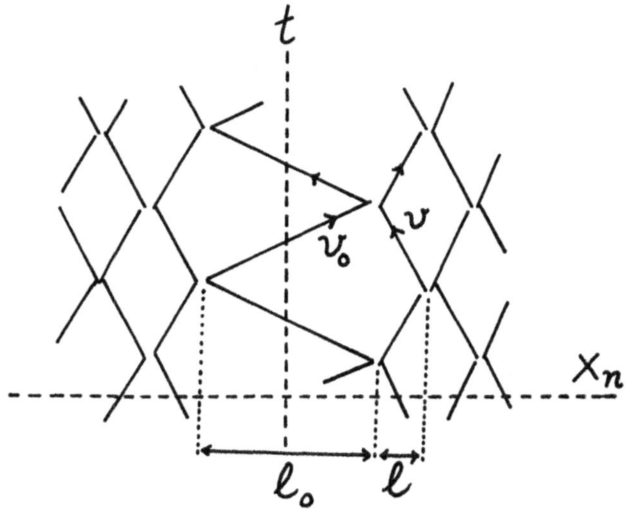

Fig.3 t~x diagram of the quasi-localized mode of a hard sphere system.

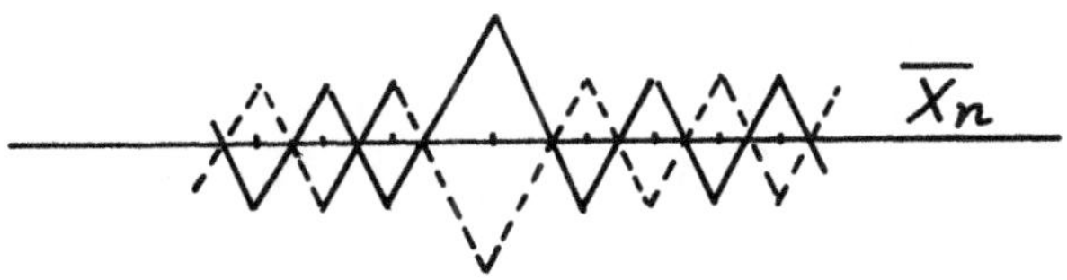

Fig.4 The quasi-localized mode of a hard sphere system.

with $\overline{x}_n$ the average position. In this diagram the diameter of each particles is ignored, and the lines indicate the motion of the particles. For each collision the conservation of the momentum

$$m_0 v_0 = mv$$

holds, and this implies an elongation of the lattice around the impurity:

$$\frac{l_0}{l} = \frac{v_0}{v} = \frac{m}{m_0} \ .$$

For a hard sphere system without impurity the cnoidal wave reduces to a $t \sim x$ diagram consisting of straight lines.

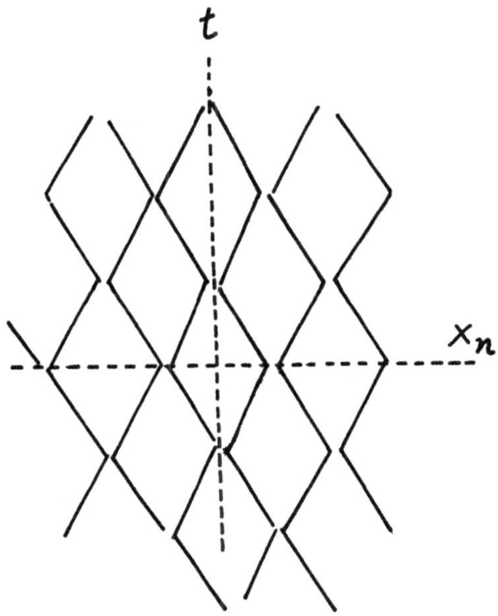

Fig.5 The maximum frequency mode of a hard sphere system.

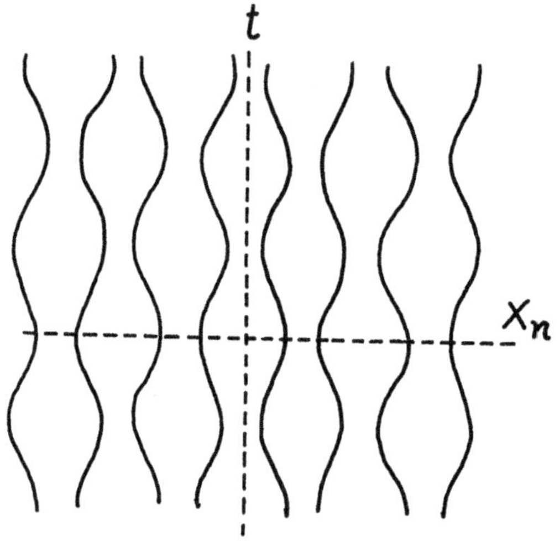

Fig.6 The maximum frequency mode of a harmonic lattice.

4. Now, let us consider the cnoidal wave in a pure nonlinear lattice with the interaction potential

$$\phi(r) = \frac{a}{b} e^{-b(r-\sigma)} + ar + \text{const.}$$

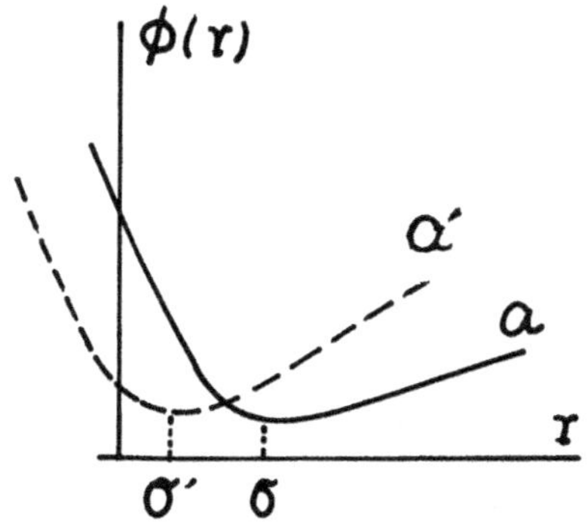

Fig.7 The exponential type of interaction
a : without pressure. a' : with an external pressure.

The equations of motion can be written as

$$m \frac{d^2 r_n}{dt^2} = a\{2e^{-b(r_n-\sigma)} - e^{-b(r_{n+1}-\sigma)} - e^{-b(r_{n-1}-\sigma)}\} .$$

Corresponding to the sinusoidal wave of the harmonic case, we have the cnoidal wave which can be written as

$$e^{-b(r_n-\sigma)} - 1 = \frac{(K\omega/2\pi)^2}{ab/m} [\mathrm{dn}^2\{2K(\frac{n}{\lambda} - \frac{\omega}{2\pi} t)\} - \frac{E}{K}]$$

with the dispersion relation

$$\omega = \frac{\pi}{K(\kappa)} \sqrt{\frac{ab}{m}} \Bigg/ \sqrt{\frac{cn^2 2K(\kappa)/L}{sn^2 2K(\kappa)/L} + \frac{E(\kappa)}{K(\kappa)}}$$

With increasing amplitude, the modulus κ approaches to unity, and ω vanishes. This is because of the expansion of the lattice due to the effect of the motion on the nonlinearity. The expansion of the lattice is given by

$$\overline{r_n - \sigma} = \frac{1}{b} \ln C$$

where

$$C = \left[\frac{\theta_0(1/\lambda)}{\theta_0(0)} \right]^2 \left(1-\left(1-\frac{E}{K}\right) sn \frac{2K}{\lambda}\right)$$

To prevent the expansion of the lattice, we have to apply a pressure. With the pressure f, the lattice can be expressed by the substitution.

$$a \to a' = a + f = aC$$

$$\sigma \to \sigma' = \sigma - \frac{1}{b} \ln C.$$

Then it can be shown that the maximum frequency of the pure lattice is given by

$$\omega_m = \omega(\lambda=2) = \frac{\pi\sqrt{ab/m}}{(1-\kappa^2)^{1/4} K(\kappa)}$$

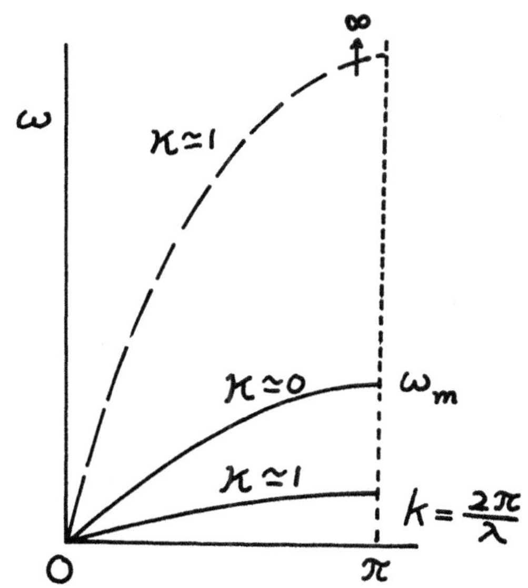

Fig.8 Dispersion relation of a nonlinear lattice.
solid curve : free expansion (no external pressure).
dashed curve : with an external force to prevent expansion.

which increases indefinitely as the amplitude or the modulus increases ($\kappa \to 1$), and approaches to the harmonic case as the modulus diminishes ($\kappa \to 0$):

$$\omega_m \to \begin{cases} \sqrt{\dfrac{ab}{m}}\, e^{K(\kappa)/2} / K(\kappa) & (\kappa \to 1) \\ 2\sqrt{\dfrac{ab}{m}} & (\kappa \to 1)\ . \end{cases}$$

It can be said therefore that there is no definite frequency band for a nonlinear lattice, and this fact will not admit the strictly localized mode in a nonlinear lattice. As we have seen in the case of the hard sphere system, the oscillation around the impurity of the quasi-localized mode will extend to infinity in the case of nonlinear lattices.

5. Some of my colleagues have examined the effect of a light impurity in an exponential lattice: H. Nagashima and S. Watanabe observed a considerably localized mode using a nonlinear LC-circuit, and recently S. Takeno has shown the same result numerically.

NONLINEAR OSCILLATION REGIMES IN SOME PHYSICAL PROBLEMS

O. I. Bogoyavlensky
L. D. Landau Institute for Theoretical Physics,
the Academy of Sciences of the USSR, Moscow, USSR.

ABSTRACT

Complex nonlinear oscillation regimes having stochastic properties and allowing detailed research of their dynamics due to presence of their separatrix approximation have been found of late in a number of problems in general theory of relativity, gas dynamics, theory of perturbations of completely integrable systems and theory of finite-dimensional approximations of hydrodynamics equations (so called systems of hydrodynamic type) [1-5]. Formulation of some of these problems is put in Sections 1, 2, 3 of the present paper, Section 4 presents the unified derivation of oscillation regimes, contained in them and the conditions sufficient for the existance of such regimes in dynamical systems of more general type. The properties of the given oscillatory regimes for concrete problems discussed in Sections 1, 2, 3 are enumerated in Section 5.

SECTION 1

Homogeneous cosmological models in general theory of relativity are defined by the condition, that four-dimensional space - time manifold M^4 with Einsteinian metric g_{ij} has a group of isometries G with space-like three-dimensional orbits (i.e. the metric on orbits is negative). It is also usually supposed that matter has a stress-energy tensor

$$T_{ij} = (p+\varepsilon)u_i u_j - pg_{ij}, \quad p = k\varepsilon, \quad 0 \leqslant k \leqslant 1 \qquad (1)$$

(where p - pressure, ε - density of energy, u^i - velocity of matter and that system of references is synchronous, that is $g_{oo} = 1$, $g_{o\alpha} = 0\,(\alpha = 1,2,3)$).

The most complicated are opened cosmological models of type VII and VIII by Bianchi and closed model of type IX with isometry group $G = SU(2)$. Let us consider this last case with additional assumption that matter doesn't move on the average, that is $u_i = \delta_{io}$.

At the given assumptions there exists a basis of the G -invariant vector fields X_1, X_2, X_3 tangent to the orbits of group SU(2), in which the metric is diagonal, that is $\langle X_i, X_j\rangle = g_{ij} = -\exp(q_i(t))\delta_{ij}$. Einstein equations

$$R_{ij} - 1/2\, g_{ij}R = \frac{8\pi k}{c^4} T_{ij} \tag{2}$$

after change of time $d\tau = dt\cdot\exp(-k(q_1+q_2+q_3)/2)$ are equivalent to Hamiltonian system in phase space $R^6(p_i, q_i)$ $(i = 1,2,3)$:

$$\dot{p}_i = -\partial H/\partial q_i,\quad \dot{q}_i = \partial H/\partial p_i,$$

$$H = \left[2\sum_{i<j}^{3} p_i p_j - \sum_{i}^{3} p_i^2 + \frac{1}{4}\left(2\sum_{i<j}^{3} \exp(q_i+q_j) - \sum_{i}^{3} \exp 2q_i\right)\right]$$

$$\times \exp(-(1-k)(q_1+q_2+q_3)/2). \tag{3}$$

The system (3) has a monotone function

$$F = 3\frac{d}{dt}\exp((q_1+q_2+q_3)/6) = (p_1+p_2+p_3)\exp(-(q_2+q_2+q_3)/3),$$

$$dF/dt \leqslant 0. \tag{4}$$

One of the most important physical problems in study of homogeneous cosmological model of type IX is a question of typical (general) behaviour regimes of metric g_{ij} at space contraction ($F\to +-\infty$) and at space expansion at the early stage ($F \gg 1$) near cosmological singularity $\det|g_{ij}| = 0$. Oscillation regime of Belinsky, Lifchits, Khalantnikov [6] is the general behaviour regime of metric at space contraction. This oscillatory regime is derived below, in Section 5, with the help of separatrix approximation of trajectories of system (3) being a special case of some system of more general type. Complete enumeration of all typical behaviour regimes of metric at space expansion at the early stage and all possible passages between them, including correct putting of problems of typical initial states for the first time in the joint work of S. P. Novikov and the author [1] on the basis of complete research of (3) by the methods of qualitative theory of differential equations.

SECTION 2

The Toda lattice is known to be an infinite system of unit mass particles, whose interaction is determined by the potential

$$V = \sum_i \exp(q_i - q_{i+1})$$

where q_i is the displacement of the i-th particle from the equilibrium. The periodic Toda lattice is determined by $q_{i+n+1} \equiv q_i$ condition and has the Hamiltonian

$$H = \frac{1}{2}\sum_{i}^{n+1} p_i^2 + \sum_{i}^{n} \exp(q_i - q_{i+1}) + \exp(q_{n+1} - q_1) \qquad (5)$$

The problems of behaviour of the Toda lattice perturbations was considered in our work [4] [1]. The systems with arbitrary masses of particles and with potentials, in which all the particles interact with each other for example

$$V_1 = \sum_i \exp\left((q_i - q_{i-1}) + \sum_k \varepsilon_{ik}(q_i - q_{i+k})\right).$$

are included into admissible perturbations.

The whole class of admissible perturbations is described by Hamiltonian systems

$$\dot{p}_i = -\partial H/\partial q_i, \quad \dot{q}_i = \partial H/\partial p_i, \qquad (6)$$

$$H = \frac{1}{2}\sum_{i,j}^{n} a_{ij} p_i p_j + \sum_{k,m}^{n+1} b_{km} \exp(\{\alpha_k, q\} + \{\alpha_m, q\}).$$

Here $\alpha_1, \ldots, \alpha_{n+1}$ are vectors in n-dimensional space R^n with coordinates $\alpha_k = (d_{k1}, \ldots, d_{kn})$, q being the vector $(q_1, \ldots, q_n)$. In R^n two scalar products (x,y) and $\{x,y\}$ are given

$$(x,y) = \sum_{i,j}^{n} a_{ij} x_i y_j, \quad \{x,y\} = \sum_{i}^{n} x_i y_i$$

Vectors α_k and quadratic forms a_{ij}, b_{km} satisfy the following

---

1. This problem was independently discussed in the work of Casati and Ford [7] as well. Dynamics of the two particles system with arbitrary masses and Toda potential has been investigated numerically in this work. The results [7] evidently point out the presence of stochasticity at H>>1. This agrees with the results of paper [4]

conditions:

1. For any vector p in R^n

$$\max_k (\alpha_k, p) > 0$$

2. For all k

$$(\alpha_k, \alpha_k)\, b_{kk} > 0$$

Under these conditions complicated oscillation regime with ergodic properties is realized, and the Toda lattice is isolated by some deep algebraic symmetry.

SECTION 3

The definition of hydrodynamic type systems was given in paper [8]. Such systems arise at finite-dimensional approximation of hydrodynamic equations and have the form

$$\dot{u}^i = \Gamma^i_{jk} u^j u^k \tag{7}$$

at constant Γ^i_{jk}. According to [8] hydrodynamic type system has the integral of energy of type $E = a_{ij} u^i u^j$ and conserves phase volume (i.e. div $u^i = 0$). The above mentioned systems (6) on mapping

$$Q_k = \exp(\{\alpha_k, q\}), \quad k = 1,2,\ldots,n+1$$

transform into hydrodynamic type systems (at $\alpha_1 + \ldots + \alpha_{n+1} = 0$).

SECTION 4

Let us define general class of dynamical systems, including systems (3), (5), (6) and some hydrodynamic type systems, for which the existance of complicated nonlinear oscillation regimes will be proved.

This class consists of systems of differential equations of the following type (variables are divided into two groups $Q^\alpha \in R^m, p^i \in R^n$):

$$\begin{aligned} \dot{Q}^\alpha &= a^\alpha_{\beta\gamma} Q^\beta Q^\gamma + Q^\alpha (b^\alpha_i p^i + \lambda^\alpha), \\ \dot{p}^i &= c^i_{\beta\gamma} Q^\beta Q^\gamma + d^i_{\beta j} Q^\beta p^j + \lambda^i_\beta\, Q^\beta \end{aligned} \tag{8}$$

(summation over repeated indices is understood here and throughout this paper).

Proposition I. The following conditions A, B, C are sufficient for the existance in system (8) nonlinear oscillation regime, admitting separatrix approximation:

A. For almost all $p^i \in R^n$

$$\max_\alpha (b^\alpha_i p^i + \lambda^\alpha) = b^\gamma_i p^i + \lambda^\gamma > 0, \quad \gamma = \gamma(p)$$

For all α vector $b^\alpha_i \neq 0$.

B. For each α matrixes of (n+1) x (n+1) order

$$S_\alpha = \begin{vmatrix} a^\alpha_{\alpha\alpha} & b^\alpha_j \\ c^i_{\alpha\alpha} & d^i_{\alpha j} \end{vmatrix}$$

have Jordan's diagonal normal form and all their eigenvalues λ_i have Re $\lambda_i = 0$.

C. For all $\beta \neq \alpha$ $a^\alpha_{\beta\beta} = 0$.

Let us introduce coordinates

$$q^\alpha = Q^\alpha/\Omega, \quad \Omega = ((Q^1)^2 + \ldots + (Q^m)^2)^{1/2}. \tag{9}$$

System (8) in coordinates q^α, p^i, Ω has the form

$$\begin{aligned} \dot{q} &= a^\alpha_{\beta\gamma} q^\beta q^\gamma \Omega - q^\alpha (b^\alpha_i p^i + \lambda^\alpha) \\ &- q^\alpha (a^\delta_{\beta\gamma} q^\delta q^\beta q^\gamma \Omega - (q^\delta)^2 (b^\delta_i p^i + \lambda^\delta), \\ \dot{\Omega} &= \Omega (a^\delta_{\beta\gamma} q^\delta q^\beta q^\gamma \Omega + (q^\delta)^2 (b^\delta_i p^i + \lambda^\delta)), \\ \dot{p}^i &= \Omega (c^i_{\beta\gamma} q^\beta q^\gamma \Omega + d^i_{\beta j} q^\beta p^j + \lambda^i_\beta q^\beta). \end{aligned} \tag{10}$$

System (10) is considered according to definition of coordinates (9) on invariant manifold $(q^1)^2 + \ldots + (q^m)^2 = 1$, $\Omega > 0$ and continues to the boundary $\Omega = 0$ being invariant manifold as well. The system has 2m sets of singular points $M_\alpha^\varepsilon = R^n (\alpha=1,\ldots,M;\ \varepsilon = \pm 1)$ with coordinates

$$q^\beta = \varepsilon\, \delta_\alpha^\beta, \quad \Omega = 0, \quad p^i \in R^n.$$

The points belonging to sets M_α^ε are marked (p,α,ε). Eigenvalues of system (10) in the singular points are the following (corresponding eigen directions are pointed out in brackets):

$$\mu_\beta = (b_i^\beta p^i + \lambda^\beta) - (b_i^\alpha p^i + \lambda^\alpha) \quad \text{(variables } q^\beta,\ \beta = 1,\ldots,m\ \beta \neq \alpha)$$

$$\mu_m = b_i^\alpha p^i + \lambda^\alpha \quad \text{(variable } \Omega),$$

$$\mu_{m+1} = \ldots = \mu_{m+n} = 0 \quad \text{(variables } p^i),$$

It follows from condition A that almost all singular points M_α^ε, except for submanifolds of less dimension, are nondegenerated and unstable (the singular point is called nondegenerated, provided the number of its zero eigenvalues is equal to the dimension of the set M_α^ε).

Define two subsets V_α^ε and W_α^ε on each M_α^ε :

on V_α^ε $(b_i^\alpha p^i + \lambda^\alpha) < 0.$

on W_α^ε $\alpha = \gamma(p)$.

Separatices going from the point $(p_o,\alpha,\varepsilon) \in V_\alpha^\varepsilon$ (at $t \to -\infty$) have form:

$$q^\beta = c^\beta \exp((b_i^\beta p_o^i + \lambda^\beta)t) \cdot (\Sigma_\delta (c^\delta)^2 \exp(2t(b_i^\delta p_o^i + \lambda^\delta)))^{-1/2},$$

$$\Omega(t) \equiv 0, \quad p^i(t) = p_o^i, \quad c^\alpha = \varepsilon, \quad -\infty<t<+\infty \tag{11}$$

(here β are indexes for which $\mu_\beta > 0$ for the rest of indexes β $q^\beta \equiv 0$). Let $\gamma = \gamma(p_o)$. If $c^\gamma>0$ all separatrices (11) at $t\to+\infty$ pass to singular point $(p_o,\gamma(p_o), +1) \in W_\gamma^{+1}$, if $c^\gamma<0$- to singular point $(p_o,\gamma(p_o), -1) \in W_\gamma^{-1}$. Exceptional separatrices for which $c^\gamma = 0$ are unstable. We shall not discuss them below while defining the map T_1.

Separatrices outgoing from singular points $(p_o,\alpha,\varepsilon) \in W_\alpha^\varepsilon$ are situated on invariant (owing to condition c) manifold $q^\beta = \varepsilon\delta_\alpha^\beta$. System (10) on this manifold after change of time $d\tau/dt = \Omega > 0$ transforms into linear system

$$\dot{X} = \varepsilon S_\alpha \cdot X + \varepsilon \cdot Y \tag{12}$$

where vectors X, Y have coordinates $X = (\varepsilon\Omega, p^i)$, $Y = (\lambda^\alpha, \lambda^i_\alpha)$. Owing to condition B each trajectory of system (12) starting on plane $\Omega = 0$ at $b^\alpha_i p^i + \lambda^\alpha > 0$ (separatrix, starting from singular points W^ε_α) crossed this plane again (at $b^\alpha_i p^i + \lambda^\alpha << 0$). Denote thus defined map by T^ε_α . Consequently the map T_1 (two-valued) transforming initial singular point into final point of separatrix which starts from the former point is determined on sets of singular points V^ε_α, W^ε_α.
The map T_1 has the form:

$$(p,\alpha,\varepsilon) \in W^\varepsilon_\alpha : \quad T_1(p,\alpha,\varepsilon) = (T^\varepsilon_\alpha(p),\alpha,\varepsilon) \in V^\varepsilon_\alpha ,$$

$$(p,\alpha,\varepsilon) \in V^\varepsilon_\alpha : \quad T_1(p,\alpha,\varepsilon) = \begin{cases} (p,\gamma(p),+1) \in W^{+1}_\gamma , \\ (p,\gamma(p),-1) \in W^{-1}_\gamma , \end{cases} \tag{13}$$

In the latter case map T_1 is two-valued.

The trajectory of system (10) starting from small enough neighborhood of singular sets will move along some succession of separatrices in diagram

$$\begin{array}{ccccccc} & & & W^{+1}_\alpha & \longrightarrow & V^{+1}_\alpha & \nearrow \cdots \\ & & \nearrow & & & & \searrow \cdots \\ \cdots \longrightarrow & V^\Sigma_\alpha & & & & & \\ & & \searrow & & & & \nearrow \cdots \\ & & & W^{-1}_\alpha & \longrightarrow & V^{-1}_\alpha & \searrow \cdots \end{array} \tag{14}$$

as long as desired (the arrows mean separatrices, passing between singular sets). That is oscillatory regime in system (8), (10) allows separatrix approximation. Later we shall point out class of systems (8) in which space Q^α, p^i is divided into invariant regions; in each of these regions all passages in diagram (14) are single-valued.

In initial coordinates oscillation regime has the following dynamics. The trajectory, starting at small enough meanings of coordinates Q^α, periodically turns out in the neighborhood of some points of plane $P(Q^1=\ldots=Q^m=0)$. These points result one from another by successive action of map T. Plane P is divided into subsets P_α: there is $\gamma(p) = \alpha$ on P_α. On each P_α the map $T = T^\varepsilon_\alpha$ (see (13)), i.e. $T(p) = T^\varepsilon_{\gamma(p)}$, $\varepsilon = \pm 1$. Coordinate $Q^{\gamma(p)}$ changes greatly while trajectory passes between two successive points of plane P while sign $Q^{\gamma(p)} = \varepsilon$, and for the rest of coordinates $|Q^{\beta}| <<<< | Q^{\gamma(p)}|$. Stochastic properties of oscillation regime are defined by map T properties.

Map T (and separatrix diagram (14)) comes to simple maps in important particular case of systems (8); if coefficients

$$a^{\alpha}_{\beta\gamma} = 0 \quad \text{at} \quad \beta,\gamma\neq\alpha \ . \tag{15}$$

In this case all planes $Q^{\alpha} = 0$ are invariant manifolds of system (8) and consequently each of 2^m regions σ, in which sign $Q^{\alpha} = \varepsilon_{\alpha}$, is also invariant. If the trajectory in region σ moves along some separatrix (11) then sign $c^{\gamma} = \varepsilon_{\gamma}$ (see (11))., this is why all passages in diagram (14) for region σ are single-valued. Corresponding map $T = T_{\sigma}$ on the set P_{α} has the form $T_{\sigma} = T_{\alpha}^{\varepsilon\alpha}$ and is also single-valued. All maps T_{σ} and corresponding oscillation regimes, generally speaking, are different.

If a system of type (8) at additional condition (15) is considered only in some one region σ (sign $Q^{\alpha} = \varepsilon_{\alpha}$) sufficient conditions of oscillation regime existance may be weakened: instead of condition B it's enough to demand, that $\mathrm{Re}\lambda_i \geqslant 0$ where λ_i are eigenvalues of matrixes $\varepsilon_{\alpha}\cdot S_{\alpha}$.

SECTION 5

The examples of systems, satisfying the conditions of proposition I are the systems described in Sections 1, 2, 3. Consider concrete aspect of oscillation regimes in these cases.

1) Hamiltonian system (3) on the level $H = 0$, describing homogeneous cosmological model of type IX in empty space after mapping $Q_i = \exp q_i$ and time change

$$d\tau_1/d\tau = \exp(-(1-k)\ (q_1+q_2+q_3)/2)$$

accepts form (8) and satisfies condition A in singular points $H = 0$, $Q_i = 0$ and conditions B, C, (15). In this case oscillation regime described above in Section 4 gives the following combinatorial model of alteration of momenta P_i.

Consider plane $\mathcal{L}$: $p_1 + p_2 + p_3 = -1$. Cone K of singular points $H = 0$, $Q_i = 0$ or $2\Sigma^3_{i<j}p_ip_j - \Sigma^3_i p_i^2$ cuts out on plane $\mathcal{L}$ circumference S^1, which inscribed into equilateral triangle Δ, cut out on plane $\mathcal{L}$ by quadrant $P_i \leqslant 0$. Circumference S^1 is divided into three arcs (1,2,3) by the point of tangency with triangle. For the system

in consideration map T acts on rays of cone K and therefore produces map τ on circumference S^1. Map τ , as it may easily be shown by integration of system (12) for this case, is a projection of arc number i (1,2,3) of circumference S^1 by the rays, outgoing from top of triangle number i on two other arcs as shown in the following figure.

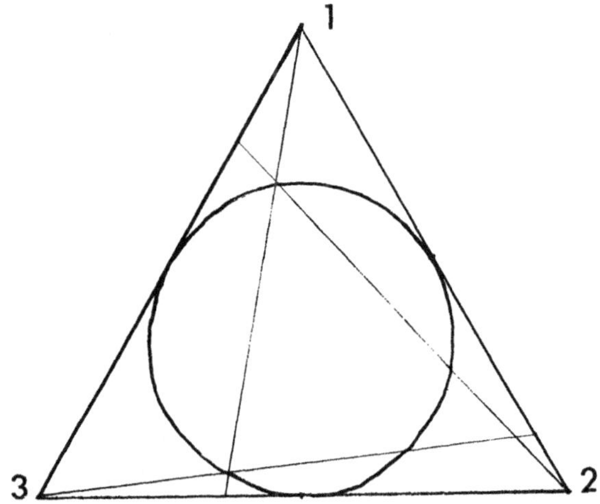

Figure 1

Map τ extends length of arc everywhere and only in points of tangency $|d\tau/dp| = 1$. There exists everywhere dense countable set of periodical points of map τ , all these points being hyperbolic (unstable). There is a piecewise smooth invariant at automorphism τ measure; complete measure of the whole of circumference turns out to be infinite. The described combinatorial model of oscillation regime may be shown to be isomorphous to well-known Belinsky-Lifshits-Khalatnikov model of changes of "Kasner exponents" of metric g_{ij} [6].

2) Hamiltonian systems with hamiltonians (6), describing perturbations of periodic Toda lattice, transform into system of type (8) at mapping $Q_k = \exp(\{\alpha_k, q\})$ and satisfy conditions A, B, C, (15). Corresponding system (10) is defined in region $Q_k \geq 0$.

Oscillation regime at a positively determined metric a_{ij} is realized at the energy $H \gg 1$. There are long sections I_N of the trajectory along which momentum p is approximately constant, momentum p changes between two successive sections I_N, I_{N+1} and

$$p(I_{N+1}) = T(p(I_N)) = \tau_k(p(I_N)) \tag{16}$$

where number k is determined by the condition

$$(\alpha_k, p(I_N)) = \max_m(\alpha_m, p(I_N)) \tag{17}$$

and map τ_k is the reflection in the plane orthogonal to vector α_k (in the metric a_{ij}):

$$\tau_k(p) = p - \frac{2(\alpha_k, p)}{(\alpha_k, \alpha_k)} \alpha_k. \tag{18}$$

The map T given in (16)-(18) is almost everywhere determined in R^n and is piecewise continuous.

The map T for the Toda lattice (5) is periodic-here vectors α_k are the roots of the simple Lie algebra of type A_n (SL(n+1)). The map T will be periodic for the Hamiltonian system (6) provided the Coxeter groups G, generated by reflections (18) is finite (if d is the order of group G, then $T^{d!+d} = T^d$, the map T, however, may be irreversible). All the finite Coxeter groups G are known and except for three special cases, they are Weyl groups of the simple Lie algebras (provided G is not decomposed into a product of two other groups). The exceptions are two Coxeter groups in three- and four-dimensional spaces and the infinite series of dyadral groups (symmetry groups of regular polygons) in two-dimensional space.

The periodicity property of the map T isolates the Toda lattice (5) among its general perturbations (6). In the general case (6) the Coxeter group G is infinite and its closure (at positively determined metric a_{ij} coincides with the orthogonal group O(n); for general vector p the set $T^k(p)$ (where k are all natural numbers) is everywhere dense on the sphere (p,p) = const. The general Hamiltonian system (6) is unlikely to have any integrals except for energy H.

However, there are several exceptional systems of the type (6) with the periodic map T and a large set of integrals. Namely, hamiltonian systems of type (6) for each simple Lie algebra are presented in paper [4], allowing as well as the Toda lattice (5) L - A pair representation and having n integrals.

Note, hamiltonian system (3) at the level H = 0 with time change $d\tau_1/d\tau = \exp(-(1-k)(q_1+q_2+q_3)/2)$ turns into system of type (6). Metric a_{ij} for this case is indefinite and Coxeter group G is infinite but G acts in R^3 discretely.

3) Systems of hydrodynamic type (7) allowing representation (8) (with $\lambda^{\alpha} = \lambda_{\beta}^{\ i} = 0$) and having energy integral E of form $E = (u^1)^2 + \ldots + (u^n)^2$ automatically satisfy conditions A, B (due to conservation of volume and due to existance of energy integral E). Realization of condition c, generally speaking, should be demanded in addition.

The simplest example of hydrodynamic type systems are Euler equations of rigid body movement, or triplet

$$\dot{v}_1 = \ell(v_2^2 - v_3^2), \quad \dot{v}_2 = -\ell v_1 v_2, \quad \dot{v}_3 = \ell v_1 v_3$$

obviously satisfying conditions A, B, C, (15). Here unstable mode v_1 is coordinates p^i, and v_2, v_3 - linear combinations of stable modes are coordinates Q^{α}.

There is a special class of hydrodynamic type systems automatically satisfying conditions A, B, C, (15). Such systems may be received by superposition of triplets according to the following rule: if two triplets are engaged by some mode, this mode in both of them is either stable or unstable. Double engagements of triplets are admissible, and each triplet may be engaged with any number of others. Hydrodynamic type systems obtained in this way have additional integral $F = Q^1 \cdot \ldots \cdot Q^m$, in oscillation regime dimensionless integral $F \cdot E^{-m/2} << 1$.

For such hydrodynamic type systems matrixes S_{α} have rank 2 and each mapping $T_{\alpha}^{+1} = T_{\alpha}^{-1}$ is reflection τ_{α} in the plane orthogonal to vector (b_i^{α}). Thus Coxeter group G, generated by reflections τ_{α} may be associated with map T, as well as for perturbations of the Toda lattice. Closure of group G is known to coincide with orthogonal group O(n) (for m>n) for general set of vectors (b_i^{α}). In general case map T is evidently ergodic on spheres of constant radius (common Euclidean measure is invariant measure for reversible map T) though a proof of this fact is already absent in two-dimensional case where the problem is reduced to shifting of segments on circumference.

Note: Analogous method of separatrix approximation of dynamical system trajectories was applied in the author's paper [3] to the research of dynamics of gravitational gaseous ellipsoid in classical gas dynamics. Using these methods the author showed that in purely gas dynamical problem arising of arbitrary final number of ellipsoid 's oscillations, which are then substituted by infinite expansion is possible. Complicated nonlinear oscillation regime of changes of ellipsoid's semi-axises arises at the presence of gravitational interaction of gas particles. This regime is accompanied by rotation of ellipsoid in space. This oscillatory regime has much in common with the described above regimes. Corresponding dynamical system, however, doesn't satisfy the conditions of proposition I, and therefore requires separate consideration.

REFERENCES

1. Bogoyavlensky, O. I., Novikov, S. P.: Soviet Zhurn. Eksp. Teor. Fiz. 64, 1475-1495 (1973).
2. Bogoyavlensky, O. I.,: Soviet Zhurn. Eksp. Teor. Fiz. 70, 361-373 (1976).
3. Bogoyavlensky, O. I.,: Soviet Appl. Math. Mech. 40, 2, 270-280 (1976).
4. Bogoyavlensky, O. I.,: Commin. Math. Phys. 51, 3, 201-209 (1976).
5. Bogoyavlensky, O. I.,: Soviet Reports of Academy of Sciences 232, 6, 1289-1292 (1977).
6. Landau, L. D., Lifshits, E. M.,: Classical Theory of Fields. Moscow: Nauka 1973.
7. Casati, G., Ford, J.: Phys. Rev. A12, 1702-1709 (1975).
8. Oboukhov, A. M.: Fluid Dynam. Trans. 5, 193-200 (1971).

METRIC UNIVERSALITY IN NONLINEAR RECURRENCE

M. J. Feigenbaum
Theoretical Division, Los Alamos Scientific Laboratory
University of California, Los Alamos, New Mexico 87545/USA

Abstract

A brief resumé of some new, universal features of recurrence relations is presented in a context of possible biological consequences.

It has been argued[1] that recursion equations of the form

$$X_{n+1} = f(X_n, \lambda_i) \qquad i = 1 \ldots, N \tag{1}$$

even in the simplest case of $N = 1$ can model certain discrete populations. In practice, f looks like Fig. 1, and is presumably chosen so as to agree with (low precision) field data. Whether there exists <u>any</u> <u>f</u> so that Eq. (1) actually affords a correct modelling is unknown.

We have recently completed[2,3] a study of equations of the form (1), with a single parameter λ which has yielded a number of remarkable results: Features emerge of a precise quantitative nature, the numerical values of which are <u>independent</u> of the specific f of (1). Accordingly, criteria exist to empirically decide if indeed Eq. (1) for <u>any</u> f can model such populations. That is, no guess as to the functional form of f is a priori necessary: if various numbers fail to mark the data, there <u>can be no f</u>.

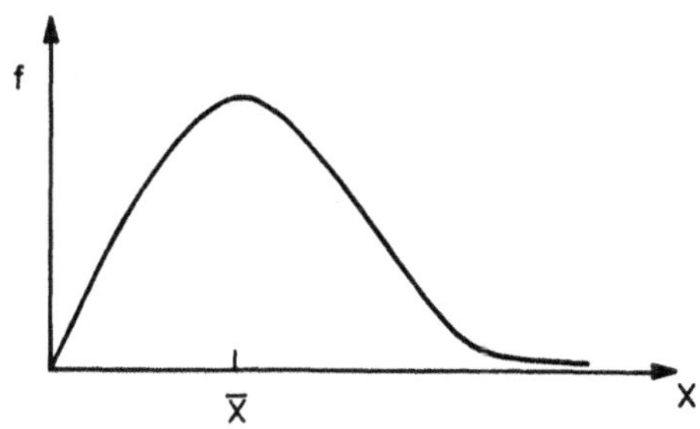

Figure 1

The key feature of Eq. (1) with f as in Fig. 1 is the ability to support limit cycles of all orders: that is, for a fixed λ for n sufficiently large, the X_n's alternate in a definite pattern among a set of r points, this set called an "r-cycle." As λ is increased, the eventual behavior remains that of an r-cycle until a "bifurcation value" is reached, at which point the r-cycle becomes unstable, going over into a 2r-cycle. Upon continued increase of λ this phenomenon recurs until an infinite set is reached (at some definite, finite, value of λ). The prototypic equation

$$X_{n+1} = \lambda X_n (1 - X_n) \tag{2}$$

is representative of such a system. This behavior has been known to be generic to a class of f's that "look like" Fig. 1. Just how large a class of functions has this behavior is unknown, although sufficient conditions are known.[4] For our purposes, any function of the class (whatever its exact extent) suffices.

Our results pertain to all highly bifurcated cycles of Eq. (1) (that is, after a given cycle has doubled in size many times). To utilize some of our results to discriminate for or against the form of Eq. (1), it is necessary to empirically follow a population through a number of bifurcations. To employ other of our results, it is merely necessary to know that the population is maintained in some fixed, highly bifurcated cycle. (For sufficiently smooth f's--such as (2)--"highly" can mean just several times.)

Let us now be specific. Our results hold independent of f for all f of the same nature of maximum--which we shall take here to be a normal, quadratic maximum (i.e., $f'(\bar{X}) = 0$, $f''(\bar{X}) < 0$), as otherwise, exact symmetries must exist to enforce $f''(\bar{X}) = 0$, etc., which is a priori unsuspected. We shall assume this throughout the rest of the discussion.

Defining λ_n to be that value of λ at which the $m\cdot 2^n$-cycle becomes unstable, bifurcating to an $m\cdot 2^{n+1}$-cycle, λ_n depends upon f, as does λ_∞: $\lambda_\infty = \lim_{n\to\infty} \lambda_n$. Our first result is that λ_n asymptomatically converges geometrically to λ_∞:

$$\lambda_\infty(f) - \lambda_n(f) \propto \delta^{-n} \qquad n \gg 1$$

where

$$\delta = 4.66920160910\ldots$$

is independent of f. Should the values λ_n be measurable, then

$$\delta_n \equiv \frac{\lambda_{n+1} - \lambda_n}{\lambda_{n+2} - \lambda_{n+1}}$$

converges to δ for large n. ($\delta_2 = 4.67\ldots$ for Eq. (2), to indicate the rapidity of onset of the asymptotic region.)

The next feature to observe (actually the most fundamental aspect of bifurcation) is a rescaling phenomenon. Define by d_n the spacing between those two elements of the $m\cdot 2^n$-cycle nearest $X = \bar{X}$ (the abscissa of the maximum of f). Then

$$\alpha_n \equiv \frac{d_{n+1}}{d_n}$$

converges to a constant α, also independent of f:

$$\alpha = 2.50229078750957\ldots\ .$$

Thus should neither α nor δ mark the data, one must reject the recursive form of Eq. (1) for any reasonable f. (The emergence of <u>different</u> values of <u>existing</u> limits from those quoted can be compatible with Eq. (1), but for an f of nonquadratic maximum--which is unlikely. Rather a failure of $\alpha_n \to \alpha$ and $\delta_n \to \delta$ should be observed if (1) is false.) Both tests depend upon following a population through several bifurcations; crude data should be sufficient at least for seeing α. (It is worthy to emphasize that α_n and δ_n are determined by <u>ratios</u> so that any consistent units may be used.)

Underlying the previous results is the fact that the locations of elements of highly bifurcated cycles are, to within a unit of scale, also universally determined within local clusters. Thus for $r \ll m \cdot 2^n$, the r elements nearest $\bar{X}$ are determined with respect to, say, the distance from $\bar{X}$ to the nearest element to $\bar{X}$. One finds an X_1 nearest $\bar{X}$; a pair of elements on the other side of $\bar{X}$ from X_1 next nearest; a quartet (actually, a pair of closer spaced pairs) of elements on the same side of $\bar{X}$ as X_1 next nearest still, etc. Crudely, the spacings of X_1 to $\bar{X}$, the centroid of the next nearest pair to $\bar{X}$; the centroid of the quartet to $\bar{X}$, etc. <u>increase each time by a factor of</u> α. (For a given kind of λ_n--i.e., whether the bifurcation value, the value of the most stable $m \cdot 2^n$-cycle, etc., the exact location of each nearby element can be computed from universal functions determined by the theory. We are simply quoting a rough result independent of more detailed experimental control that should be strikingly present.)

Our thrust so far has been the delineation of some decisive experiments on the validity of recursive modelling. However arduous the performance of these experiments might prove, it must be realized that an infinite bifurcation sequence is the very hallmark of such models and the <u>minimal</u> structure to be empirically established to justify <u>any</u> reference to equations of the form (1) in the context of population-dynamics. And even crude measurements of this structure should bear out our result. That is, if the form (1) applies, the numbers α and δ for example, must emerge as absolute numbers of nature.

Beyond this, though, it should also be realized that a certain philosophical orientation has emerged. We are saying that Fig. 1, dictated by the mechanisms of linearity in the dilute region and strong competition in the dense region fix not merely the qualitative form of the theory, but strikingly, a bulk of the exact numerical predictions: knowledge of the "true" recurrence functions for a large variety of exact predictions becomes irrelevant, leaving its very "meaning" in question.

Finally, we have spoken only in a language of population-dynamics. Certainly this is unnecessarily restrictive: these results evidently apply as well to economic, meteorological, social, etc., realizations of the same phenomenon. Of course, should such modelling have arisen only through naive approximation and devoid of real contact, this too could be exposed by our results in their discriminative role. And then, in the realms of turbulence, field theories and other matters pertaining to infinite self-couplings it is not impossible that the behavior of even so simple a nonlinear system as (1), might serve in a propaedeutic vein.

This work was performed under the auspices of the U. S. Department of Energy.

1. R. May, "Simple mathematical models with very complicated dynamics," Nature, Vol. 261, June 10, 1976, p. 459.
2. M. Feigenbaum, "Quantitative Universality for a Class of Nonlinear Transformations," J. Stat. Phys., Vol. 19, No. 1, 1978.
3. M. Feigenbaum, "The Formal Development of Recursive Universality," Los Alamos preprint LA-UR-78-1155.
4. N. Metropolis, M. L. Stein, and P. R. Stein, "On Finite Limit Sets for Transformations on the Unit Interval," J. Comb. Theory, July 1973, Vol. 15, No. 1, p. 25.

MAGNETIC FLUX ANNIHILATION IN A LARGE JOSEPHSON JUNCTION

Alwyn C. Scott
Department of Electrical and Computer Engineering
University of Wisconsin
Madison, Wisconsin 53706/USA

1. Introduction

The physical phenomenon to be discussed in this paper is the "displaced linear branch" in the volt-ampere characteristics of Josephson tunnel junctions having dimensions large compared with the Josephson penetration length (~0.1 mm). This effect was reported several years ago [1-3] and an explanation based on a continuous flow of magnetic flux quanta (fluxons) was suggested. Since that time, several striking advances have been made in the understanding of the analytic behavior exhibited by solutions of the underlying nonlinear partial differential equation, the sine-Gordon equation

$$\phi_{xx} - \phi_{tt} = \sin \phi \ . \tag{1.1}$$

Particularly important have been i) the development of an inverse scattering transform method (ISTM) for (1.1) [3-6], ii) the application of Whitham's method (WM), of slow perturbation of a single phase (periodic traveling wave) solution, to a dissipative version of (1.1) [7,8]

$$\phi_{xx} - \phi_{tt} - \alpha\phi_t = \sin \phi \ , \tag{1.2}$$

iii) Ablowitz's discussions of the extension of Whitham's method to multiple phase (multiply periodic) solutions [9-12], and iv) demonstrations that appropriate multiply periodic solutions assumed by Ablowitz do in fact exist for (1.1) [13-15]. The present discussion should be considered in context with previous efforts to apply the fruits of soliton theory to the problems of fluxon dynamics in large Josephson junctions [16,17]. Such studies are of interest not only as an aid to understanding the volt-ampere characteristics of Josephson junctions [1-3,18,19], but also because the soliton properties of fluxons may make them useful as carriers of information [20-25].

The displaced linear branch was first observed on 1 mm. × 1 mm. crossed strip Josephson junctions as volt-ampere characteristics of the sort displayed in Fig. 1. Above a critical current (I_c) for zero voltage (~1/2 ampere), the voltage (V) was found to increase linearly with the difference between current and critical current. Thus

$$\boxed{V = K(I-I_c)} \tag{1.3}$$

where the constant of proportionality (K) is much less sensitive to temperature than the normal electron component of tunneling current. It was suggested [1] that the input power (VI) implied by (1.3) is carried by moving fluxons from the edges of the junction toward the center where it is given up in events of fluxon-antifluxon annihilation.

Although the two dimensional effects which appear in the square junction geometry of Fig. 1 are of both theoretical and practical interest, the present discussion will be confined to the one dimensional geometry indicated in Fig. 2. In this case a

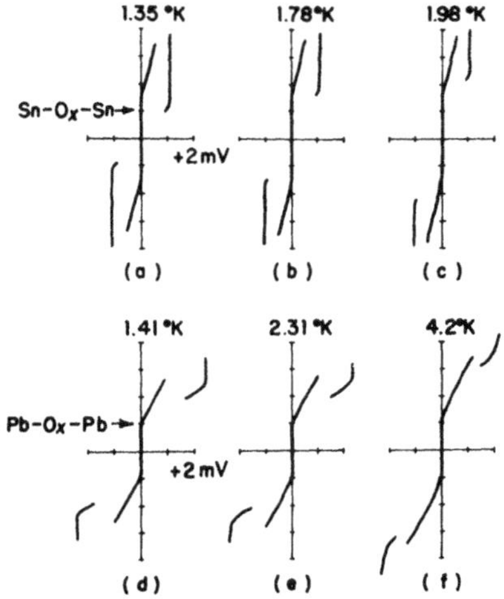

Figure 1. Volt-ampere characteristics of typical experimental junctions at various temperatures. (a), (b), and (c) are Sn-O_X-Sn, vertical: 200 mA/div, horizontal: 2 mV/div. (d), (e), and (f) P_b-O_X-Pb, vertical 500 mA/div, horizontal: 2mV/div.

mechanical analog of (1.2) [26] can be used to demonstrate the existence of flux flow domains near the edges and a flux annihilation domain near the center as shown in Fig. 3.

The paper is organized in the following way. The linearized version of (1.2) which is obtained by neglecting the sinϕ term is discussed in Section II in order to provide a perspective for considering the nonlinear problem. Then using results obtained by Pelinovskii and Shavratzkii [8] and in [17], the flux flow domain is described in Section III. The nonlinear solution in the flux flow domain is based upon the single phase (periodic traveling wave) solution described in detail by Whitham [7]. The flux annihilation domain considered in Section IV involves the propagation of fluxons in both directions and requires therefore a double phase (doubly periodic) solution [9-12]. In the flux annihilation domain, fluxons and antifluxons combine to form "breather solitons" which give up their energy to the dissipation represented by the $-\alpha\phi_t$ term in (1.2). To understand the annihilation domain, it is important to appreciate the nature of breather solitons; thus two appendices are included which discuss (A) the structure of a breather, and (B) the use of the inverse scattering transform method (ISTM) to study the generation of breathers through the interaction of double phase components. A third appendix (C) gives a summary account of single phase and double phase solutions for (1.1). The material in these appendices should be appreciated in order to follow the discussion of the main text. Finally we conclude with a review of the results which appear to be useful in interpreting currently available experimental data and some suggestions for future experiments.

The unnormalized form of (1.2) is

$$\Phi_{XX} - \ell g \Phi_{TT} - g \ell \Phi_T = j_0 \ell \sin(2\pi\Phi/\Phi_0) \quad (1.4)$$

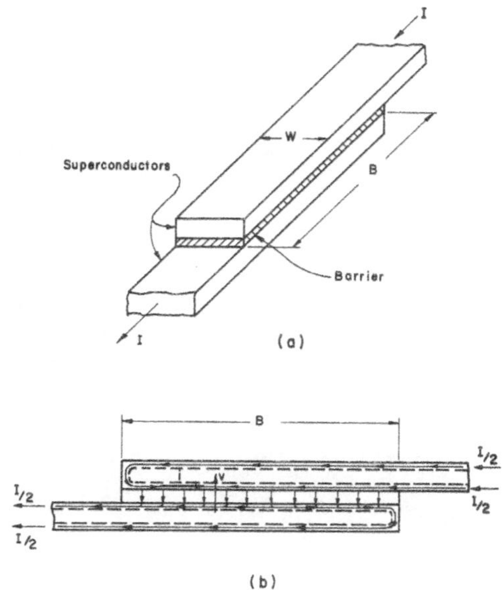

Figure 2. One dimensional geometry used for the theoretical analysis (not to scale). (a) perspective view, (b) section showing voltage v across insulating barrier and current i parallel to the barrier.

where

X is distance in the laboratory coordinate system,

T is time in the laboratory coordinate system,

Φ_T = V(X,T) is the voltage across the barrier,

Φ_X = -I(X,T) is the superconducting surface current flowing parallel to the barrier,

g is the shunt conductance through the barrier carried by (Giaever type [27]) normal electron tunneling per unit length,

ℓ is the inductance presented to the superconducting surface current per unit length,

c is the shunt capacitance of the barrier per unit length,

and

j_0 is the maximum (Josephson type [28]) superconducting electron tunneling through the barrier per unit length.

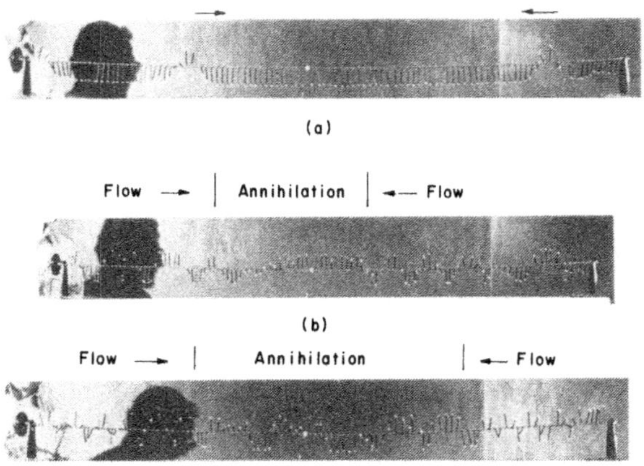

Figure 3. Kink flow domains and the kink annihilation (breather) domain on a mechanical analog of (1.2). (a) The crank is turned very slowly, (b) it is turned faster, (c) faster still.

Thus

$$\Phi(X, T) = \int^{T} V(X, T')dT' \tag{1.5}$$

is the magnetic flux which has passed point X at time T. It is convenient to measure this in units of the magnetic flux quantum or fluxon

$$\begin{aligned} \Phi_0 &= h/2e \\ &= 2.0678 \times 10^{-15} \text{ volt-seconds.} \end{aligned} \tag{1.6}$$

If we define a Josephson length

$$\lambda_J \equiv \left[\frac{\Phi_0}{2\pi j_0 \ell}\right]^{1/2} \tag{1.7}$$

and note that the characteristic velocities of (1.4) are u_0 where

$$u_0 = [\ell c]^{-1/2} \quad , \tag{1.8}$$

then (1.4) reduces to (1.2) with the normalizations

$$x = X/\lambda_J \tag{1.9}$$

$$t = u_0 T/\lambda_J \tag{1.10}$$

$$\phi = 2\pi\Phi/\Phi_0 \tag{1.11}$$

$$\alpha = g\ell u_0 \lambda_J \quad . \tag{1.12}$$

Furthermore, if the shunt voltage across the barrier is normalized as

$$v = \frac{V}{\left[\frac{\Phi_0 j_0}{2\pi c}\right]^{1/2}} = \frac{V}{V_N} \tag{1.13}$$

and superconducting surface current as

$$i = \frac{I}{\left[\frac{\Phi_0 j_0}{2\pi\ell}\right]^{1/2}} = \frac{I}{I_N} \quad , \tag{1.14}$$

then

$$\phi_t = v$$
$$\phi_x = -i \quad . \tag{1.15a,b}$$

In this paper, we will be primarily concerned with an analysis of the normalized equation (1.2). A more detailed discussion of the approximations involved in obtaining (1.4) can be found in reference [17] together with some tables of experimentally measured values of the parameters.

2. The Linear Approximation

We begin with a brief discussion of the equation

$$\phi_{xx} - \phi_{tt} - \alpha\phi_t = 0 \tag{2.1}$$

which is a linearization of (1.2) by neglecting the sine term. From (1.4) it can be seen that this is physically equivalent to assuming the coefficient of Josephson tunneling current (j_0) to be small. Eq. (2.1) can also be written as the first order pde's

$$v_t + i_x = -\alpha v$$
$$i_t + v_x = 0 \tag{2.2a,b}$$

where v and i are normalized shunt voltage and series current as defined in (1.13) through (1.15). The experimental situation indicated in Fig. 2b can then be represented as in Fig. 4a where the dimension $b = B/\lambda_J$. Physical symmetry and the boundary conditions require the current (i) to be zero at the center in Fig. 4a. Thus we need consider only the half junction shown in Fig. 4b with the boundary conditions

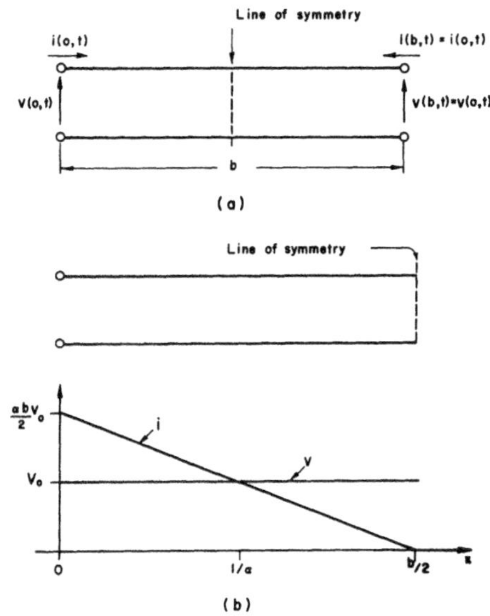

Figure 4. A linear approximation to the problem.

$$v(0, t) = v_0 \qquad i(b/2, t) = 0 \quad . \tag{2.3a,b}$$

If we assume v_0 = const. and seek steady state solutions, (2.2) implies

$$v(x) = v_0 \qquad i(x) = \frac{\alpha b v_0}{2}\left(1 - \frac{2x}{b}\right) \quad . \tag{2.4a,b}$$

Now return to (1.15) and write

$$\phi = vt - ix \tag{2.5}$$

which can be viewed as a traveling wave with phase velocity

$$u = v/i \quad . \tag{2.6}$$

From (2.4) we see that

$$\frac{\alpha b}{2} < 1 \Rightarrow u > 1 \text{ for } 0 \le x \le b/2 \tag{2.7a}$$

and

$$\frac{\alpha b}{2} > 1 \Rightarrow u < 1 \text{ for } 0 \le x \le 1/\alpha$$
$$u > 1 \text{ for } 1/\alpha < x \le b/2 \ . \qquad (2.7b)$$

Thus for sufficiently small α, the phase velocity of a single traveling wave (2.5) will exceed unity for almost all x.

We can, if we wish, resolve ϕ into the sum of two traveling waves, one carrying flux to the right and the other carrying flux to the left. Then

$$\phi = (v_1 t - i_1 x) + (v_2 t + i_2 x) \ . \qquad (2.8)$$

where $v_1 + v_2 = v_0$ and $i_1 + i_2 = i(x)$. This resolution is not unique, but in anticipation of results to be obtained below for the nonlinear problem we suggest the following for the case $\alpha b/2 > 1$.

For $0 \le x < 1/\alpha$ set $v_2 = 0$ and $i_2 = 0$ and take v_1 and i_1 to be the values given in (2.4). For $1/\alpha \le x \le b/2$ take

$$i_1 = v_1 = \frac{v_0}{2}\left(\frac{b-1/\alpha-x}{b/2-1/\alpha}\right)$$
$$-i_2 = v_2 = \frac{v_0}{2}\left(\frac{x-1/\alpha}{b/2-1/\alpha}\right) \ . \qquad (2.9a,b)$$

Thus for $0 \le x < 1/\alpha$ there is only a single traveling wave component and it has a phase velocity less than unity. For $1/\alpha < x \le b/2$ there are two components and the phase velocity of each has a magnitude equal to unity.

3. The Flux Flow Domain

In this section we address ourselves to the nonlinear equation (1.2). Consider first a steady state solution. Then $\phi_{xx} = \sin\phi$ which is readily integrated to

$$\frac{1}{2}\phi_x^2 = E - \cos\phi \qquad (3.1)$$

where E is a constant of integration. The conditions $\phi \to 0$ and $\phi_x \to 0$ at large x requires $E = 1$ whereupon

$$\phi_x = \pm 2 \sin \phi/2 \ .$$

From (1.15) i has maximum magnitude of $i_c = 2$ when $\phi(0) = \pi$. Thus the critical current (into <u>both</u> ends)

$$\boxed{I_c = 4\left[\frac{\Phi_0 j_0}{2\pi\ell}\right]^{1/2}} \qquad (3.2)$$

is established below which the voltage $(v = \phi_t)$ is zero and above which flux must flow. This effect has been carefully investigated [29].

Suppose now that $i(0) > i_c$ $(I > I_c)$ so a steady state solution for (1.2) does not exist. As has been shown in reference [17], it is convenient to turn to Whitham's method of averaging for which it is supposed that the solution appears locally as a traveling wave of the form

$$\phi = \phi(\theta)$$
$$\theta = \omega t - k \qquad (3.3a,b)$$

where ω and k are allowed to be slowly varying functions of x and t. Then $\phi(\theta)$ is defined locally as the elliptic integral

$$\theta = \sqrt{k^2 - \omega^2} \int^{\phi} \frac{d\phi'}{\sqrt{2(E - \cos\phi')}} \quad , \qquad (3.4)$$

where we assume $E > 1$. Dynamic equations for the slow variation of k, ω and E with x and t are [7,8,17]

$$[\omega F''(E)F(E)]_t + [kF'(E)F(E)]_x = -\alpha\omega F'(E)F(E)$$
$$k_t + \omega_x = 0 \quad . \qquad (3.5a,b)$$

The functions F'(E) and F(E) are complete elliptic integrals of the first and second kind respectively and defined as

$$F'(E) \equiv \frac{1}{2\pi} \int_0^{2\pi} \frac{d\phi}{\sqrt{2(E - \cos\phi)}}$$
$$F(E) = \frac{1}{2\pi} \int_0^{2\pi} \sqrt{2(E - \cos\phi)}\, d\phi \qquad (3.6a,b)$$

Note that in the <u>large amplitude linear limit</u>, $E \to \infty$,

$$F' \to (2E)^{-1/2}$$
$$F \to (2E)^{1/2} \quad . \qquad (3.7a,b)$$

It is convenient to require that ϕ increase by 2π when θ does, whereupon k, ω and E are related by the <u>nonlinear dispersion equation</u>

$$k^2 - \omega^2 = [F'(E)]^{-2} \quad . \qquad (3.8)$$

In the large amplitude linear limit ($E >> 1$), the product

$$F'(E)F(E) \to 1$$

as is shown in Figure 5. Furthermore (3.4) and (3.8) imply

$$\phi \to \theta$$

so

$$v = \phi_t \to \theta_t = \omega$$
$$i = -\phi_x \to -\theta_x = k \qquad (3.9a,b)$$

and (3.5) approach the linear equations (2.2) which were considered in the previous section. Figure 5 indicates that this large amplitude linear limit is approached

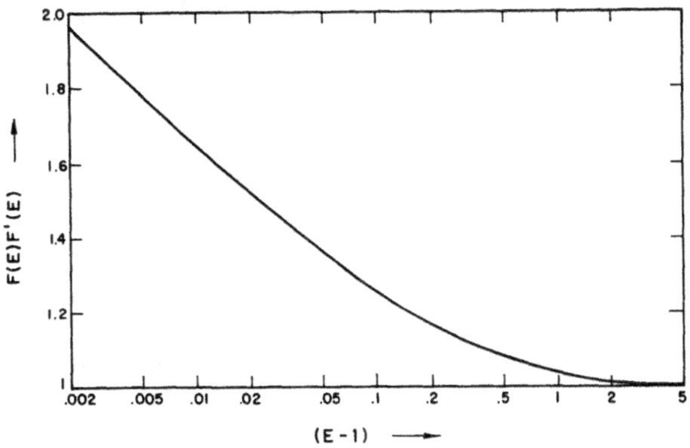

Figure 5. $F(E)\,F'(E)$ vs. $(E-1)$.

within a few percent for $E > 2$.

Suppose now that the current (i) is increased to the maximum zero voltage ($v = 0$) level. As i goes from 0^- to 0^+, the steady state current, $i(x)$, will jump from the solution of (3.1) (with $E = 1$) to a solution of (3.8) with $\omega = 0$ or

$$k = k_c = [F'(E_c)]^{-1} \tag{3.10}$$

where E_c is the critical value of the integration constant E for a traveling wave solution carrying the critical current. Since $k_c \approx i_c = 2$, (3.10) is satisfied for $F'(E_c) \approx 0.5$ or

$$E_c \approx 2.237 \quad .$$

For positive values of voltage ($\omega > 0$), this periodic solution will move.

To determine the relation between terminal current, $i(0)$, and voltage, $\omega(0)$, when the flux is moving, we must eliminate E from the nonlinear dispersion equation (3.8). This can be done by noting [7,17] that traveling wave velocity (or nonlinear phase velocity)

$$u = \omega/k \tag{3.11}$$

and E are related as

$$\int_0^u \frac{du'}{1-(u')^2} = \int_{E_c}^{E} \left[-\frac{F'(E)}{F(E)}\right]^{1/2} dE' \tag{3.12}$$

where the lower limits have been chosen to satisfy the condition

$u = 0$ when $E = E_c$.

Since $E_c \approx 2$, the high amplitude linear limit $[F(E) \approx (2E)^{1/2}$ and $F'(E) \approx (2E)^{-1/2}]$ is appropriate for evaluating the right hand side of (3.12). Then

$$E(0) = E_c \left(\frac{1+u}{1-u}\right) \tag{3.13}$$

which upon substitution into (3.8) gives

$$\boxed{\omega(0) = k(0) - \sqrt{2E_c}} \quad . \tag{3.14a}$$

This is the equation for the displaced linear branch. Taking account of (3.9) and the normalizations in (1.13) and (1.14), it can be written

$$\boxed{V = \frac{1}{2}\sqrt{\frac{\ell}{c}}\,(I - I_c)} \quad . \tag{3.14b}$$

Thus the constant slope (K) which appears in (1.3) is equal to half the characteristic impedance of the linear operator in (1.2) [1], and the weak temperature dependence of this slope (see Figure 1) is readily appreciated.

Now let us consider how far such a steady state traveling wave solution can be extended into the junction. The assumption of steady state in (3.5b) implies

$$\omega = \text{const} \tag{3.15a}$$

and from (3.5a), k(x) must satisfy

$$[kFF']_x = -\alpha\omega FF' \quad . \tag{3.15b}$$

Pelinovskii and Shavratzkii [8] have shown that there is a definite limit in x beyond which a single phase (periodic traveling wave) solution of (3.15) does not exist. To see this consider the quantity kFF'. Using (3.8) it is seen to be

$$kFF' = F\sqrt{1 + \omega^2 (F')^2} \tag{3.16}$$

which, for fixed ω, is a function of E with a minimum value ($E = E_{min}$) satisfying the condition

$$\omega^2 \{[F'(E_{min})]^2 + F(E_{min})F''(E_{min})\} + 1 = 0 \quad . \tag{3.17}$$

The character of kFF' as a function of E and ω is displayed in Fig. 6 and E_{min} vs. ω is plotted in Fig. 7. When kFF' is equal to its minimum value, (3.15b) can no longer be satisfied because the right hand side requires that kFF' continue to decrease. It is only for the range of x for which $E > E_{min}$ that the steady state single phase solution can exist.

The value of x (say x_0) at which kFF' has fallen to its minimum value is an upper estimate for the boundary of the flux flow region. Equation (3.15) can be integrated with the boundary conditions

$$E = E(0) \quad \text{at } x = 0$$

$$E = E_{min} \quad \text{at } x = x_0$$

to obtain

$$x_0 = \frac{1}{\alpha} \int_{E_{min}}^{E(0)} \left[\frac{1}{F} \sqrt{\frac{1}{\omega^2} + (F')^2} + \frac{\omega F''}{\sqrt{1 + \omega^2 (F')^2}} \right] dE \quad . \tag{3.18}$$

where from (3.13) and (3.14)

$$E(0) = \omega\sqrt{2E_c} + E_c \quad . \tag{3.19}$$

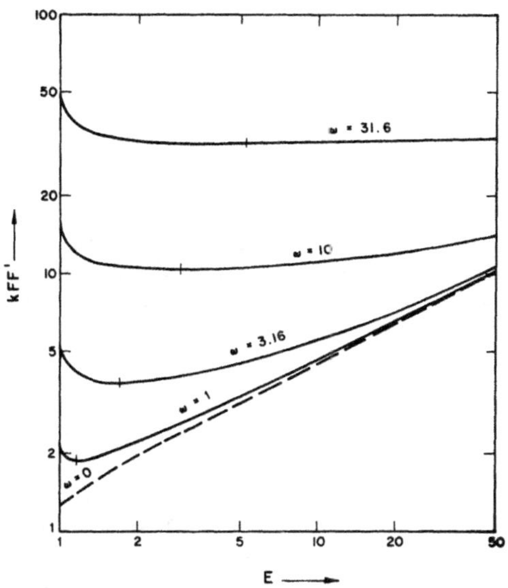

Figure 6. k F(E)F'(E) vs. E.

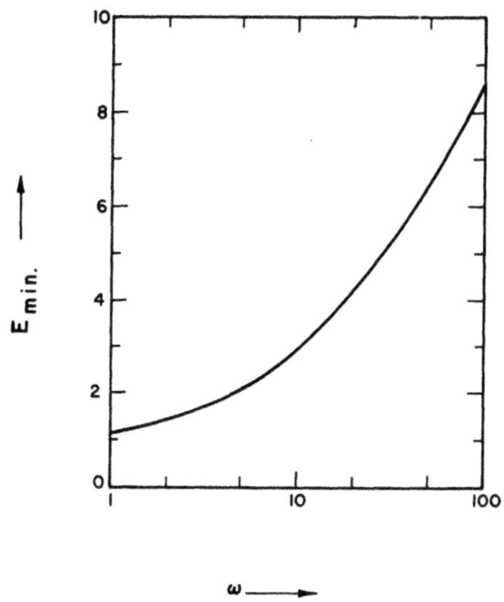

Figure 7. E_{min} vs. ω.

In the large amplitude linear limit (3.18) implies that $x_0 \to (2E_c)^{1/2}/\alpha\omega$ as $\omega \to \infty$, and $x_0 \to .607/\alpha\omega$ as $\omega \to 0$.

We are now prepared to calculate the power flowing across the boundaries of the flux flow region. In general $P(x) = \langle v(x)i(x)\rangle \approx \omega k(x)$ where the approximation improves for large E as indicated in (3.9). Taking the boundaries of the flux flow region as $x = 0$ and $x = x_0$ (see Fig. 8), the power flowing in at $x = 0$ is

$$P(0) \doteq \omega k(0)$$

$$\doteq \omega^2 + \omega\sqrt{2E_c} \qquad (3.20a,b)$$

where (3.14a) has been used to evaluate k(0). In a similar way, the power flowing out at $x = x_0$ is

$$P(x_0) \doteq \omega k(x_0)$$

$$\doteq \omega\sqrt{\omega^2 + 2E_{min}} \qquad (3.21a,b)$$

where (3.8) has been used to evaluate $k(x_0)$.

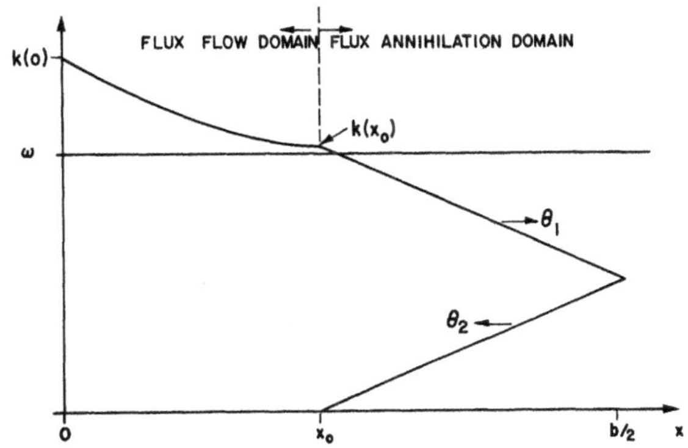

Figure 8. Steady state components of voltage (ω) and current (k) in the flux flow domains. In the flux annihilation domain, an approximation to the double phase solution is sketched.

If the assumption of steady state in the flux flow domain is correct, the difference between these incoming and outgoing powers must equal the d.c. dissipation. Thus we expect

$$P(0) - P(x_0) = \alpha\omega^2 x_0 \tag{3.22}$$

whereupon a second estimate for x_0 is

$$x_0 \doteq \frac{1}{\alpha}\left[1 + \frac{\sqrt{2E_c}}{\omega} - \sqrt{1 + \frac{2E_{min}}{\omega^2}}\right] \quad . \tag{3.23}$$

We do not expect precise agreement between the values of x_0 calculated from (3.18) and (3.23) because the power flow estimates in (3.20a) and (3.21a) are not exact for smaller values of E.

However, as Fig. 9 shows, the agreement is rather good. Equation (3.23) gives values for x_0 which are 7-10% higher than those calculated from (3.18) when ω lies between 1 and 2. Equation (3.23) is, of course, much more convenient than (3.18). For $\omega > 10$, the asymptotic expression

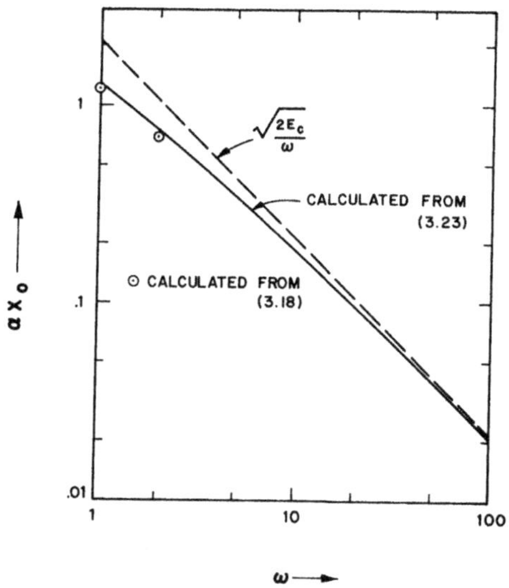

Figure 9. Maximum length of the flux flow domain (x_0) vs. voltage (ω).

$$x_0 \sim \frac{\sqrt{2E_c}}{\alpha\omega} \tag{3.24}$$

is a reasonable approximation.

4. The Flux Annihilation Domain

Let us suppose that we have a large Josephson junction which exhibits a displaced linear branch (Fig. 1) in the volt-ampere characteristics as is implied by (3.14). Assume further that α, b, ω and E_c are adjusted such that the relation

$$\frac{\alpha b}{2}\omega^2 = \omega(\omega + \sqrt{2E_c}) \tag{4.1}$$

holds. This relation says that the d.c. power into the junction (RHS) equals the d.c. power dissipated on the junction (LHS). If ω is increased above the value which satisfied (4.1), the LHS will be greater than the RHS and the displaced linear branch will not supply power to the junction at the rate it would be dissipated. Thus the value of ω which satisfies (4.1) is the "break voltage" ω_B (see Fig. 10) above which the displaced linear branch will not continue. Figure 1 shows that such a break voltage is a characteristic feature of the displaced linear branch.

For $\omega_B >> 1$, (4.1) implies that a necessary condition for observing a break in the displaced linear branch is

$$\frac{\alpha b}{2} \sim 1 \tag{4.2a}$$

or, in unnormalized units,

$$gB \sim 2 \sqrt{\frac{c}{\ell}} \tag{4.2b}$$

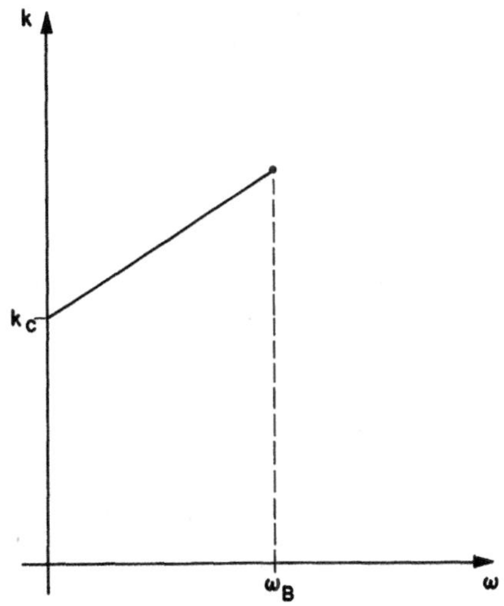

Figure 10. The break voltage (ω_B) in the displaced linear branch.

Now consider the difference between the power flowing into the flux annihilation domain, $P(x_0)$, and the d.c. power being dissipated in this domain

$$P_{dc} = \alpha\omega^2 \left(\frac{b}{2} - x_0\right) \quad . \tag{4.3}$$

Thus, from (3.21) and (3.23)

$$P(x_0) - P_{dc} = \omega^2 + \omega\sqrt{2E_c} - \alpha\omega^2 \frac{b}{2} \quad . \tag{4.4}$$

From (4.1) this difference is zero when $\omega = \omega_B$. For $\omega < \omega_B$, (4.4) indicates that the d.c. power flow into the flux annihilation domain is greater than the d.c. power dissipation in that domain. Thus for $\omega < \omega_B$ it is not possible to have a steady state (ω = const) in the flux annihilation domain. There must be an additional a.c. component of voltage to dissipate the power difference given by (4.4). The source for this a.c. voltage is the continuous creation of breather solitons near the center of the junction.

The dynamic activity in the flux annihilation domain is considerably more complex than in the flux flow domains. To get some qualitative appreciation for this activity we turn to the ideas developed in Section II and in the appendices. Consider first the problem of representing the d.c. components or the time averages of v and i. Following Appendix C we assume a double phase solution

$$\phi(x, t) = \Phi(\theta_1, \theta_2) \tag{4.5}$$

with

$$\begin{aligned} \theta_1 &\equiv \omega_1 t - k_1 x \\ \theta_2 &= \omega_2 t - k_2 x \quad . \end{aligned} \tag{4.6a,b}$$

Then, motivated by (2.8), we try the simple form

$$\Phi(\theta_1, \theta_2) = \theta_1 + \theta_2 \quad . \tag{4.7}$$

Equations (C-14) both become

$$(k_1 + k_2)_x = -\alpha(\omega_1 + \omega_2) \tag{4.8a}$$

and (C-6c, d) imply

$$\omega_1 + \omega_2 = \omega \quad \text{(a constant).} \tag{4.8b}$$

The boundary condition $k_1(x_0) + k_2(x_0) = k(x_0)$ and (4.8a) implies

$$k_1 + k_2 = k(x_0) - \alpha\omega(x - x_0) \quad . \tag{4.9}$$

The symmetry condition discussed in connection with Fig. 4 requires

$k_1(\frac{b}{2}) + k_2(\frac{b}{2}) = 0$, but from (4.9)

$$\begin{aligned} k_1(\tfrac{b}{2}) + k_2(\tfrac{b}{2}) &= k(x_0) - \alpha\omega(\tfrac{b}{2} - x_0) \\ &= \omega + \sqrt{2E_c} - \frac{\alpha b}{2}\,\omega \quad . \end{aligned} \tag{4.10}$$

This is equal to zero only at the break voltage $\omega = \omega_B$. For $\omega < \omega_B$, the simple double phase function assumed in (4.7) cannot correctly represent the d.c. components. Some more versatile expression (probably using Riemann theta functions) will be required [13-15]. In the special case $\omega = \omega_B$, however, we can use (4.5) - (4.7) where

$$k_1(x) = k(x_0) - \omega(x-x_0) + \frac{k(x_0)}{2}\left(\frac{x-x_0}{b/2-x_0}\right)$$

$$k_2(x) = -\frac{k(x_0)}{2}\left(\frac{x-x_0}{b/2-x_0}\right) \quad . \qquad (4.11a,b)$$

For $\omega_B - \omega << \omega_B$, (4.11) should be approximately correct. Furthermore from (3.24) and (4.2a)

$$x_0 \sim \frac{\sqrt{2E_c}}{\omega}\frac{b}{2}$$

$$<< \frac{b}{2}$$

so most of the junction will be occupied by the flux annihilation domain. Also the rate at which kinks enter this domain at $x = x_0$ will be $\omega/2\pi$. Since each breather forms by coupling a kink entering from the left and an antikink from the right, the steady state rate of breather formation must equal $\omega/2\pi$.

Where in the flux annihilation domain will these breathers form? To answer this question we can use the results of Appendix B as a qualitative guide. Just at the center of the junction, $\phi_x = 0$ and $\phi_t = \omega$. Thus, from (B-22), we expect the number of breathers formed per unit time and distance to be $\alpha\omega/2\pi$. Since the rate of rotation falls from a maximum at the center to zero at the edges of the flux annihilation domain, the net rate of breather formation should equal $\alpha b\omega/4\pi$. Then the condition (4.2a) implies a breather formation rate $\sim\omega/2\pi$.

Thus we have a fairly self consistent picture of breather formation for the case $\omega - \omega_B << \omega_B$. We expect the breathers to remain near the center of the junction because they are formed near the center of the junction as <u>stationary</u> breathers.

The case $\omega << \omega_B$ is much less well understood for several reasons.

i) The qualitative nature of the d.c. solution is not clear.

ii) From (4.4) the a.c. voltage is no longer a small fraction of the d.c. voltage.

iii) For $\omega \sim 1$, $x_0 \sim b/2$ and the flux annihilation domain becomes very narrow. In this case, the a.c. voltage is probably not confined to the flux annihilation domain.

Numerical studies of the case $\omega << \omega_B$ would be helpful in understanding these uncertain aspects. As far as the external observation of a displaced linear branch is concerned, however, all that is required is for a single phase (traveling wave) solution to be established near the edges of the junction.

5. Conclusions

Here an attempt will be made to draw together the somewhat tattered strands of the above discussion by indicating which results may have experimental significance and by suggesting some directions for future research.

a) Critical Current

From (3.2), the critical current (I_c) above which flux will begin to move is $4[\Phi_0 j_0/2\pi\ell]^{1/2}$. Since it is practical to compute j_0 and ℓ from independent measurements on a small junction [17], the ideal value for I_c can be compared with that actually observed on a large junction. Such a comparison is important because irregularities in the barrier thickness (d in Fig. 2) may cause flux pinning and increase I_c above the value given by (3.2).

b) Slope of the Displaced Linear Branch

Equation (3.14b) indicates that the slope of the displaced linear branch should equal $2\sqrt{c/\ell}$ where the factor of two accounts for current flow into both ends of a long junction (see Fig. 2b). This result confirms the suggestion made in [1]. Since c and ℓ can be independently measured on a small junction, this striking feature of the above theory for the displaced linear branch is always subject to direct experimental check.

c) Conditions for Observing the Displaced Linear Branch

In (4.2) it was noted that the condition for observing a break in the displaced linear branch is $\alpha b \approx 2$ or, in unnormalized units, $gB \approx 2\sqrt{c/\ell}$. If $\alpha b \gg 2$, the Josephson current becomes a small portion of the total current, and the situation reduces to the linear problem studied in Section II. If, on the other hand, $\alpha b \ll 2$, the junction becomes a resonant electromagnetic structure, and the step structure observed by Fiske [30-33] will dominate the volt-ampere characteristic. Thus the condition

$$\frac{\alpha b}{2} \sim 1$$

or (5.1a,b)

$$gB \sim 2\sqrt{c/\ell}$$

also determines the range of experimental parameters for which the displaced linear branch should be unambiguously observed.

d) Numerical Studies

It was assumed, in the discussions of Sections III and IV, that the a.c. voltage associated with the decay of breather solitons is confined to the flux annihilation domain and does not "leak out" into the flux flow domain. This assumption seems plausible because stationary breathers are generated at the center of the junction and it is confirmed in a rough way by observations on the mechanical model as shown in Fig. 3. But it would be comforting to have a direct numerical test of this picture. An appropriate problem would be integration of (1.2) with end conditions

$$\omega(0, t) = \omega(b, t) = 0 \text{ for } t < 0$$

$$= v \text{ for } t > 0$$

and with $\alpha b = 2$, $\alpha = 0.1, 0.01, 0.001$, and $v = 1, 10, 100$. An asymptotic form of the dynamic activity should be sought as $t \to \infty$.

e) Experiments with Light Sensitive Junctions

Barone et al. [34] have recently demonstrated that properly prepared Pb - CdS - In junctions can be changed from "small" ($b < 1$) to "large" ($b > 1$) by ordinary optical illumination. This change occurs because the Josephson tunneling is increased and therefore λ_J is decreased under illumination. An increase in the Josephson tunneling should correspond to an increase in the normal electron (Giaever) tunneling and therefore an increase in α. Thus αb should be an even more sensitive function of illumination than b. Indeed, from (5.16), αb is directly proportional to the Giaever tunneling density.

Thus it may be possible to experimentally observe a transition from "Fiske steps" on a "small" junction to flux flow on a "large" junction as the illumination level is increased.

APPENDIX A: STRUCTURE OF A BREATHER

1. Classical Structure

Consider the sine-Gordon equation in characteristic coordinates

$$\phi_{\xi\tau} = \sin\phi \tag{A-1}$$

which is equivalent to (1.1) under the independent variable transformation

$$\begin{aligned} \xi &= \frac{1}{2}(x-t) \\ \tau &= \frac{1}{2}(x+t) \end{aligned} \tag{A-2a,b}$$

If $\phi_{n-1}(\xi,\tau)$ is a solution of (A-1), it is easily shown that $\phi_n(\xi,\tau)$ which satisfies

$$\begin{aligned} (\phi_n-\phi_{n-1})_\xi &= 4i\gamma \sin\left(\frac{\phi_n+\phi_{n-1}}{2}\right) \\ (\phi_n+\phi_{n-1})_\tau &= -\frac{i}{\gamma}\sin\left(\frac{\phi_n-\phi_{n-1}}{2}\right) \end{aligned} \tag{A-3a,b}$$

is also a solution [35]. This is called a Bäcklund transform which generates a new solution from a known (old) solution; by repeated application a hierarchy of solutions can be obtained. If $i\gamma_j$ is real, an additional kink or soliton will be introduced with the asymptotic form

$$\phi_j = 4\tan^{-1}[\exp(2i\gamma_j\xi - i\tau/2\gamma_j)] \tag{A-4}$$

and velocity in the laboratory (x, t) frame [17]

$$u_j = \frac{4\gamma_j+1}{4\gamma_j-1} \quad . \tag{A-5}$$

As George Lamb has suggested [36], a convenient way to generate a breather solution to (1.1) is to perform two Bäcklund transforms with complex γ's and

$$\gamma_1 = -\gamma_2^* \quad . \tag{A-6}$$

Such a solution, ϕ_B, is given by [16,35,36]

$$\tan\frac{\phi_B}{4} = \left(\frac{\gamma-\gamma^*}{\gamma+\gamma^*}\right)\tan\left(\frac{\phi_1-\phi_2}{4}\right) \tag{A-7}$$

where ϕ_1 and ϕ_2 have the Gudermannian form indicated in (A-4). Thus in the laboratory (x, t) frame a <u>breather</u> <u>soliton</u> has the form

$$\phi_B = 4\tan^{-1}\left\{\left(\frac{\gamma-\gamma^*}{\gamma+\gamma^*}\right)\frac{\exp[i(\gamma-\frac{1}{4\gamma})x-i(\gamma+\frac{1}{4\gamma})t]-\exp[-i(\gamma^*-\frac{1}{4\gamma^*})x+i(\gamma^*+\frac{1}{4\gamma^*})t]}{1+\exp[i(\gamma-\gamma^*-\frac{1}{4\gamma}+\frac{1}{4\gamma^*})x+i(-\gamma+\gamma^*-\frac{1}{4\gamma}+\frac{1}{4\gamma^*})t]}\right\} \quad . \tag{A-8}$$

Upon examination of (A-8), the following points will be noted.

i) As $\gamma \to$ a pure imaginary #, $\phi_B \to 2\pi$ and the breather disassociates into a kink-antikink pair.

ii) As $\gamma \to$ a real #, $\phi_B \to 0$ and is proportional to $\sin(kx-\omega t)$ where $k = \gamma - 1/4\gamma$, $\omega = \gamma + 1/4\gamma$ and $\omega^2 = 1 + k^2$.

iii) The envelope of the breather is determined by the denominator in (A-8). The envelope velocity

$$u_e = \frac{\gamma + \frac{1}{4\gamma} - \gamma^* - \frac{1}{4\gamma^*}}{\gamma - \frac{1}{4\gamma} - \gamma^* + \frac{1}{4\gamma^*}} \tag{A-9}$$

is positive for $|\gamma| > 1/2$ and negative for $|\gamma| < 1/2$. Thus the condition for a stationary breather is

$$|\gamma| = \frac{1}{2} \ . \tag{A-10}$$

In terms of an angle parameter (β) which relaxes from $\frac{\pi}{2} \to 0$ as two kinds decay into a stationary breather of zero energy, (see Fig. 11) a stationary breather takes the form

$$\phi_B = 4 \tan^{-1} \{\tan\beta \sin[(\cos\beta)t] \operatorname{sech}[(\sin\beta)x]\} \ . \tag{A-11}$$

The rest energy of a breather can be calculated as the total energy of a stationary breather from the Hamiltonian density

$$H = \frac{1}{2} \phi_x^2 + \frac{1}{2} \phi_t^2 + 1 - \cos\phi$$

corresponding to (1.1). At $t = 0$, (A-11) reduces to $\phi_B(x,0) = 0$; so $H = \frac{1}{2} \phi_{B,t}^2(x,0)$ and the total energy

$$U_B = \int_{-\infty}^{\infty} H dx$$

or

$$U_B = 16 \sin\beta \ . \tag{A-12}$$

When β reaches $\pi/2$, the breather disassociates into two kinks of the form (A-4), so the energy of a kink is

$$U_k = 8 \ .$$

2. Quantum structure

Fadeev [37,38] and Dashen, Hasslacher and Neveu [39] have carried through an exact quantization of the breather soliton described by (A-11). The discrete energy levels were found to be

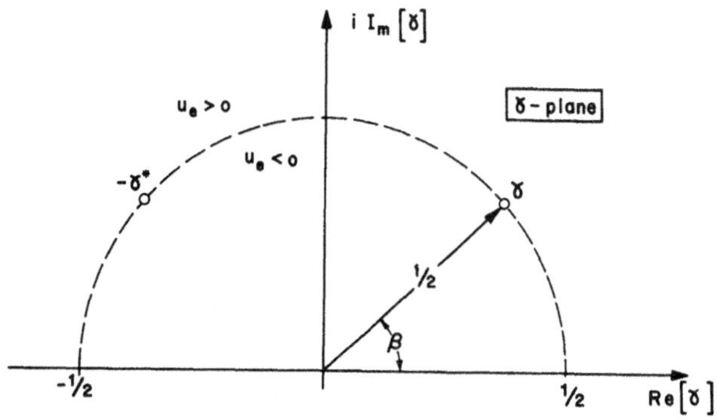

Figure 11. Locus of eigenvalues for stationary breathers.

$$U_n = 2\tilde{U}_k \sin\left[\frac{\hbar\omega_J}{2\tilde{U}_k}\right](n + 1/2) \tag{A-13}$$

where $\tilde{U}_k$ is the unnormalized energy of a kink (fluxon), $\hbar$ is Planck's constant divided by 2π and

$$\omega_J = u_0/\lambda_J \tag{A-14}$$

is the Josephson frequency which was used in (1.10) to normalize the time scale. There is correspondence between the classical frequency ($\cos\beta$) in (A-11) and the frequency for emission or absorption in (A-13) if

$$\beta = \frac{\omega_J}{2\tilde{U}_k}(n + 1/2) \quad . \tag{A-15}$$

Thus the allowed quantum states for a stationary breather are evenly spaced on the circle $|\gamma| = 1/2$ (see Fig. 11). Quantum effects should become important in predicting the dynamics of a breather if the number of these states is not large compared with unity. To evaluate this situation we note that in unnormalized units

$$\tilde{U}_k = 8cV_N^2 \lambda_J = \frac{4}{\pi} \Phi_0 j_0 \lambda_J \tag{A-16}$$

and turn to the parameters measured on the Josephson transmission lines discussed in [17]. From (A-15) the total number of levels is

$$N = \frac{4\tilde{U}_k}{\pi \hbar \omega_J} . \tag{A-17}$$

Table I Estimates of quantum structure for breathers

	N53C	N25L	
λ_j	2.63×10^{-4}	1.27×10^{-3}	meter
ω_J	6.69×10^{10}	1.81×10^{10}	rad/sec.
j_0	1.9	.097	amp/meter
J_0	1.22	.062	amp/cm ^{2}
U_k	13.1×10^{-19}	3.24×10^{-19}	joules
$N(=4\tilde{U}_k/\pi\hbar\omega_J)$	2.37×10^5	2.16×10^5	----

From the last entry in Table I it is clear that the number of quantum levels is so large that classical dynamics should be sufficiently accurate for the description of these particular Josephson transmission lines. As for other structures it should be noted that $N \propto j_0 \lambda_J^2$ or

$$N \propto W \tag{A-18}$$

where W is the width of the transmission line (see Fig. 2a). If W were decreased by two orders of magnitude (from 64 microns to 0.64 micron), quantum corrections would still be of minor importance. This is in accord with a previous estimate of the need for quantum corrections in a point contact Josephson junction [40].

APPENDIX B: A COHERENT BREATHER GAS FROM ROTATING INITIAL CONDITIONS

Here we use the inverse scattering transform method for the sine-Gordon equation (1.1), which was discovered by Ablowitz, Kamp, Newell and Segur [4], to see how a breather gas can be generated. In laboratory coordinates the equations for evolution of the scattering variable $\psi = (\psi_1, \psi_2)$ are

$$\psi_{1,x} = -\frac{i}{2}\left(\gamma - \frac{1}{4\gamma}\cos\phi\right)\psi_1 + \frac{1}{2}\left[\frac{i}{4\gamma}\sin\phi - \frac{1}{2}(\phi_x - \phi_t)\right]\psi_2$$

$$\psi_{2,x} = \frac{1}{2}\left[\frac{i}{4\gamma}\sin\phi + \frac{1}{2}(\phi_x - \phi_t)\right]\psi_1 + \frac{i}{2}\left(\gamma - \frac{i}{4\gamma}\cos\phi\right)\psi_2 \qquad \text{(B-1a,b)}$$

and

$$\psi_{1,t} = \frac{i}{2}\left(\gamma + \frac{1}{4\gamma}\cos\phi\right)\psi_1 + \frac{1}{2}\left[\frac{i}{4\gamma}\sin\phi + \frac{1}{2}(\phi_x - \phi_t)\right]\psi_2 \qquad \text{(B-2a,b)}$$

$$\psi_{2,t} = \frac{1}{2}\left[\frac{i}{4\gamma}\sin\gamma - \frac{1}{2}(\phi_x - \phi_t)\right]\psi_1 - \frac{i}{2}\left(\gamma + \frac{1}{4\gamma}\cos\phi\right)\psi_2 .$$

These equations are closely related to the Bäcklund transform equations (A-3) [5, 41] for which we noted that a purely imaginary γ introduced a kink of soliton and a pair $(\gamma, -\gamma^*)$ introduced a breather into the total solution.

The application of such scattering equations to compute the evoluation of nonlinear waves from specified initial conditions has been discussed in detail [5, 17]. Here I will merely sketch a particular calculation of interest to our study of the flux annihilation domain in Section IV.

Refer to Figure 12 and consider the following initial conditions on $\phi(x, t)$ which is to be a solution of (1.1)

$$\phi(x, 0) = 0$$

$$\phi_t(x, 0) = 0 \quad \text{for} \quad |x| > p \qquad \text{(B-3a,b)}$$

$$= V \quad \text{for} \quad |x| \leq p.$$

A scattering problem is defined as indicated where ψ_i and ψ_r are the asymptotic incident and reflected waves in Region ③ and ψ_t is the asymptotic transmitted wave in Region ①. Since ψ obeys (B-1,2) in all regions, the asymptotic forms indicated in Figure 12 follow directly from assuming $\phi \to 0$ as $|x| \to \infty$ and writing

$$\gamma' = \gamma - \frac{1}{4\gamma} . \qquad \text{(B-4)}$$

Bound states of the scattering problem require $a(\gamma') = 0$ and $\mathrm{Im}(\gamma') > 0$. The corresponding bound state eigenvalues represent the kinks and breathers to be found in the evolution of $\phi(x,t)$. It is easily shown that under (B-4) the upper (lower) half of the γ plane maps to the upper (lower) half of the γ'-plane; thus the condition for a bound state is the same in laboratory (x,t) coordinates as in the characteristic (ξ,τ) coordinates usually used to discuss the ISTM for the sine-Gordon equation.

Since the bound state eigenvalues are independent of time, they can be determined from the initial conditions (B-3) at t = 0.

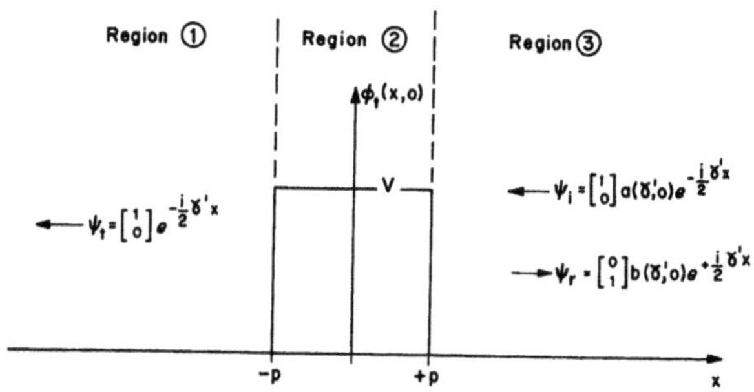

Figure 12. The scattering problem at t = 0 for rotating initial conditions.

In Region ①

$$\psi_1 = e^{-\frac{i}{2}\gamma' x}$$
$$\psi_2 = 0 \; . \tag{B-5a,b}$$

In Region ② (proceeding as in [17])

$$\psi = \begin{bmatrix} (\cos mx - \frac{i\gamma'}{2m} \sin mx) & \frac{V}{4m} \sin mx \\ -\frac{V}{4m} \sin mx & (\cos mx + \frac{i\gamma'}{2m} \sin mx) \end{bmatrix} \times \begin{bmatrix} \psi_{10} \\ \psi_{20} \end{bmatrix} \tag{B-6}$$

where

$$(2m)^2 \equiv (\gamma')^2 + (\tfrac{V}{2})^2 \; . \tag{B-7}$$

The boundary conditions at $x = -p$ determine the constants ψ_{10} and ψ_{20} as

$$\begin{aligned} \psi_{10} &= (\cos mp - \frac{i\gamma'}{m} \sin mp) e^{\frac{i}{2}\gamma' p} \\ \psi_{20} &= (- \frac{V}{4m} \sin mp) e^{\frac{i}{2}\gamma' p} . \end{aligned} \qquad \text{(B-8a,b)}$$

Then the boundary conditions at $x = +p$ determine

$$a(\gamma', 0) = e^{i\gamma' p}[(\cos mp - \frac{i\gamma'}{2m} \sin mp)^2 - (\frac{V}{4m})^2 \sin^2 mp] \qquad \text{(B-9)}$$

as the incident wave amplitude at time $t = 0$.

The bound states are determined by the condition $a(\gamma', 0) = 0$ which can be expressed as

$$\cot 2mp = \frac{i\gamma'}{2m} . \qquad \text{(B-10)}$$

This condition together with (B-7) is just the pair of equations studied previously in connection with the problem of fluxon propagation [17]. The only difference is that γ has been replaced by γ' which is related to γ by (B-4). Thus we can immediately state that all the roots of (B-10) which lie in the upper half of the γ'-plane lie on the imaginary axis, and the construction of Figure 11 in reference [17] will find them. In particular we note that $2m_n p \approx n\pi$ for most of the roots, so from (B-7)

$$\gamma'_n \approx \frac{i}{p} \sqrt{(\frac{Vp}{2})^2 - (n\pi)^2} . \qquad \text{(B-11)}$$

To see which of these roots correspond to breathers we must return to the γ-plane and use the results of the preceding appendix. First we write (B-4) in the form

$$\gamma = \frac{1}{2} \left[\gamma' \pm \sqrt{(\gamma')^2 + 1} \right] . \qquad \text{(B-12)}$$

As V is increased from zero, threshold levels (V_n) are reached at which new zeros of $a(\gamma', 0)$ appear in the upper half of the γ'-plane. Each enters at the origin and moves up the imaginary axis of the γ'-plane. From (B-12) a corresponding pair of zeros enters the γ-plane at $\gamma = \pm 1/2$ and move up the circle $|\gamma| = 1/2$ indicated in Figure 11. Thus stationary breathers are generated just above threshold by the initial conditions (B-3). From the construction in Figure 11 of [17], it is easily seen that the threshold condition for the nth breather pair is

$$V_n = (2n-1)\pi/p. \qquad \text{(B-13)}$$

From (B-11) we note that as $p \to \infty$ most of the zeros lie at $\gamma' = iV/2$. For $V < 2$, (B-12) indicates that they lie on the circle $|\gamma| = 1/2$ in Figure 11 and cluster at

$$\beta = \sin^{-1}(V/2) . \qquad \text{(B-14)}$$

As $p \to \infty$ and $v < 2$ in Figure 12, we have $\phi_{xx} \to 0$ everyhwere whereupon (1.1) reduces to the pendulum equation

$$\phi_{tt} = \sin \phi. \tag{B-15}$$

This extended oscillation may be viewed as "gas" of a large number of synchronized or coherent breathers.

It is interesting to compare the energy of this coherent state with the total input energy U_0. From the Hamiltonian density (A-12) and the initial conditions (B-3), this energy is

$$U_0 = V^2 p \tag{B-16}$$

which can be portioned between i) kinds (of which there are none when $V < 2$), ii) breathers and iii) radiation. Since these three energy components are positive definite [38], the energy of the N breathers

$$U_{NB} \leq U_0. \tag{B-17}$$

An upper bound on U_{NB} is obtained by noting that the approximation (B-11) underestimates $|\gamma'_n|$ so

$$|\gamma'_n| \geq \sqrt{1 - \left(\frac{2n\pi}{Vp}\right)^2} \tag{B-18}$$

and from (B-14) the energy of the nth breather

$$U_{NB} > 8V \sqrt{1 - \left(\frac{2n\pi}{Vp}\right)^2} . \tag{B-19}$$

From (B-13) the number of breathers (N) is the largest integer less than

$$\frac{Vp+\pi}{2\pi} .$$

Thus there are no breathers, and no breather energy, when

$$Vp < \pi .$$

In general the total breather energy

$$U_{NB} > \sum_{n=1}^{N} 8V\sqrt{1 - \left(\frac{2n\pi}{Vp}\right)^2} . \tag{B-20}$$

For

$$Vp >> 2\pi$$

the right hand side of (B-20) can be approximated by the integral

$$\frac{4}{\pi} V^2 p \int_0^1 \sqrt{1-y^2}\, dy = V^2 p \ .$$

Thus (B-20) and (B-17) together imply

$$U_{NB} \to U_0$$

as $p \to \infty$ with $V < 2$. For a number of breathers large compared with unity, essentially all the input energy goes into the breathers. If $Vp = 2\pi N$, (B-20) implies U_{NB}/U_0 is greater than 55% for $N = 2$, greater than 71% for $N = 3$ and greater than 79% for $N = 4$.

Finally let us suppose that

$$V >> 2$$

and we are interested in the behavior of the dissipative sine-Gordon equation (1.2). The time average of sin ϕ will be small, and, for sufficiently large p, the x derivatives can be neglected. Then (1.2) reduces to

$$\frac{dV}{dt} \approx -\alpha V \ . \qquad \text{(B-21)}$$

From (B-13), a breather forms every time V decreases by $2\pi/p$. Thus a rough estimate for the number of breathers formed per unit time and per unit distance is given by

$$N = \frac{\alpha V}{2\pi} \ . \qquad \text{(B-22)}$$

APPENDIX C: DOUBLE PHASE SOLUTIONS

This appendix is a summary account of the theory of multiphase solutions for nonlinear wave equations which has been developed over the past few years by Ablowitz [9 - 12].

Suppose we have a nonlinear wave equation for $\phi(x, t)$

$$N(\phi) = 0 \tag{C-1}$$

which can be obtained from an Euler variation of the Lagrangian density. Thus

$$N(\phi) = \frac{\delta}{\delta\phi} L(\phi_t, -\phi_x, \phi) \; . \tag{C-2}$$

Suppose further that $\phi(x, t)$ can be written as a double phase function

$$\phi(x, t) = \Phi(\theta_1, \theta_2) \tag{C-3}$$

such that L is a doubly periodic function of θ_1 and θ_2. Then choose

$$\begin{aligned} \theta_1 &= \omega_1 t - k_1 x \\ \theta_2 &= \omega_2 t - k_2 x \end{aligned} \tag{C-4a,b}$$

and average L over both periods to

$$\begin{aligned} \mathcal{L} &\equiv \frac{1}{4\pi^2} \int_0^{2\pi} \int_0^{2\pi} L(\omega_1\Phi_{\theta_1} + \omega_2\Phi_{\theta_2}, k_1\Phi_{\theta_1} + k_2\Phi_{\theta_2}, \Phi) d\theta_1 d\theta_2 \\ &= \mathcal{L}(\omega_1, \omega_2, k_1, k_2). \end{aligned} \tag{C-5}$$

If ω_1, ,ω_2, k_1 and k_2 are not constant but allowed to vary <u>slowly</u> with x and t, two dynamic equations for this slow evolution can be obtained from Euler variations of $\mathcal{L}$ with θ_1 and θ_2, Thus

$$\frac{\delta}{\delta\theta_1} \mathcal{L} = 0 \Longrightarrow \left(\frac{\partial\mathcal{L}}{\partial\omega_1}\right)_t - \left(\frac{\partial\mathcal{L}}{\partial k_1}\right)_x = 0 \tag{C-6a}$$

$$\frac{\delta}{\delta\theta_2} \mathcal{L} = 0 \Longrightarrow \left(\frac{\partial\mathcal{L}}{\partial\omega_2}\right)_t - \left(\frac{\partial\mathcal{L}}{\partial k_2}\right)_x = 0. \tag{C-6b}$$

Two additional equations are conservation of periods for the two components or

$$\begin{aligned} (\omega_1)_x + (k_1)_t &= 0 \\ (\omega_2)_x + (k_2)_t &= 0 \; . \end{aligned} \tag{C-6c,d}$$

A necessary condition for this description to be valid is that the original equation (C-1) have the double phase solutions indicated in (C-3) for which L is doubly periodic. It has recently been demonstrated that multiple phase solutions [$\phi(x, t) = \Phi(\theta_1, \theta_2, \ldots, \theta_N)$ for any finite integer N] exist for those particular nonlinear wave equations which display soliton behavior [13-15]. This class includes the sine-Gordon equation (1.1).

A Lagrangian density for (1.1) is

$$L = \frac{1}{2}\phi_x^2 - \frac{1}{2}\phi_t^2 - \cos\phi. \tag{C-7}$$

If $\phi(x, t)$ has the doubly periodic form in (C-3)

$$L = \frac{1}{2}(k_1\Phi_{\theta_1} + k_2\Phi_{\theta_2})^2 - \frac{1}{2}(\omega_1\Phi_{\theta_1} + \omega_2\Phi_{\theta_2})^2 - \cos\Phi \tag{C-8}$$

so (C-6a,b) become

$$\begin{aligned} &\frac{\partial}{\partial t}\int_0^{2\pi}\int_0^{2\pi}(\omega_1\Phi_{\theta_1}^2 + \omega_2\Phi_{\theta_1}\Phi_{\theta_2})d\theta_1 d\theta_2 + \frac{\partial}{\partial x}\int_0^{2\pi}\int_0^{2\pi}(k_1\Phi_{\theta_1}^2 + k_2\Phi_{\theta_1}\Phi_{\theta_2})d\theta_1 d\theta_2 = 0 \\ &\frac{\partial}{\partial t}\int_0^{2\pi}\int_0^{2\pi}(\omega_2\Phi_{\theta_2}^2 + \omega_1\Phi_{\theta_1}\Phi_{\theta_2})d\theta_1 d\theta_2 + \frac{\partial}{\partial x}\int_0^{2\pi}\int_0^{2\pi}(k_2\Phi_{\theta_2}^2 + k_1\Phi_{\theta_1}\Phi_{\theta_2})d\theta_1 d\theta_2 = 0. \end{aligned} \tag{C-9a,b}$$

These are (3.22) of reference [9].

Now consider (1.2) in which the sine-Gordon equation is made dissipative through addition of the small loss term $-\alpha\phi_t$. Following the discussion in Whitham [7, p. 510], we note that (C-9a) can be interpreted as

$$\langle\phi_t\rangle_{1t} - \langle\phi_x\rangle_{1x} + \langle\sin\phi\rangle_1 = 0 \tag{C-10}$$

where

$$\langle\cdot\rangle \equiv \int_0^{2\pi}\int_0^{2\pi}(\cdot)\Phi_{\theta_1}d\theta_1 d\theta_2 . \tag{C-11}$$

Noting the correspondence between (C-10) and (1.1) in the form $\phi_{tt} - \phi_{xx} + \sin\phi = 0$, we can perform a similar averaging of (1.2) to obtain

$$\langle\phi_t\rangle_{1t} - \langle\phi_x\rangle_{1x} + \langle\sin\phi\rangle_1 = -\alpha\langle\phi_t\rangle_1 . \tag{C-12}$$

Similar treatment with respect to the averaging

$$\langle\cdot\rangle_2 \equiv \int_0^{\pi}\int_0^{\pi}(\cdot)\Phi_{\theta_2}d\theta_1 d\theta_2 \tag{C-13}$$

leads to the pair of dynamic equations

$$\frac{\partial}{\partial t}\int_0^{2\pi}\int_0^{2\pi}(\omega_1\Phi_{\theta_1}^2+\omega_2\Phi_{\theta_1}\Phi_{\theta_2})d\theta_1 d\theta_2+\frac{\partial}{\partial x}\int_0^{2\pi}\int_0^{2\pi}(k_1\Phi_{\theta_1}^2+k\Phi_{\theta_1}\Phi_{\theta_2})d\theta_1 d\theta_2=$$

$$=-\alpha\int_0^{2\pi}\int_0^{2\pi}(\omega_1\Phi_{\theta_1}^2+\omega_2\Phi_{\theta_1}\Phi_{\theta_2})d\theta_1 d\theta_2$$

(C-14a,b)

$$\frac{\partial}{\partial t}\int_0^{2\pi}\int_0^{2\pi}(\omega_2\Phi_{\theta_2}^2+\omega_1\Phi_{\theta_1}\Phi_{\theta_2})d\theta_1 d\theta_2+\frac{\partial}{\partial x}\int_0^{2\pi}\int_0^{2\pi}(k_2\Phi_{\theta_2}^2+k_1\Phi_{\theta_1}\Phi_{\theta_2})d\theta_1 d\theta_2=$$

$$=-\alpha\int_0^{2\pi}\int_0^{2\pi}(\omega_1\Phi_{\theta_1}^2+\omega_2\Phi_{\theta_1}\Phi_{\theta_2})d\theta_1 d\theta_2 \ .$$

These, together with the conservation equations (C-6c,d) are the dynamic equations used in this paper. In Section III, only a single phase is assumed (3.3) so (C-14a) reduces to (3.5a). In Section IV, both phases are employed for the d.c. components. It is necessary in either case to assume

$$\alpha << 1$$

so that the dissipative variation over a cycle is small and the averaging indicated in (C-5) is well defined.

Ablowitz [9] has suggested a way to get some integral constraints on the function $\Phi(\theta_1, \theta_2)$. Considering (C-3) as a change of independent variables, (1.1) becomes

$$g_1\Phi_{\theta_1\theta_1}+2h\Phi_{\theta_1\theta_2}+g_2\Phi_{\theta_2\theta_2}+\sin\Phi=0 \qquad \text{(C-15)}$$

where $g_i \equiv \omega_i^2 - k_i^2$ and $h \equiv \omega_1\omega_2 - k_1k_2$. This equation can be written in the form of a conservation law as

$$\frac{\partial}{\partial\theta_1}(\tfrac{1}{2}g_1\Phi_{\theta_1}^2-\tfrac{1}{2}g_2\Phi_{\theta_2}^2-\cos\Phi)+\frac{\partial}{\partial\theta_2}(h\Phi_{\theta_1}^2+g_2\Phi_{\theta_1}\Phi_{\theta_1})=0$$

or as (C-16a,b)

$$\frac{\partial}{\partial\theta_1}(h\Phi_{\theta_2}^2+g_1\Phi_{\theta_1}\Phi_{\theta_2})+\frac{\partial}{\partial\theta_2}(\tfrac{1}{2}g_2\Phi_{\theta_2}^2-\tfrac{1}{2}g_1\Phi_{\theta_1}^2-\cos\Phi)=0.$$

From these it follows that

$$\int_0^{2\pi}(\tfrac{1}{2}g_1\Phi_{\theta_1}^2-\tfrac{1}{2}g_2\Phi_{\theta_2}^2-\cos\Phi)d\theta_2=-E_1$$

and (C-17a,b)

$$\int_0^{2\pi}(\tfrac{1}{2}g_2\Phi_{\theta_2}^2-\tfrac{1}{2}g_1\Phi_{\theta_1}^2-\cos\Phi)d\theta_1=-E_2 \ .$$

If Φ is independent of θ_2, (C-17a) is satisfied by the elliptic function in (3.4).

Under the assumption h = 0, (C-15) becomes

$$g_1\Phi_{\theta_1\theta_1} + g_2\Phi_{\theta_2\theta_2} + \sin\Phi = 0 \qquad \text{(C-18)}$$

which Lamb [36] has shown to have solutions of the form

$$\Phi = 4\tan^{-1}[f(\theta_1)g(\theta_2)] \qquad \text{(C-19)}$$

where f and g are elliptic functions. This is the doubly periodic function suggested recently by Ben-Abraham [42]. For our purposes, however, the assumption h = 0 is too severe. It requires

$$\left(\frac{\omega_1}{k_1}\right)\cdot\left(\frac{\omega_2}{k_2}\right) = 1 \qquad \text{(C-20)}$$

which means that the two component traveling waves must have phase velocities in the same direction with one going faster than the characteristic velocity of (1.1) and the other slower.

From references [13-15] it seems that a simple prescription for constructing a more general double phase solution to (1.1) would be as follows: i) Express the elliptic function defined in (3.4) in terms of theta functions, ii) Augment the arguments of these theta functions from θ to $\theta_1 + \theta_2$, and iii) Check that the resulting expression is a solution of (1.1).

Acknolwedgements

It is my pleasure to acknowledge helpful discussions with J. G. Berryman, D. J. Kaup and D. W. McLaughlin.

REFERENCES

1. A. C. Scott and W. J. Johnson, Internal flux motion in large Josephson junctions, Appl. Phys. Lett. 14 (1969), 316-318.

2. A. Barone, Flux-flow effect in Josephson tunnel junctions, J. Appl. Phys. 42 (1971), 2747-2751.

3. A. Barone, Thermal effect in U-I characteristic of large Josephson junctions, Phys. Status Solidi 13 (1972), K93-K95.

4. M. J. Ablowitz, D. J. Kaup, A. C. Newell and H. Segur, Method for solving the sine-Gordon equation, Phys. Rev. Lett. 30 (1973), 1262-1264.

5. M. J. Ablowitz, D. J. Kaup, A. C. Newell and H. Segur, The inverse scattering transform-Fourier analysis for nonlinear problems, Studies in Appl. Math. 53 (1974), 249-315.

6. D. J. Kaup, Method for solving the sine-Gordon equation in laboratory coordinates, Studies in Appl. Math. 54 (1975), 165-179.

7. G. B. Whitham, Linear and nonlinear waves, Wiley-Interscience, New York (1974).

8. E. N. Pelinovskii and S. X. Shavratzkii, On the damping of stationary waves in a system described by a nonlinear Klein-Gordon equation, Zhurnal Prikladnoi Mekhaniki i Tekkincheskoi Fisiki 5 (1974), 55-59, (in Russian; translation available from The Librarian, Math. Res. Center, University of Wisconsin, Madison).

9. M. J. Ablowitz and D. J. Benney, The evolution of multi-phase modes for nonlinear dispersive waves, Studies in Appl. Math. 49 (1970), 225-238.

10. M. J. Ablowitz, Applications of slowly varying nonlinear dispersive wave theories, Studies in Appl. Math. 50 (1971), 329-344.

11. M. J. Ablowitz, Approximate methods for obtaining multiphase modes in nonlinear dispersive wave problems, Studies in Appl. Math. 51 (1972), 17-55.

12. M. J. Ablowitz, A note on resonance and nonlinear dispersive waves, Studies in Appl. Math. 54 (1975), 61-70.

13. B. A. Dubrovin and S. P. Novikov, Periodic and conditionally periodic analogs of the many-soliton solutions of the Kortweg-de Vries equation, Sov. Phys. JETP, 40 (1975), 1058-1063.

14. A. R. Its and V. B. Mateev, Schrödinger operators with finite-gap spectrum and N-soliton solutions of the Korteweg-de Vries equation, Teoreticheskaya i Matematicheskaya Fisika, 23 (1975), 51-68. Translated by Plenum Publishing Corporation (1976).

15. E. Date and S. Tanaka, Analogue of inverse scattering theory for discrete Hill's equation and exact solutions for the periodic Toda lattice (preprint, 1976).

16. A. C. Scott, Propagation of magnetic flux on a long Josephson tunnel junction, Il Nuovo Cimento 69B (1970), 241-261.

17. A. C. Scott, F. Y. F. Chu and S. A. Reible, Magnetic flux propagation on a Josephson transmission line, J. Appl. Phys. 47 (1976) 3272-3286.

18. J. R. Waldram, A. B. Pippard and J. Clarke, Theory of the current-voltage characteristics of SNS junctions and other superconducting weak links, Philos. Trans. Roy. Soc. London 268 (1970), 265-287.

19. T. A. Fulton and R. C. Dynes, Single vortex propagation in Josephson tunnel junctions, Solid State Communications 12 (1973), 57-61.

20. T. A. Fulton, R. C. Dynes and P. W. Anderson, The flux shuttle-a Josephson junction shift register employing single flux quanta, Proc. IEEE 61 (1973), 28-35.

21. T. A. Fulton and L. N. Dunkleberger, Experimental flux shuttle, Appl. Phys. Lett. 22 (1973), 232-233.

22. K. Nakajima, T. Yamashita and Y. Onodera, Mechanical analog of active Josephson transmission line, J. Appl..Phys. 45 (1974), 3141-3145.

23. K. Nakajima, Y. Onodera, T. Nakamura and R. Sato, Numerical analysis of vortex motion on Josephson structures, J. Appl. Phys. 45 (1974), 4095-4099.

24. K. Nakajima, Y. Sawada and Y. Onodera, Nonequilibrium stationary coupling of solitons, J. Appl. Phys. 46 (1975), 5272-5279.

25. K. Nakajima, Y. Onodera and Y. Ogawa, Logic design of Josephson network, J. Appl. Phys. 47 (1976), 1620-1627.

26. A. C. Scott, Active and nonlinear wave propagation in electronics, Wiley-Interscience, New York (1970).

27. I. Giaever, Metal-insulator-metal tunneling phenomena in solids, Plenum Press, New York (1969), 19-30.

28. B. D. Josephson, Supercurrents through barriers, Advances in Phys. 14 (1965), 419-451.

29. W. J. Johnson and A. Barone, Effect of junction geometry on maximum zero-voltage Josephson current, J. Appl. Phys. 41 (1970), 2958=2960. JAP 41 (1970), 2958.

30. M. D. Fiske, Temperature and magnetic field dependencies of the Josephson tunneling current, Rev. Mod. Phys. 36 (1964), 221-222.

31. I. M. Dmitrenko, I. K. Yanson and V. M. Svistunov, Interaction of the alternating Josephson current with resonant modes in a superconducting tunnel structure, JETP Lett. 2 (1965), 10-13.

32. I. O. Kulik, Theory of "steps" of voltage-current characteristic of the Josephson tunnel current, JETP Lett. 2 (1965), 84-87.

33. D. N. Langenberg, D. J. Scalapino and B. N. Taylor, Josephson-type superconducting tunnel junctions as generators of microwave and submillimeter wave radiation, Proc. IEEE 54 (1966), 560-575.

34. A. Barone, G. Modernò, M. Russo and R. Vaglio, Light-induced transition from "small" to "large" Josephson junctions, Phys. Lett. 53A (1975), 393-394.

35. D. W. McLaughlin and A. C. Scott, A restricted Bäcklund transformation, J. Math. Phys. 14 (1973), 1817-1828.

36. G. L. Lamb, Jr., Analytical descriptions of ultrashort optical pulse propagation in a resonant medium, Rev. Mod. Phys. 43 (1971), 99-124.

37. L. D. Faddeev, Hadrons from leptons? JETP Lett. 21 (1975), 64-65.

38. V. E. Korepin, P. P. Kulish, and L. D. Faddeev, Soliton quantization, JETP Lett. 21 (1975), 138-139.

39. R. F. Dashen, B. Hasslacher and A. Neveu, Particle spectrum in model field theories from semiclassical functional integral techniques, Phys. Rev. D 11 (1975), 3424-3450.

40. A. C. Scott, Quantum theory of the small Josephson junction, Phys. Lett. 25A (1967), 132-133.

41. F. Y. F. Chu and A. C. Scott, Bäcklund transformations and the inverse method, Phys. Lett. 47A (1974), 303-304.

42. S. I. Ben-Abraham, Exact solution to the sine-Gordon equation in two space dimensions, Phys. Lett. 55A (1976), 383-384.

SOME NON-LINEAR PHYSICS IN CRYSTALLOGRAPHIC STRUCTURES

Serge AUBRY*
Brookhaven National Laboratory+, Upton, New York 11973

*Permanent Address : *Laboratoire Léon Brillouin, C.E.N. Saclay, BP n°2, 91190 Gif-sur-Yvette, France*

1. INTRODUCTION

During the past ten years, the development of neutron scattering techniques and the improvement of the other experimental methods allowed considerable progress in the knowledge of crystallographic structures. Although most crystals can be understood on the basis of an harmonic model with small perturbation terms, many crystals which undergo a change of structure at some temperature, exhibit unexpected features, and sometimes in a rather large region around the transition temperature.

The central peak phenomenon first discovered in 1971 in strontium titanate[1)] is one of the most intriguing. The structural phase transition of this crystal at 105 K, was believed to be of the displacive type, which means that the elastic forces between the atoms are temperature dependent and some of them vanich at the transition temperature, then allowing the structure of the crystal to change. The softening of the elastic atomic constants is experimentally observable by looking (for example, by neutron scattering) at the variation of the phonon peak frequencies of the dynamical response function. In strontium titanate, the softening of one phonon peak was indeed observed but turned out to be incomplete. Close to the transition temperature, instead of vanishing, the soft phonon frequency ceases to decrease and then a zero frequency peak (the central peak) grows with a divergent intensity at the transition. Further experiments did not succeed in determining the width of this central peak. It might be zero as this has been recently shown in KDP [2)], another compound which exhibits a central peak. However, there are probably several kinds of central peak[3)] which may have a zero width or not and which require several interpretations. Some theories[4)] interpret the central peak phenomena by a strong anharmonicity of the atomic forces, the others assume the existence of defects or impurities.[5)] These theories might be complementary but until now, none of them has been convincingly shown to explain any given experiment.

Incommensurate structures in crystals are also surprising phenomena. For example, in potassium selenate[6)] which exhibits a structrural phase transition at 93 K, the structure of this crystal above the transition temperature (between 93 K and 129.5 K) is modulated by a static periodic distortion. Its period varies with the temperature and so is generally incommensurate with the lattice period.

Other various unexpected features have been observed both in compounds which generally undergo structural phase transitions, and also in magnetic structures. They are currently being intensively studied and their list grows month after month. For brevity's sake, we will not continue their description but this brief survey demonstrates the ignorance we have of strongly non-linear systems in which the standard approximations may fail drastically.

The aim of this talk is not to present a review of the different theories which have been developed to interpret these phenomena but to present a summary of some of these studies on simple but strongly non-linear models which make use of some methods in stochasticity. I hope this will show the potential applications which might exist in solid state physics for any progress in this field.

Two one-dimensional models will be described. One has been studied to understand some aspects of the non-linear dynamics in crystals when close to the transition temperature. The second one is for commensurability and incommensurability problems.

2. *PERIODIC ORBITS AND THE DYNAMICS OF A ONE-DIMENSIONAL COUPLED DOUBLE-WELL CHAIN.*

The model that we consider is a chain of atoms i, the energy of which is

$$\phi\left(\{u_i,p_i\}\right) = \sum_i \left(\frac{1}{2}p_i^2 + \frac{1}{2}C(u_{i+1}-u_i)^2 + (u_i^2-1)^2\right) \quad (1)$$

where p_i is the momentum of the atom i with mass unity.

u_i is its displacement with respect its mean position.

C is the elastic coupling constant between two consecutive atoms.

$(u^2-1)^2$ is a double-well potential (assumed to be created by a rigid sublattice of different atoms).

It is an oversimplified model[4)] to understand the dynamics of structural phase transitions, which contains some of the essential characteristics of the real crystals which undergo a structural phase transition. Its transition temperature is zero ; then all the atoms are either in $u_i = +1$ or in $u_i = -1$.

Numerical calculations have been performed[7)] to obtain the space time Fourier transform of the atomic correlation function

$$\mathcal{S}(q,\omega) = \int e^{i\omega t}\left(\sum_n e^{iqn} \langle u_i(t)\, u_{i+n}(0)\rangle\right) dt \quad (2)$$

which is an experimentally measurable quantity. Of course, $\mathcal{S}(q,\omega)$ depends on the temperature of the chain (numerically on its initial energy).

Quasi harmonic theories predict , in this model with one atom per unit cell, that $\mathcal{S}(q,\omega)$ must contain (for $\omega \geq 0$) only one phonon peak at a finite frequency $\omega_{ph}(q)$ with some broadening. In fact, numerical calculation shows that there are (mainly) three peaks, the relative intensity of which varies with the temperature. In addition, one of them is at zero frequency (central peak). We call the two others quasi-phonon peaks. This central peak and one quasi-phonon peak only exist at low temperature. When increasing the temperature, the other quasi-phonon peak appears and develops at the expense of the two first ones which disappear at high temperature. There is a finite region in temperature, where these three peaks coexist.

The central peak is very narrow close to the order-disorder limit $C \ll 1$ while there is almost no softening of the two quasi-phonon frequencies (they are almost independent of the temperature) the spectrum evolves continuously until the displacive limit ($C \gg 1$), where the central peak is then rather broad and where there is an appreciable softening of the two non-zero frequency peak in the quasi-critical region.

This striking behavior can be understood by looking to particular periodic solutions of the non-linear motion equations

$$\ddot{u}_i + 4\, u_i(u_i^2 - 1) + C(2u_i - u_{i-1} - u_{i+1}) = 0 \qquad (3)$$

When $C \gg 1$ (displacive limit) a continuous field can be introduced because u_i varies very slowly versus its index i. Considering it as a continuous variable x gives the motion equation

$$\frac{\partial^2 u}{\partial t^2} - C\,\frac{\partial^2 u}{\partial x^2} + 4u\,(u^2 - 1) = 0 \qquad (4)$$

Propagating solutions with the form $u(qx - \omega t)$ satisfy the equation

$$-(Cq^2 - \omega^2)\, u'' + 4u(u^2 - 1) = 0 \qquad (5)$$

It can be easily integrated, using the three kinds of Jacobi elliptic functions[8)]: dn and cn when $Cq^2 < \omega^2$, sn when $Cq^2 > \omega^2$. These three kinds of solutions are periodic orbits of the whole system.

Considering their energy and their amplitude versus their period in space and time, a semi-quantitative interpretation[4b,7)] of the variation of the dynamical correlation function $\mathcal{S}(q,\omega)$ versus the temperature can be given close to the displacive limit.

The family of solutions dn corresponds to small amplitude motions of the atoms in one side of their double-well, and the family cn corresponds to large amplitude symmetric motions from one side to the other side. They interpret the two quasi-phonon peaks with a non zero frequency.

The third peak, at zero frequency (central peak) is interpreted by the family of solutions sn. Some of these solutions are stationary (time-independent). They represent equidistant walls which may or not be moving uniformly. (A single wall separates the chain in two regions with different ordering e.g. $u_i > 0$ for $i > i_o$ and $u_i < 0$ for $i < i_o$. When there are several walls, it is an intuitive concept if they are distinguishable).

When the wave vector of a solution of this family goes to zero, one obtains a single moving wall (a moving kink). Due to the Lorenz invariance on x and t of the continuous motion equations (4), its energy is a function of its velocity v by the standard relation :

$$E(v) = \frac{1}{\sqrt{1-\frac{v^2}{c^2}}} E(0) \tag{6}$$

with $c = \sqrt{C}$ the limiting velocity of these kinks (physically, it is a sound velocity). Although these kinks are not true solitons (they are not kept when they interact) this particle-like behavior allows us to suppose that they behave independently if they are widely separated[4]. Then it can be assumed that they are well described by a perfect gas (which at high enough temperature might be relativistic) ; this assumption allows a quantitative calculation of the shape of the central peak of $\mathcal{S}(q,\omega)$ and of its width.

3. *LATTICE LOCKING AND STOCHASTICITY.*

The above interpretation holds only in the displacive limit (C >> 1). However, it now becomes clear that the central peak of this model is related to the existence of a family of solutions of the motion equations, which contains stationary solutions. These last solutions can be studied in the discrete model far from the displacive limit by considering a non-linear planar mapping. Eq. (3) with $\ddot{u}_i = 0$ yields :

$$C(u_{i+1} + u_{i-1} - 2u_i) - 4u_i\,(u_i^2 - 1) = 0 \tag{7}$$

The point $\underset{\sim}{P}_{i+1}$ of coordinates (u_{i+1},u_i) is then a function $T(\underset{\sim}{P}_i)$ of the point $\underset{\sim}{P}_i = (u_i,u_{i+1})$. This planar transformation T

$$\begin{pmatrix} x' \\ \\ y' \end{pmatrix} = T \begin{pmatrix} x \\ \\ y \end{pmatrix} = \begin{pmatrix} 2x + \frac{4}{C}\,x(x^2 - 1) - y \\ \\ y \end{pmatrix} \tag{8}$$

is area preserving and has similar properties to those which have been already studied[9]. Given an arbitrary initial point $\underset{\sim}{P}_1 = (u_1, u_0)$ the iterated sequence $T^n(\underset{\sim}{P}_1)$ for $-\infty < n < +\infty$ determine a solution of (7).

The essential feature of the T-mapping, which has physical consequences, is that the sequence of points which do not go to infinity are mapped into either analytical curves, or appear to be erratic (in that case, it is a Cantor set which is mapped).

In an exceptional case, it can also be mapped into a set of n, n^{th} order fixed points $\underset{\sim}{F}_1$, $\underset{\sim}{F}_2 = T(\underset{\sim}{F}_1)\ldots\underset{\sim}{F}_n = T^n(\underset{\sim}{F}_1) = \underset{\sim}{F}_1$, of the transformation T^n. However, the properties of these fixed points, determine also many properties of the T-mapping. Their iterated sequences represent commensurate configurations in model (1). (We mean that their period is commensurate with the lattice spacing). The simplest fixed points of our transformation (8) are first order. They are $\underset{\sim}{H}_1 = (1,1)$ $\underset{\sim}{H}_2 = (-1,-1)$ and $\underset{\sim}{F}_o = (0,0)$. $\underset{\sim}{H}_1$ and $\underset{\sim}{H}_2$ are hyperbolic. In other words, the eigenvalues of the linearized form of T, in $\underset{\sim}{H}_1$ or $\underset{\sim}{H}_2$ are real and their product is unity. The corresponding eigendirections are also real, one is attractive, the other is repulsive. Consequently, these fixed points are unstable with respect to the transformation T but their iterated sequences represent the two classical ground-states of model (1). The fixed point $\underset{\sim}{F}_o$ can be either elliptic when $C > 1$ (the eigenvalues of the linearized form of T in $\underset{\sim}{F}_o$ are complex with modulus unity and the transformation T is locally equivalent to a rotation on an ellipse) or hyperbolic when $C < 1$. When it is elliptic, is it stable with respect to the transformation T but its iterated sequences represent the dynamically unstable state $u_i \equiv 0$ of model (1).

The Kolmogorov-Arnold-Moser theorem (K.A.M. theorem)[10)] predicts that (generally) there exist invariant analytical curves with respect to the non-linear transformation T, around its nth order elliptic fixed points. On these curves, the transformation T^n is topologically equivalent to a rotation incommensurate with 2π. For example, when $\underset{\sim}{F}_o$ is elliptic, there exist connected closed analytical curves around it. A consequence is that the atomic position of the configurations of the model (1) generated by the points of one of these curves can be written as :

$$u_i = f(i + \alpha) \tag{9}$$

α is an arbitrary phase, and f is an analytical periodic function. It is clear that (9) represents an incommensurate periodic configuration of model (1). The perturbation spectrum of this configuration is obtained by linearizing (3) in ε_i after setting

$$u_i(t) = f(i + \alpha) + \varepsilon_i \exp(i\omega t) \tag{10}$$

where ω is the frequency of the perturbation mode. It yields

$$-\omega^2\varepsilon_i + C(2\varepsilon_i - \varepsilon_{i+1} - \varepsilon_{i-1}) + 4\varepsilon_i(3f^2(i + \alpha) - 1) = 0 \tag{11}$$

Differentiating (7) with respect to i and comparing with (11) shows that $\omega = 0$ is an eigenfrequency and that $\varepsilon_i = f'(i + \alpha)$ is the eigenvector. It is the translation mode. Consequently, this configuration is not locked.

The generation of the erratic sequences of the transformation T can be understood by first looking at the wall configuration of model (1) in the discrete case. A stationary configuration with a single wall (with for example $u_{-\infty} = -1$ and $u_\infty = +1$) is represented in the T-mapping by a sequence of points $T^n(\underset{\sim}{P}_o)$ which go from $\underset{\sim}{H}_2 = (-1,-1)$ for $n = -\infty$ to $\underset{\sim}{H}_1 = (1,1)$ for $n = +\infty$. These points belong to the set of points, the successive transformations of which, by T go to $\underset{\sim}{H}_1$. (this set is a line which is called the contracting sheet of $\underset{\sim}{H}_1$) and to the set of points, the successive transformations of which by T^{-1} go to $\underset{\sim}{H}_2$, this being the dilating sheet of $\underset{\sim}{H}_2$.

For a linear transformation, the contracting and dilating sheets of its unique fixed point exist only in the hyperbolic case and are straight lines parallel at the fixed point to the real eigendirections. In the non-linear cases, these sheets only exist for hyperbolic fixed points but may have extremely complex shapes (see Ref. 11). In addition, contracting and dilating sheets may intersect. The intersection points are called homoclinic if the two involved fixed points are the same, and heteroclinic if they are not. Then, the single wall configurations are represented by heteroclinic points associated with the couple of hyperbolic fixed points $\underset{\sim}{H}_1$ and $\underset{\sim}{H}_2$. They do exist as this can be checked numerically[11].

A theorem[10b], chapter III theorem 3-7 states that if one homoclinic point exists, there also exists an infinite number of such points. Their closure is a Cantor set Λ which is stable by T and in which this transformation is a Bernouilli shift. As a result, T is ergodic and mixing in Λ . Physically, this means that one cannot make a continuous change of the corresponding stationnary configuration. This implies that the perturbation spectrum cannot possess any zero frequency mode.

The same property is true for the heteroclinic points which represent the single walls walls. They are (generally) locked for any value of the coupling constant C. Random distribution of such locked walls (provided that they are not too close to each other) are the physical representation of the erratic trajectories of the transformation T. At high concentration, the walls cannot be considered as independent because they overlap one another ; the resulting interaction competes with the lattice locking and an unlocked configuration may result. This is exactly what is proven above by the K.A.M. theorem.

It turns out from these remarks that we can make the physical equivalence in model (1) configuration : representative curve in the T-mapping

unlocked $\leftrightarrow$ analytic (non-stochastic)
locked $\leftrightarrow$ non-analytic or stochastic

We refer the reader to (11) for many more details which also concern the other kind of excitations.

When C is large, most of the trajectories of T are analytical curves. This is consistent with the displacive limit where the continuous model (4) applies and where the stationary configurations are described by elliptic functions sn which are periodic and analytic. Close to this limit, the lattice locking effects become very weak, the wall dynamics are fast and the central peak broad. When C decreases, these stationary configurations one after the other become locked, the wall dynamics slower and the central peak narrower. At C = 0 the walls are perfectly locked because the chain is composed of independent particles and the central peak has a zero width.

These results are consistent with the numerical results[7)] and demonstrate the important role of the lattice effects on the dynamics associated with the central peak of this model.

4. *COMMENSURABILITY AND INCOMMENSURABILITY IN CRYSTALLOGRAPHIC STRUCTURES*[12)]

In the above model (1), the classical ground-state is trivial. However, this is not the case in many models describing the structure of new compounds as for example K_2SeO_4 mentioned in the introduction. One of the more simple models where such a situation does occur describes a one-dimensional epitaxial monolayer of a gas absorbed on a crystal surface. Its potential energy is

$$\phi\left[\{u_i\}\right] = \sum_i \left[V(u_i) + W(u_{i+1} - u_i)\right] \tag{12}$$

where u_i is the position of the ith atom of the monolayer, $V(u_i)$ is a periodic potential with period 2a which represents the crystal surface, and $W(u_{i+1} - u_i)$ is the interacting potential between neightboring atoms of the chain. Its minimum is obtained for $u_{i+1} - u_i = b$ which is generally not commensurate with 2a. The potentials are assumed to be smooth enough (say analytical) function (see ref. 12)

Finding the classical ground-state requires minimization of the energy (12). If ϕ in (12) is not considered as an energy but as the action of a single particle with coordinates u_i, with a discrete time i, the equations of extremalization of (12) describe a formal canonical system (see ref. 10, appendices 32,33). This remark is generalizable to all classical model with short-range interactions for crystallographic or magnetic structures. When V vanishes in (12), this canonical system has a motion invariant

$$u_{i+1} - u_i = \ell = \text{constant} \tag{13}$$

This system is integrable. Considering $V(u_i)$ as a small perturbation, the K.A.M. theorem (Ref. 10a, theorem 21.11 and b, theorem 2.11) is applicable. (for details see ref. 12) and yields the following conclusion.

For most ℓ incommensurate with the period 2a of the potential V (except a zero measure set) and for V small enough, there exist solutions of the extremalization equation of (12) which can be written as

$$u_i = (i\ell + \alpha) + f(i\ell + \alpha) \tag{14}$$

where f is an analytical function with period 2a, which goes to zero when V goes to zero, and with zero mean value. α is an arbitrary phase.

Now, if we consider an infinite epitaxial monolayer with fixed ends such that for N going to infinity

$$u_N - u_o = N\ell \tag{15}$$

with $\ell/2a$ irrational. The analytical solution (14) is the classical ground-state. Since it is represented by an analytical function of i, it is not locked by the substrate potential (see the previous section). It possesses a zero frequency translation mode.

For large enough potential V(u), we can prove (using some assumption (18) on V and W mentioned in the following) that the ground-state can no longer be represented by (14). Mathematically, the invariant torus with period ℓ of the K.A.M. theorem which represents the ground-state disappears. This is a transition from a non-stochastic regime to a stochastic regime. Physically, the monolayer becomes locked by the subtract potential V, for some critical value of its amplitude. This transition can be interpeted by saying that when the potential V becomes strong enough, and the particles are obliged to stay in bounded intervals close to the minima. Then, the u_i (mod 2a) cannot be dense on the interval (0,2a) which means that u_i cannot be represented by formula (14) with f periodic analytic.

The Peierls system[13] in which the Frölich mode (zero frequency translation mode) is generally assumed to exist, is very similar to this epitaxial model and might exhibit in some cases, the same kind of locking transition simultaneously with the appearance of an insulating phase. This question is to be examined in further studies.

When condition (15) (fixed volume) is relaxed, the model has free ends. It is convenient in studying it to introduce an additive parameter by changing $W(u_{i+1} - u_i)$ in (12) into

$$\tilde{W}(u_{i+1} - u_i) = W(u_{i+1} - u_i) - p(u_{i+1} - u_i) \tag{16}$$

We then get in the energy and additional term

$$-p(u_N - u_o) = - \sum_{i=0}^{N-1} p(u_{i+1} - u_i) \tag{17}$$

showing that p is the conjugate parameter of the volume $(u_N - u_o)$ of the chain and is then a pressure.

To get more results, we now need to make some assumption about V and W. We assume that

$$\begin{array}{ll} W & \text{is convex function} \\ V & \text{is symmetric in addition to being periodic.} \end{array} \tag{18}$$

This hypothesis is not unreasonable for many physical applications. The ground-state for any boundary conditions can then be proven[12] to have the form

$$u_i = (i\ell + \alpha) + \varepsilon_i \tag{19}$$

where ε_i has a zero mean value. Its modulus is bounded by 2a, but it is not necessarily an analytic function of i. The equality (19) defines ℓ as the mean distance between neighboring atoms and α as a phase. A consequence is that the existence of two phases, in the ground-state with different atomic mean distance is impossible with or without any boundary conditions. Therefore, no first-order transition can exist in this model by varying the parameters. This sharply contradicts the results which have been obtained by standard variational methods[14)] on a particular model (12) which satisfies (18).

The energy of the ground-state with fixed ends (15) versus ℓ is

$$\psi(\ell) = \psi_o(\ell) - p\ell \tag{20}$$

where $\psi_o(\ell)$ is the energy without pressure of the model with fixed ends (15). This function is shown to be convex[(12)] and the ground-state at constant pressure is given by solving

$$\psi_o'(\ell) - p = 0 \tag{21}$$

which defines ℓ versus p.

We show in ref. (12) that when ℓ is commensurate with (2a) the left and right derivatives of ψ are generally different and that

$$\psi_-' (\ell) < \psi_+'(\ell) \tag{22}$$

This function $\psi'(\ell)$ makes a jump at any rational value of $\ell/2a$. This jump is shown to be related to the smallest energy of the stationary phase excitations in this commensurate ground-state with period ℓ.[12)] Then solving (20) to get $\ell(p)$ the atomic mean distance of the ground-state for a given pressure, shows that this function is constant each time that $\ell/2a$ is a rational number and for p satisfying

$$\psi_-'(\ell) < p < \psi_+'(\ell) \quad \text{and is continuous.} \tag{23}$$

This monotonous increasing function $\ell(p)$ has infinitely many steps and is called a devil's stair. We can go further in choosing particular potentials for V and W in (12)

$$W(u_{i+1} - u_i) = \frac{1}{2} C(u_{i+1} - u_i)^2 \tag{24a}$$

C is a coupling constant.

$$V(u_i) = \frac{1}{2} \omega_o^2 (u_i - 2m_i a)^2 \tag{24b}$$

ω_o^2 is a positive constant and

$$m_i = \text{Int}\left[\frac{u_i}{2a} + \frac{1}{2}\right] \qquad (24c)$$

make V(u) periodic with period 2a. However, this potential (24b) is not well-behaved since it is not differentiable for $u_i/2a$ and integer. All the configurations of this model are readily shown to be locked because this model is locally harmonic. This is consistent with the hypothesis of the version of Moser (Ref. 10, theorem 2.11) of the K.A.M. theorem which requires, when applied to our problem, that V must be at least more than three times differentiable[12).

Model (12) - (24) is exactly soluble. The atomic mean distance of the ground state of this model at constant pressure is proven to be always commesurate with the period 2a of the subtract potential V (except perhaps for a zero measure set of pressure values).

Then the function $\ell(p)$ defined by (20) is entirely composed of infinitely many constant steps (the width of them is calculable at any order). When varying the parameters of this model or the pressure, the atomic mean distance of the ground-state varies with infinitely many jumps but without first-order transition.

If such a transition exists in physical compounds as we believe, most of these jumps cannot be observed experimentally except perhaps the largest one which correspond to simple order of commensurability of ℓ with 2a and which might appear to be first-order transitions but are not. The important feature of these transitions is that for V large enough, the locking forces are practically the same for commensurate (or incommensurate) configurations at any order. Then these sequences of transitions are expected to have a global hysteresis which can be simply described by assuming a solid friction between the monolayer and the subtract. When V is smaller, this friction disappears first for incommensurate configurations. Such phenomena are very similar to the unexplained hysteresis of the apparently continuous variation of the wave vector distorsion of some crystal versus the temperature (for example, see ref. (15)).

5. *CONCLUDING REMARKS*

These exact results obtained on very simple models show that the real behavior of simple non-linear systems might be very different from what is obtained by standard approximations. This occurs because non-linear systems can exhibit qualitatively different behavior from linear systems. These results of course cannot be obtained as done usually, by approximating the non-linear system by a linear one. (Solitons, kinks, and stochasticity are examples.) In more complicated models, we may except that the results are more varied. For example, it might be possible that the ground-state is apparently random although its Hamiltonian is not. Indeed, it turns out that the assumption (18) that we made on potential V and W are essential to get the devil's stair transition that we just described. If they are relaxed, we ignore what can be the real structure of the ground-state of this model, whether or not it is periodic.

Another problem is to find the effect of the temperature and of the quantum fluctuations. They are well studied by renormalization group techniques[18]. These methods make use of a recursive transformation which transforms a Hamiltonian into a renormalized Hamiltonian. This transformation has usually several kinds of fixed points The fixed points which are attractive represent the phases of the system. Unstable fixed points with attractive and repulsive directions represent the critical points. The critical surface is simply their contracting sheet (see Sec. 3). The behavior of the mapping of the renormalization group transformation has always been found to exhibit such a simple behavior but never stochasticity or only limit cycles. We believe that the explanation of this fact is simply that there are only a few possible structures in the considered models and therefore of possible transitions and fixed points. In systems as simple as the epitaxial model (12) considered in a several dimension version, and in many other models which might be involved in crystal structure, infinitely many phases are possible and therefore infinitely many fixed points in the recursive renormalization transformation.

In addition, it is now known that new possible features, which are not believe to be exceptional, like a strange (or fractal[16,17]) attractor may exist in recursive transformations. For example, if a strange attractor exists for the recursive renormalization transformation of some model, it has a finite basin (the set of points which converge to the strange attractor by the recursive transformation) which in some sense plays the same role as the critical surface in usual cases. The corresponding transition should be not at a critical temperature, but extended on a finite interval of temperature $T_1 < T < T_2$. We think that in this range of temperature the crystal structure be composed of static macro-domains providing then an alternative explanation[5] for a zero width central peak phenomenon. In other cases, we may have limit cycles for the recursive renormalization transformation which may describe transition with modulated structures like those of K_2SeO_4 [6]. We believe that models with conflicting interactions between the atoms like the one we considered here in (12) might provide examples of such unexpected behavior of the renormalization group.

These remarks remain still speculative and need detailed studies on particular models. But for this reason, we believe that mathematical progress in non-linear problems should be very useful to help in the understanding of these complex situations and therefore to get explanations of some recent experiments on unusual crystallographic and magnetic structures and perhaps also on amorphous structures.

REFERENCES

1. T. Riste, E.J. Samuelsen, K. Otnes, J. Feder, Solid State Commun. 9, 1455 (1971) ; S.M. Shapiro, J.D. Axe, G. Shirane, T. Riste, Phys. Rev. B 6, 4332 (1972).
2. L.N. Dursvala and R.W. Gammon, Phys. Rev. Lett. 38, 1081 (1977).
3. K.B. Lyons and P.A. Fleury, Phys. Rev. Lett. 37, 161 (1976).

4. a) J.A. Krumhansl and R.J. Schrieffer, Phys. Rev. B 11, 3535 (1975).
b) S. Aubry, J. Chem. Phys. 64, 3392 (1976).
c) T. Schneider and E. Stoll, Phys. Rev. B 13, 1216 (1976).
5. T. I. Halperin and C.M. Varma, Phys. Rev. B 14, 4030 (1976).
6. M. Iizumi, J.D. Axe, G. Shirane, and K. Shimaoka, Phys. Rev. B 15, 4392 (1977).
7. S. Aubry, Thesis, Université Pierre et Marie Curie (Paris) 1975.
8. I.S. Gradshteyn and I.M. Ryshik, Table of Integral Series and Products, Academic Press, New York and London, p. 904.
9. M. Henon, Quat. Appl. Math. 27, 291 (1970).
10. a) V.I. Arnold and A. Aveze, *Ergodic Problems in Classical Mechanics*, W.A. Benjamin, Inc. (1968).
b) J. Moser, *Stable and Random Motions in Dynamical Systems*, Princeton University Press (1973).
11. S. Aubry, to be published in J. Math. Phys.
12. S. Aubry, Manuscript in preparation.
13. See, for example, W. Dieterich, Adv. in Phys. 25, 615 and references therein.
14. S.C. Ying, Phys. Rev. B 3, 4160 (1971).
15. W.D. Ellenson, S.M. Shapiro, G. Shirane, A.F. Garito, to be published in Phys. Rev. B (1977).
16. M. Henon, Comm. Math. Phys. 50, 69 (1976).
17. Y. Pomeau, Private communication.
18. M.E. Fisher, Rev. Mod. Phys. 46, 597 (1974) ; D. Nelson and M.E. Fisher, Ann. Phys. 91, 226 (1975).

+ Work at Brookhaven supported by U.S. Department of Energy under Contract n° EY-76-C-02-0016 .

LASER INSTABILITIES - AN EXAMPLE FROM SYNERGETICS

H. Haken and A. Wunderlin
Institut für Theoretische Physik
der Universität Stuttgart

Abstract

Some important laser instabilities are reviewed by applying an order parameter concept for nonequilibrium systems developed by one of us (H.H.). The first laser threshold, where a transition from a completely disordered state to a state ordered on a macroscopic scale occurs, is discussed under various viewpoints. The second threshold, where the c.w. solution becomes unstable, is considered for two different cases: Depending on certain parameters, either regular pulses are obtained as a new ordered state or a seemingly irregular spiking occurs. In the latter case contact is made with the Lorenz model of turbulence of fluids.

Introduction

Over the past few years a new scientific discipline named synergetics [1 - 6] is developing rapidly. Synergetics deals with the cooperative behavior and self-organization of systems which are composed of many subsystems. Phenomena like the spontaneous formation of new structures or self-organization occur when such systems are driven by external influences far away from thermal equilibrium.

It is important to note that the concepts and mathematical methods developed in this new field are not restricted to physical systems [1-5] but can be applied to quite different disciplines like non-equilibrium chemical reactions, biological systems, morphogenesis and even sociological models.

We shall discuss in the following an example from physics, the laser which is one of the most striking and best understood examples of a synergetic system. But before going through the details we will briefly discuss some basic ideas and mathematical methods developed in synergetics [1,7,8] .

We consider a system composed of many subsystems which can be exhaustively described by a set of variables $u_1, u_2 \ldots u_N$ which we combine to a state vector $\underset{\sim}{u} = (u_1, u_2 .. u_N)$. The dynamics of the system under consideration may then be governed by an equation of motion of the general type

$$\dot{\underset{\sim}{u}} = \underset{\sim}{G}(\underset{\sim}{u}, \underset{\sim}{\sigma}) + D \nabla^2 \underset{\sim}{u} + \underset{\sim}{F}(t). \tag{1}$$

The first term $\underset{\sim}{G}(\underset{\sim}{u}, \underset{\sim}{\sigma})$ contains the in general nonlinear kinetics of the system, the second allows for spatial diffusion with a diffusion matrix D , and $\underset{\sim}{F}(t)$ represents the fluctuating forces. The external parameters which measure the external influences are combined to a vector $\underset{\sim}{\sigma}$.

Implicitly there are three different time scales inherent in the system (1). τ_1 an extreemly large scale in comparison to all the other time scales is associated with the external parameters and allows us to treat them as quasi static quantities. The second time scale τ_2 is a measure for the time variation of the state vector $\underset{\sim}{u}$ and the third τ_3 corresponds to the short scale of the fluctuations:

$$\tau_1 \gg \tau_2 \gg \tau_3 . \tag{2}$$

We assume that we can find a stationary state to Eq. (1) called $\underset{\sim}{U}_0$ which is stable for a certain set of values of the external parameters. For simplicity we may assume $\underset{\sim}{U}_0$ as a homogeneous state in time and space but note that the formalism remains applicable if already is a spatially or time ordered state [7] . The stability of $\underset{\sim}{U}_0$ may be checked by linear stability analysis.

When we change the external parameters some of the eigenmodes of the linear problem may become unstable, i.e., the state $\underset{\sim}{U}_0$ looses its stability and a new state will emerge. The aim of synergetics is not only to find the instability points but also to predict the new state beyond the instability point. Therefore our mathematical formalism must be able to describe the spontaneous ordering out of a homogeneous state, i.e., a disorder - order transition as well as transitions from an ordered state to a new, e.g., more complex ordered state.

To describe spontaneous ordering mathematically we note that linear stability analysis already yields an important hint, i.e., it splits the modes into two classes near the instability point: Stable modes and unstable modes. If we denote the amplitudes of the modes by $\{\xi\}$ we may write schematically

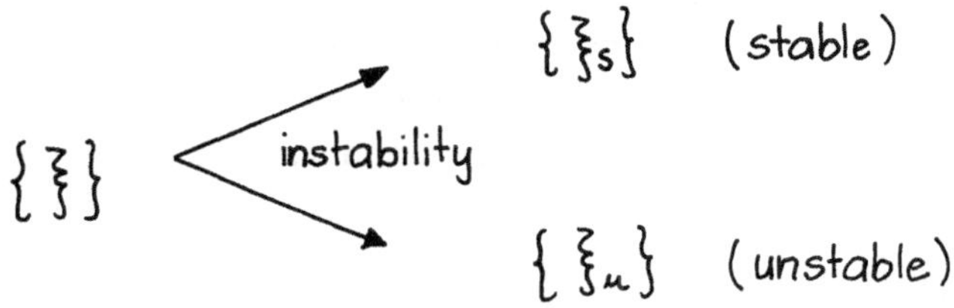

(3)

Both types of modes are assumed well seperated by their different time scales. Whereas the stable modes remain strongly damped also beyond the instability point the unstable ones grow and vary slowly on a much larger time scale, i.e., they dominate the macroscopic behavior of the whole complex system. It is just this fact which allows for adiabatic elimination of the stable modes from Eq. (1) and leaves us with a set of equations for the amplitudes of the unstable modes alone:

$$\dot{\xi}_u = \lambda_u \xi_u + \nabla^2 \xi_u + H(\{\xi_u\}) + F_u(t) .$$

(4)

$H(\{\xi_\mu\})$ contains all nonlinear terms in ξ_μ, $F_\mu(t)$ is the projection of the fluctuating forces on the critical mode. It turns out that the unstable modes are stabilized beyond the instability point in two ways: By the nonlinearities itself and secondly via the coupling to the stable modes which again is prescribed by the nonlinearities.

In conclusion we find a new state vector of the system beyond the instability point $\{\xi_\mu\}$ spanned by the amplitudes of the unstable modes only. Thus the original space is drastically reduced and the unstable modes prescibe the behavior of the whole system, i.e., they act as order parameters on the system and slave all the other modes. It is just this concept of order parameters which in many cases makes complex cooperative behavior of systems far from thermal equilibrium again tractable. In the following we shall apply this concept to instabilities of the laser.

The laser model [9]

We shall consider the following set up for a laser (compare Fig. 1). The laser active material is embedded inbetween two mirrors which serve for mode selection and on the other hand constitute a feedback mechanism between atoms and light field. Usually one of the mirrors is partly transparent. The laser active material consists of a host material where N laser active atoms

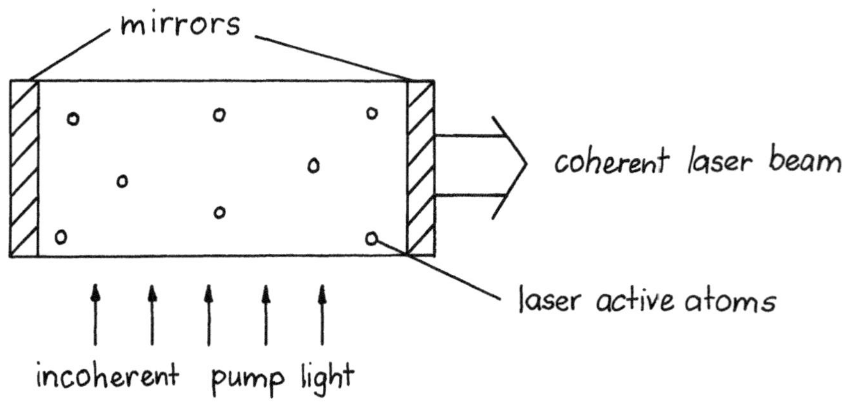

Fig. 1: Typical setup of a laser

are distributed at random. These laser active atoms will be treated as two level atoms which are far away from each other, i.e., they interact only via the electromagnetic field. The state of the laser is controlled by an external pump mechanism, e.g., by incoherent pump light.

Obviously this system is composed of many independent subsystems, the single laser active atoms.

It is a well known task to derive the Hamiltonian [9] of the coherent interaction between the two level atoms and the light field in the dipole approximation. The light field is described by photon creation and annihilation operators b_λ^+ and b_λ, respectively, which obey the usual Bose commutation relations. The index λ distinguishes the different modes prescribed by the cavity. We furthermore will assume that they are polarized in one direction only.

The atoms are described by raising and lowering operators

$$\alpha_\mu^+ = (a_2^+ a_1)_\mu \quad ; \quad \alpha_\mu = (a_1^+ a_2)_\mu \, , \tag{5}$$

where the index μ labels the different atoms. α_μ^+ raises the electron of atom μ from the ground state 1 to the upper level 2. In the order to get a complete description of the atomic state we furthermore introduce the operator

$$\sigma_\mu = (a_2^+ a_2 - a_1^+ a_1)_\mu \, , \tag{6}$$

which measures the difference in occupation of the two levels 2,1.

The Hamiltonian now reads

$$H = \frac{1}{2}\hbar\nu \sum_\mu \sigma_\mu + \hbar \sum_\lambda \omega_\lambda b_\lambda^+ b_\lambda + \sum_{\mu,\nu} \hbar \left(g_{\mu\lambda}^* \alpha_\mu^+ b_\lambda + h.c. \right) . \tag{7}$$

The first two terms correspond to the electrons and light field respectively, ν is the transition frequency between the atomic states 2 and 1, and ω_λ the frequency of the different modes λ of the electromagnetic field. The third term represents the dipole interaction between the two level atoms and the light field in the rotating wave approximation and $g_{\mu\lambda}$ denotes the coupling constant. Because the laser device is an open system we have to couple all the atoms as well as the electromagnetic field modes to external heat baths. These heat baths take into account the external pump, a homogeneous broadening of the atomic line by a line width γ , and losses of the electromagnetic field by damping constants κ_λ . The bath variables are eliminated via the Heisenberg equations of motion and one is finally left with a set of equations for the field modes and the atomic variables alone:

$$\dot{b}_\lambda^+ = (i\omega_\lambda - \kappa_\lambda) b_\lambda^+ + i \sum_\mu g_{\mu\lambda}^* \alpha_\mu^+ + F_\lambda^+(t) , \tag{8}$$

$$\dot{\alpha}_\mu^+ = (i\nu - \gamma)\alpha_\mu^+ - i \sum_\lambda g_{\mu\lambda} b_\lambda^+ \sigma_\mu + \Gamma_\mu^+(t) , \tag{9}$$

$$\dot{\sigma}_\mu = \gamma_\parallel (d_0 - \sigma_\mu) + 2i \sum_\lambda (g_{\mu\lambda}\alpha_\mu b_\lambda^+ - h.c.) + \Gamma_{\sigma_\mu}(t) , \tag{10}$$

where d_0 measures the pump of the single atoms. The quantities $F_\lambda^+(t)$, $\Gamma_\mu^+(t)$, and $\Gamma_{\sigma_\mu}(t)$ are operators corresponding to fluctuating forces. Their correlation functions can be calculated from quantum mechanical consistency arguments: The commutation relations of the operators to the dynamic variables must be fulfilled for all times t.

It is rather difficult to solve these operator equations (compare [9]). Fortunately it turns out that already mean field theory yields fairly good agreement with experimental results. The mean field approximation can simply be achived by the replacement

$$b_\lambda^+ \to b_\lambda^* \; ; \quad \alpha_\mu^+ \to \alpha_\mu^* \; ; \quad \sigma_\mu \to \sigma_\mu \; ; \quad \text{etc.} \tag{11}$$

in the basic equations of motion (8-1o), where b_λ^* , α_μ^* , σ_μ are now c - number variables. The resulting set of equations constitute the basis for the following considerations.

Single mode operation

To make contact with the formalism described in the introduction we note that the components of the state vector $\underset{\sim}{u}$ are in the case of a laser the amplitudes of the electromagnetic field modes, the polarization of the single atoms ($\propto \alpha_\mu^*$) and their inversion σ_μ . If we neglect fluctuations the stationary state $\underset{\sim}{u}_0$ of the laser for low pump is given by

$$b_\lambda^* = \alpha_\mu^* = 0 \quad ; \quad \sigma_\mu = d_0 . \tag{12}$$

Physically (12) means that the laser is acting as a usual lamp and is therefore only driven by the fluctuations. The transitions of the single atoms are caused by spontaneous emission and one observes independent single wave tracks. This is also reflected in the statistics of the light field which is built up from many independent events and therefore must have a Gaussian distribution. The state $\underset{\sim}{u}_0$ of the laser may be considered as completely disordered.

The stability of the state (12) can now be tested by linear stability analysis. We expect that a certain subset of the eigenvalues of the linear problem will depend on the index λ as well as on d_0 . It turns out that for low pump (12) is stable. By increasing the pump more and more the real part of one of these eigenvalues say with index λ_0 will first change its sign and λ_0 should be the mode which is closest to resonance with the atomic transition frequency ν . For simplicity we shall assume exact resonance: $\omega_{\lambda_0} = \nu$.

From the general considerations in the introduction we know that near the unstable point we have to take into account the unstable mode only. Eqs. (8-1o)[+] can now be simplified considerably by introducing a rotating frame

$$b_{\lambda_0}^* = \tilde{b}_{\lambda_0}^* \exp(i\omega_{\lambda_0} t) \; ; \; \alpha_\mu^* = \tilde{\alpha}_\mu^* \exp(i\nu t) ; \; \text{etc.} \tag{13}$$

[+] We will always make use of the simplification (11)

(We shall drop the tilde and the index λ_0 in the following.) Furthermore we note the possibility to introduce collective atomic variables

$$S^* = \sum_\mu \exp(ikx_\mu)\,\alpha_\mu^* \quad ; \quad D = \sum_\mu \sigma_\mu \,, \tag{14}$$

where we assumed plane waves for the cavity modes. After some obvious calculations we find the following new set of equations

$$\dot{b}^* = -\kappa b^* + igS^* + F^*(t)\,, \tag{15}$$

$$\dot{S}^* = -\gamma S^* - igb^* D + \Gamma_S^*(t)\,, \tag{16}$$

$$\dot{D} = \gamma_\parallel (D_0 - D) + 2ig(Sb^* - c.c.) + \Gamma_D(t)\,, \tag{17}$$

with

$$D_0 = Nd_0 \quad \text{and} \quad g_\mu^* = g\exp(ikx_\mu)$$

$$\Gamma_S^* = \sum_\mu \Gamma_\mu^* \exp(ikx_\mu)\; ; \quad \Gamma_D = \sum_\mu \Gamma_{\sigma_\mu}(t)\,. \tag{18}$$

Eqs. (15-17) show most clearly the feedback mechanism inherent in the original set (8-1o): The atomic dipole moments act via a sum (compare the definition of S^* (14)) on the light field mode λ_0 . The light field with the macroscopic dipole moment S^* drives the inversion (17) and light field and inversion act back on the polarization again. Whether this feedback mechanism yields a new state depends on the relation between pump and the damping and coupling constants.

In order to calculate the new state beyond the instability point we will apply the formalism of generalized Ginzburg-Landau equations.

The result is an equation for the amplitude of the unstable mode alone, where ξ^* plays the role of an order parameter

$$\dot{\xi}^* = \kappa \Lambda \xi^* - \beta |\xi|^2 \xi^* + \hat{F}(t), \tag{19}$$

where we put

$$\Lambda = (D_0 - D_c)/D_c . \tag{20}$$

D_c denotes the value of the pump parameter at the critical point and is given by

$$D_c = \frac{\kappa \gamma}{g^2} , \tag{21}$$

and β is an abbreviation for the expression

$$\beta = 4 g^2 \kappa \, (\gamma \cdot \gamma_{\parallel})^{-1} . \tag{22}$$

$\hat{F}(t)$ is the projection of the fluctuating forces on the critical mode. If $\Lambda < 1$ and if the atomic dipoles are heavily damped in comparison to the field mode ($\kappa \ll \gamma$ is always fulfilled in lasers) [10] we may replace

$$\xi^* \longrightarrow b^* . \tag{23}$$

Thus the amplitude of critical light field mode b^* plays the role of an order parameter in the laser. The equation of motion of the order parameter just has the form of a time dependent Ginzburg-Landau equation with fluctuating forces. For a further discussion of (19) we proceed to the corresponding Fokker-Planck equation and seek for its stationary solution which can be written in the form

$$f_0 = \mathcal{N} \exp(-\Phi/2Q), \tag{24}$$

where $\mathcal{N}$ is a normalization constant and Q denotes the intensitiy coefficient of the fluctuating force $\hat{F}(t)$. Φ is a potential which has all properties of a Ljapunov function and therefore can be discussed along the same lines as the Ginzburg-Landau free energy for continuous equilibrium phase transitions:

$$\phi = -\kappa \Lambda |b|^2 + \beta/2 \, |b|^4 . \qquad (25)$$

Eq. (25) constitutes the analogies between the laser transition which takes place far from thermal equilibrium, and continuous phase transitions in equilibrium (if we confine ourselves to the mean field approximation). The transition is connected with a broken symmetry, the fluctuations show a critical slowing down etc. The state beyond threshold shows a pronounced ordering of the single atomic dipole moments which now act in a cooperative fashion. This is reflected by

$$\langle |b^*| \rangle \neq 0, \qquad (26)$$

i.e., the mean of the modulus of the order parameter is unequal zero with a fixed otherwise arbitrary phase. The rotational symmetry is restored by a slow diffusion of the phase along the minimum of $\emptyset$. The statistics of the laser light indicate that the laser indeed is in a coherent state: It can be described by a Poisson distribution.

We conclude: At threshold the laser undergoes a transition from a completely disordered state to a new state which is ordered on a macroscopic scale. In addition one can show that beyond threshold a mode selection process takes place which may yield a situation where just one mode survives. The mode selection process can be understood qualitatively in the following way. If the inverse of the correlation length is much smaller than the spacing between different modes prescribed by the cavity only single mode operation should be possible [19] .

First order transition

We shall discuss the threshold of the single mode laser in further analogy to phase transitions. We expect that additional external influences may change the character of the transition drastically. An example is provided by the injection of a coherent external signal [11] which is in resonance with the atomic transition frequency ν and has a definite phase φ_0 . This signal can be represented as

$$E = E_0 \exp(i\nu t - ikx + i\varphi_0) + c.c. \qquad (27)$$

Because of the externally prescribed phase φ_0 the degeneracy in phase of the ordinary transition should be lifted. Indeed the result again can be written as stationary solution of a Fokker-Planck equation (24) with a potential $\emptyset_1$

$$\phi_1 = \phi - 2\hat{E}_0 \cdot r \cos(\hat{\varphi}_0 - \varphi), \tag{28}$$

where $\emptyset$ is given in (25), r and φ are defined through $b^* = r \cdot \exp(i\varphi)$. $\hat{E}_0$ and $\hat{\varphi}_0$ are proportional to E_0 and φ_0, respectively. The rotational symmetry of $\emptyset$ is removed: $\emptyset_1$ yields a pinning of φ to the minimum $\varphi = \hat{\varphi}_0$ which means that the phase can no longer diffuse in an undamped manner.

A second example where the rotational symmetry is preserved but the threshold of the laser has features of a first order transition can be obtained by inserting a saturable absorber between the laser material and the mirrors [12] . The absorber can be taken into account by replacing the loss constant of the cavity mode by

$$K \rightarrow K_0 + K_s\left(1 + |b|^2/I_s\right)^{-1}, \tag{29}$$

where I_s is a constant. The corresponding potential solution then reads

$$\phi_2 = (G - K_0)|b|^2 - I_s K_s \ln\left(1 + \frac{|b|^2}{I_s}\right) - \frac{1}{2} G |b|^4 / I_{as},$$
$$\left(I_{as} = \gamma \cdot \gamma_{\parallel} \cdot (4g^2)^{-1}\right) \tag{3o}$$

with $G = g^2/\gamma \, D_0$. Changing G we observe a set of curves $\emptyset\,(b, G)$ which are deformed in a similar way if we increase the pump as it is known from first order phase transitions in thermal equilibrium, especially a hysteresis effect occurs.

To summerize: All these calculations to the single mode laser show the importance of the order parameter concept. By the method of generalized Ginzburg-Landau equations we could reduce the state vector of a complex many body problem to one variable only. We shall now discuss the situation where several order parameters will be present.

Multi mode operation [1, 3, 9]

We briefly consider the case when many modes are present. For this end we introduce the frequency difference

$$\Omega_\lambda = \omega_\lambda - \nu, \tag{31}$$

which measures the distance of the mode frequency ω_λ from the atomic line center. We then have to distinguish between two different situations:

a) If the separation of the frequencies corresponding to different modes is comparably large, i.e.,

$$\gamma_\parallel < |\Omega_{\lambda_1} - \Omega_{\lambda_2}|, \tag{32}$$

we are allowed to average over the phases

$$\overline{b_\lambda^* b_{\lambda'}} = n_\lambda \delta_{\lambda\lambda'}. \tag{33}$$

Here n_λ is the number of photons present in the mode λ which is proportional to the intensity of mode λ. Again it is possible to eliminate the atomic variables by the formalism described in the introduction along similar lines as in the single mode case. We finally find a potential corresponding to the stationary solution of the Fokker-Planck equation for multimode operation

$$\phi_3 = \frac{1}{2}\sum_\lambda |b_\lambda|^2 (2\kappa_\lambda - w_\lambda D_0) + \frac{D_c}{\gamma_\parallel}\sum_{\lambda,\lambda'} w_\lambda w_{\lambda'} |b_\lambda|^2 |b_{\lambda'}|^2, \tag{34}$$

where w_λ is defined as

$$w_\lambda = 2\gamma g^2 (\Omega_\lambda^2 + \gamma^2)^{-1}. \tag{35}$$

The possible stable stationary configurations of the many modes can be determined by calculating the minima of ϕ_3. The different modes are coupled via the fourth order term in (34) and one may verify the mode selection process from this potential where only the mode closest to resonance will survive. On the other hand if some of the modes are degenerated, ϕ_3 determines their amplitudes and gives an answer on the behavior of the fluctuations near the critical points.

b) If we find in contrast to (32) the situation

$$|\Omega_{\lambda_1} - \Omega_{\lambda_2}| \lesssim \gamma_{\parallel} \quad , \tag{36}$$

the phases of the different modes play an important role and a phase average is no longer possible. The above described procedure remains applicable but the calculations are more lengthy and complicated. The result yields now a potential which essentially depends on the phases of the different modes. The minima of the potential also prescribe the phases of the coexisting modes, i.e., a phase locking occurs.

Continuously many modes [1,3,9]

We start with a continuum of modes running in one direction through a homogeneously pumped laser material. As it is known from previous sections only modes near resonance are important. We therefore confine ourselves to a quasi monochromatic wave, i.e.,

$$|\Omega_\lambda| \ll \gamma \quad ; \quad |\Omega_\lambda - \Omega_{\lambda'}| \ll \gamma_{\parallel} \; . \tag{37}$$

We may now eliminate the unstable modes as before and finally form a wave packet like

$$\Psi(x,t) = \int_{-\infty}^{+\infty} b_k \exp(ikx - i\vartheta|k|t) \tag{38}$$

in the equation of the order parameters. In (38) we have replaced the index λ by the wave vector k and made use of the relations

$$\omega_\lambda = c|k_\lambda| \quad ; \quad \Omega_\lambda = \vartheta|k_\lambda| . \tag{39}$$

The final equation for the order parameter $\Psi(x,t)$ then reads ($\alpha = -k\Lambda$)

$$\dot{\Psi} = -\alpha\Psi + c\left(i\vartheta\frac{d}{dx} + v\right)^2\Psi - 2\tilde{\beta}|\Psi|^2\Psi + \hat{F}(x,t) \tag{40}$$

and the potential to the stationary of the corresponding Fokker-Planck equation is given by

$$\phi_4 = \int dx\left\{\alpha|\Psi|^2 + \beta|\Psi|^4 + c\left|\left(i\vartheta\frac{d}{dx} - v\right)\Psi\right|^2\right\} . \tag{41}$$

(4o) is exactly of the form of the time dependent Ginzburg-Landau equation in superconductivity and the potential corresponds to the free energy if we replace v by the x-component of the vector potential, Q by $2k_B T$ (k_B is the Boltzmann constant) and if D is identified with temperature T. Up to this point we have discussed the first laser threshold from various aspects. We now shall analyze the stability of this ordered state in order to show how transitions from the ordered state to a new more complex ordered state may happen as well as a situation where one proceeds for higher pump to a chaotic state of the laser.

Second threshold, many modes [13]

To discuss the second threshold of a laser where the c.w. solution becomes unstable we consider a ring laser with periodic boundary conditions. For simplicity we neglect the fluctuating forces. The basic equations (8-1o) can be transformed to space dependent macroscopic variables by the transformation

$$
\begin{aligned}
E(x,t) &= \sum_{\lambda} b_\lambda \exp(-ik_\lambda x), \\
P(x,t) &= \sum_{\mu} \delta(x-x_\mu)\, \vartheta^{\cdot} \alpha_\mu, \\
D(x,t) &= \sum_{\mu} \delta(x-x_\mu)\, \sigma_\mu
\end{aligned}
\tag{42}
$$

(where ϑ is the atomic dipole moment). A detailed calculation shows that we may confine ourselves to real quantities E, P which we normalize as well as D to the c.w. solutions in a rotating frame (compare (13)). The resulting equations have the form

$$
\begin{aligned}
(\partial/\partial t + \kappa + c\,\partial/\partial x)\, E &= \kappa P, \\
(\partial/\partial t + \gamma)\, P &= \gamma E D, \\
(\partial/\partial t + \gamma_\parallel)\, D &= \gamma_\parallel(\Lambda+1) - \gamma_\parallel \Lambda E P,
\end{aligned}
\tag{43}
$$

where we made the approximation $\kappa_\lambda \approx \kappa$. Linear stability analysis indicates an instability of the c.w. solutions if the pump becomes higher than Λ_c :

$$\Lambda_c = 4 + 3\varepsilon + 2\left(2(1+\varepsilon)(2+\varepsilon)\right)^{1/2}, \tag{44}$$

with $\varepsilon = \gamma_{\parallel}/\gamma$. The mode which first becomes unstable has a wave number k_c unequal to zero

$$k_c = -i\left(\lambda_u \frac{\kappa\gamma\gamma_{\parallel}}{2(\gamma+\gamma_{\parallel})}\left(\Lambda_c+\varepsilon\right)\right); \tag{45}$$

$$\left(\lambda_u^2 = \frac{1}{2}(\varepsilon - 3\Lambda_c)\,\gamma_{\parallel}\cdot\gamma\right).$$

The application of the formalism of generalized Ginzburg-Landau equations yields an equation for the amplitude of critical mode alone

$$\dot{\xi}^* + c\,\partial/\partial x\,\xi^* = \tilde{\lambda}\xi^* + \tilde{a}|\xi|^2\xi^* - \tilde{b}|\xi|^4\xi^* \tag{46}$$

and the coefficients $\tilde{a}$, $\tilde{b}$ have the properties

$Re\,\tilde{a} > 0$; $Re\,\tilde{b} > 0$.

If the pump Λ is higher than the critical Λ_c, $\tilde{\lambda}$ is positive and if $\Lambda < \Lambda_c$, $\tilde{\lambda}$ is negative.

The stationary solutions to (46) again can be discussed from a potential which shows that the transition at second threshold is discontinuous.[+]

Above threshold the electric field strength E can be written as

$$E = (1+E_0) + E_1\cos(\omega_c\tilde{\tau} + \varphi_1) + E_2\cos(2\omega_c + \varphi_2) + \\ + E_3\cos(3\omega_c\tilde{\tau} + \varphi_3) + \ldots, \tag{47}$$

where $\tilde{\tau}$ is defined through

$$\tilde{\tau} = t - (\omega_c\cdot c)^{-1}k_c x. \tag{48}$$

[+]Numerical calculations show that the character of the transition essentially depends on the length of the ring laser [13] .

Second threshold of the single mode laser

As was already shown in the foregoing example the c.w. action of a laser may become unstable. To show how this instability may arise in the case of a single mode laser we start from the equations (15-17) and normalize the equations to the c.w. solutions again. To be as clear as possible we again neglect the fluctuating forces and put the phases inherent in (15-17) equal to zero. Introducing

$$E = b/b_{c.w.} \, , \quad P = S/S_{c.w.} \, ; \quad D = D/D_{c.w.} \, ; \tag{49}$$

we obtain

$$\begin{aligned} \dot{E} &= -\kappa\,(E-P) \\ \dot{P} &= -\gamma\,(P-ED) \\ \dot{D} &= \gamma_{\parallel}\,(\Lambda+1) - \gamma_{\parallel} D - \gamma_{\parallel}\,\Lambda E P . \end{aligned} \tag{5o}$$

E again measures the electric field strength, P the over all polarization and D the total inversion of the atoms. Again we obtain the threshold condition by linearizing (5o) around the stationary state $E = P = D = 1$ and have to fix the values of Λ where the eigenmodes of the linear problem become unstable. We find two modes becoming unstable simultaneously if [15]

$$\Lambda > \Lambda_c = \frac{(\gamma+\gamma_{\parallel}+\kappa)(\gamma+\kappa)}{\gamma\,(\kappa-\gamma-\gamma_{\parallel})} \, . \tag{51}$$

From (51) we note the following condition for the instability to be present

$$\kappa > \gamma + \gamma_{\parallel} \tag{52}$$

This means that the second threshold of the single mode laser can only be observed in cavities with high losses. To clarify the nature of the state beyond threshold we perform the following shift and scaling operations [15] to the variables E,P,D:

$$E = \frac{1}{(\gamma_{\parallel}\Lambda)^{1/2}} x \; ; \quad P = \frac{1}{(\gamma_{\parallel}\Lambda_c)^{1/2}} y \; ; \quad D = r - z \; ; \quad t = t'/\gamma \tag{53}$$

where

$$r = (\Lambda + 1). \tag{54}$$

If we furthermore introduce

$$\sigma = \frac{\kappa}{\gamma} \; ; \quad b = \frac{\gamma_{\parallel}}{\gamma} \; ; \tag{55}$$

we finally arrive at the equations

$$\begin{aligned} \dot{x} &= \sigma(y - x), \\ \dot{y} &= rx - y - xz, \\ \dot{z} &= -bz + xy. \end{aligned} \tag{56}$$

These are the Lorenz equations [16] well known from the Bénard problem in hydrodynamics. The transition of the single mode laser at second threshold therefore is equivalent to the Lorenz instability. This instability should be observable in lasers where as the corresponding conditions are not met in realistic fluids.
From machine calculations [16] we know that beyond threshold a seemingly random spiking of the laser will occur despite of the fact that (56) are purly deterministic equations.

It should be noted that the trajectories of the system beyond the instability can be visualized as cuts between a breathing sphere and cylinder in a somewhat different frame [17] . It is then possible to estimate the maximal radius of the sphere as well as of the cylinder in the stationary state.

Finally we note that contact to the Lorenz instability can also be made for a quasimonochromatic field propagating in one direction through a homogeneously pumped laser material with a uniform velocity [18] .

In conclusion we may say that higher and higher pumping does not necessarily yield a whole sequence of order - order transitions. The Lorenz instability, e.g.,marks a boarder beyond which seemingly chaotic behavior is observed. Thus the situation in the laser is analogous to that of turbulence in the Bênard problem for high Prandtl numbers. In other models the turbulent state may occur after a whole sequence of bifurcations (Landau-Hopf-model) or a few bifurcation steps only [2o] , [21].

Summary

We treated instabilities of the laser to demonstrate the powerfulness of methods developed in synergetics. Especially we showed how spontaneous ordering out of a completely disordered state may arise, discussed a condition how to proceed from an ordered state to a new more complex ordered state and finally considered the possibility to come from an ordered state to a chaotic state if we are far enough away from thermal equilibrium.

References

[1] H. Haken: "Synergetics, an Introduction,"Springer-Verlag Berlin, Heidelberg, New York (1977). (In this book many further references can be found)

[2] H. Haken, ed.: Synergetics (Proceedings of a Symposium on Synergetics, Elmau 1972), Teubner-Verlag, Stuttgart (1973)

[3] H. Haken: "Cooperative effects in systems far from thermal equilibrium and in nonphysical systems." Rev. Mod. Phys. 47, 67 (1975)

[4] H. Haken, ed.: "Cooperative effects, Progress in synergetics" North Holland, Amsterdam (1974)

[5] H. Haken, ed.: "Synergetics - A Workshop" Springer-Verlag Berlin, Heidelberg, New York (1977)

[6] A. Wunderlin: "Some aspects of synergetics" In: F.C. Auluck, ed.: Frontiers of theoretical physics, Delhi (1977)

[7] H. Haken: Z. Physik B21, 1o5 (1975); B22, 69 (1975); B 23, 388 (1975) and to be published.

[8] A. Wunderlin, H. Haken: Z. Physik B21, 393 (1975)

[9] H. Haken in: Encyclopedia of Physics, S. Flügge ed., vol. 25/2c "Laser Theory" Springer-Verlag, Berlin, Heidelberg, New York (197o) with many further references

[1o] W. Hauger, Dissertation Stuttgart (1977)

[11] J.F. Scott, M Sargent III, C.D. Cantrell: Opt. Commun. 15,13 (1975)

[12] W.W. Chow, M.O. Scully, E.W. van Stryland: Opt. Commun. 15 , 6 (1975)

[13] H. Haken, H. Ohno: Opt. Commun. 16, 2o5 (1976) and to be published

[14] H. Risken, K. Nummedal: Phys Lett. 26 A, 275 (1968); J. appl. Phys. 39 , 4662 (1968)

[15] H. Haken Phys. Lett. 53A, 77 (1975)

[16] E.N. Lorenz: J. Atmos.Sci. 2o, 13o, (1963)

[17] H. Haken, A. Wunderlin: Phys. Lett. 62A , 133 (1977)

[18] R. Graham: Phys. Lett. 58A , 44o (1976)

[19] R. Graham: Progress in Optics XII, 233 (1974)

[2o] D. Ruelle, F. Takens: Commun. math. Phys 2o, 167 (1971)

[21] H.L. Swinney, P.R. Fenstermacher, J.P. Gollup: in [5] , p.6o

DYNAMICS AND ERGODICITY OF THE INFINITE HARMONIC CRYSTAL

A REVIEW OF SOME SALIENT FEATURES

J.L. van HEMMEN * †

Institut des Hautes Etudes Scientifiques, 91440 Bures-sur-Yvette, France.

Abstract. We review some recent developments and rigorous results concerning the dynamics and ergodicity of the a priori infinite harmonic crystal. We present several ways of constructing the dynamics, indicate the relevance of the technique of fourier transforms of measures in studying the classical states, and discuss the result that (nearly) every perfect infinite harmonic crystal in thermal equilibrium is a Bernoulli system, so is at the top of the ergodic hierarchy. Finally we exemplify the classical KMS condition.

INTRODUCTION

Since Hamilton [1] wrote his analysis of the dynamics of an infinite array of particles with harmonic nearest neighbour interactions - in one of his note books that were of course not published during his lifetime - and posed the dynamical problem, there have been very few investigations on the a priori infinite vibrating crystal. We mention only Schrödinger's description [2] of the time dependent behaviour of the same one-dimensional system in terms of his now well-known coordinates, and Hemmer's work [3]. Nevertheless one learns [4, ch.II] that the natural context for lattice dynamics in the harmonic approximation is the infinite system. Indeed, it is. In this note we want to indicate how one is able to construct the dynamics of an a priori infinite harmonic crystal, solving the dynamical problem in its full generality: in ν dimensions ($\nu = 1,2$, or 3) and with f particles per unit cell (nonprimitive crystals; $f = 1,2,\ldots$). All this however is only a part of a larger project: solving the ergodic problem for this type of system. So let us turn to ergodicity.

Equilibrium statistical mechanics is concerned with the calculation of observable quantities when the system is in thermal equilibrium, say at inverse temperature β (ensemble, statistical or phase averages). Idealizing to some extent the measurement process, one can say that the outcomes of experiments are given by (infinite) time averages of the same observable quantities. A system is called ergodic if both prescriptions coincide, *i.e.*:

If f is such a quantity, so a function on the phase space Ω, and T_t represents the dynamical evolution in Ω while $\langle f \rangle = \int_\Omega f(\omega)d\mu_\beta(\omega)$ is the phase average where the integral is taken over Ω with respect to the equilibrium state (measure) μ_β, then

* Supported by the Netherlands Organization for the Advancement of Pure Research (Z.W.O.).

† Present Address: Math. Dept., Duke University, Durham, NC 27706, U.S.A.

for almost every x

$$\lim_{T\to\infty} \frac{1}{T}\int_0^T dt\ f(T_t x) = \int_\Omega f(\omega)d\mu_\beta(\omega). \tag{1}$$

There is a well-known candidate for μ_β when the system is finite, namely the micro-canonical ensemble which lives on a certain energy surface (Ergodic hypothesis, (1) holds).

A finite harmonic crystal is not ergodic at all, the system being equivalent to a finite number of non-interacting harmonic oscillators or normal modes. Compare nevertheless this statement with [4, p. 313] where one predicts the outcomes of X-ray and neutron scattering experiments by using implicitly the ergodic hypothesis, and remember that the system considered there is finite (though very large).

But it may still come as a surprise to discover [5,6] that (nearly) every perfect (translation invariant) infinite harmonic crystal in thermal equilibrium is not ergodic but even at the top of the ergodic hierarchy, that is: Bernoulli. We now return to the essential ingredients of the proof which are relevant as well. Further details can be found in [5] unless mentioned otherwise.

(1) Solve the equations of motion of the a priori infinite system, *i.e.* specify a phase space Ω such that for each point ω in Ω there is a unique solution $T_t\omega$ to the equations of motion with $T_t\omega$ remaining in Ω for all times t. Distribution theory techniques will be very useful [5,7]; they generalize the idea of normal modes to the infinite system in a natural way.

(2) Describe the thermal equilibrium state μ_β of the infinite system; μ_β is a measure that lives on Ω, so $\mu_\beta(\Omega) = 1$. The techniques of fourier transforms of measures as introduced in [5], is quite helpful, enabling one to prove - often in a few lines - results for (non) equilibrium situations which were very difficult or even impossible to be obtained otherwise. We mention: the characterization of μ_β itself either through the classical KMS condition or through the thermodynamic limit (μ_β is unique and gaussian), the equivalence of the micro-canonical and the canonical ensemble when the number of particles $N \to \infty$, and the description [8] of the approach to a stationary, non-equilibrium state in a harmonic chain where all particles to the left have temperature T_1 and all particles to the right have temperature $T_2 \neq T_1$ at time $t = 0$.

(3) Study the ergodic behaviour. For a perfect infinite harmonic crystal (lattice) with finite range interaction one has [5,6] the following striking alternative: when the system is in thermal equilibrium, it is either not even ergodic, or it is Bernoulli, so at the top of the ergodic hierarchy and loosely speaking, "equivalent" to an infinitely repeated roulette wheel experiment à la Monte Carlo. In practice only the latter situation can occur. Interpretation: "localized disturbances steam off to infinity where they are no longer visible", an observation of Lanford [9] which is valid also here, though now in a non-trivial context.

Summarizing the discussion in a succinct way we can say

Conclusion: Thus the natural context for lattice dynamics in the harmonic approximation is the infinite system indeed. □

In the subsequent four sections we give some more, technical information.

1. DYNAMICS

We take the one-dimensional harmonic chain with nearest-neighbour interaction ($\nu = f = 1$) as an example. First we have to specify a suitable set of initial conditions for this infinite array of particles labelled by the integers $n \in \mathbb{Z}$, *i.e.* a suitable set of coordinates $x(n)$ and momenta $p(n)$ which form together the phase space Ω.

Definition [5]: Let m be an integer

$$s_m \equiv \left\{ \xi = (\xi(n)) \mid \|\xi\|_m^2 = \sum_{n=-\infty}^{+\infty} (1+n^2)^m \xi(n)^2 < +\infty \right\} . \tag{3}$$

When $m \leq -1$ the sequence $\xi(n)$ may grow, when $m \geq 0$ it has to fall off. Define moreover

$$s = \bigcap_{m=0}^{+\infty} s_m \quad , \quad s' = \bigcup_{m=0}^{+\infty} s_{-m} . \tag{4}$$

When $x \in s'$ and $\xi \in s$ the "inner product"

$$< x,\xi > = \sum_{n=-\infty}^{+\infty} x(n)\ \xi(n) \tag{5}$$

is well-defined. We write for fixed ξ : $< x,\xi > = \phi(\xi)(x)$; $\phi(\xi)$ is a continuous function of the variable x in s'. One can view on a linear transformation T in s' as a "transposed" action of an S in s by means of

$$< Tx,\xi > = < x,S\xi > \tag{6}$$

and write $T = S^*$. Moreover S determines T uniquely. Let us finally remark that analogous considerations hold true for $s' \oplus s'$, etc. We are now prepared to consider the dynamics of our example.

In suitable units $p(n) = \dot{x}(n)$. Though the notion of hamiltonian is meaningless the equations of motion do make sense

$$\ddot{x}(n) = x(n+1) + x(n-1) - 2x(n) \equiv -(\Phi x)(n),\ n \in \mathbb{Z} . \tag{7}$$

Rewrite $\ddot{x} = -\Phi x$ where Φ is the interaction matrix, in the form

$$\frac{d}{dt}\begin{pmatrix} x \\ \dot{x} \end{pmatrix} = \begin{pmatrix} 0 & \mathbb{1} \\ -\Phi & 0 \end{pmatrix}\begin{pmatrix} x \\ \dot{x} \end{pmatrix} \equiv A\begin{pmatrix} x \\ \dot{x} \end{pmatrix} \rightarrow \begin{pmatrix} x(t) \\ \dot{x}(t) \end{pmatrix} = \exp\ [tA] \begin{pmatrix} x(o) \\ \dot{x}(o) \end{pmatrix} \tag{8}$$

When $|\Phi^{n,m}| \leq \varphi(n-m)$ for a certain $\varphi \in s(\mathbb{Z})$ and $(x(o),\ \dot{x}(o)) \in s' \oplus s'$, then (8) makes sense, $(x(t),\ \dot{x}(t)) \in s' \oplus s'$ again and the solution is unique [5]. That is: $\Omega = s' \oplus s'$ is a good candidate for a phase space. We now specialize to the perfect crystal and consider two methods, A and B, in turn.

A. Distribution theory. Suppose, as in (7), $\phi^{n,m} = \varphi(n-m)$, *i.e.* translation invariance. We know $x(n;t) = \mathrm{Re}[\exp - i(kn - \omega(k)t)]$ with $\omega(k)^2 = 2[1 - \cos k]$, is a solution to (7). This suggests that we should take a linear superposition of running waves as the general solution. But then we arrive precisely at Schwartz' theory of distributions [7]. Suppose we start at $t = 0$ with

$$x(n;0) = \frac{1}{2\pi} < D_0, e^{-ikn} > \ , \ \dot{x}(n;0) = \frac{1}{2\pi} < \dot{D}_0, e^{-ikn} > \tag{9}$$

where D_0 and $\dot{D}_0$ are given distributions. Such a representation is possible precisely when x and x' are in s'. Symbolically

$$x(n) \in s' \leftrightarrow x(n) = \frac{1}{2\pi} < D, e^{-ikn} > . \tag{10}$$

Suppose now that we can represent $x(n;t)$ by

$$x(n;t) = \frac{1}{2\pi} < D_t, e^{-ikn} > \ , \ \dot{x}(n;t) = \frac{1}{2\pi} < \dot{D}_t, e^{-ikn} > \tag{11}$$

then

$$\ddot{D}_t = -\omega(k)^2 D_t$$

and we arrive at

$$\begin{pmatrix} D_t \\ \dot{D}_t \end{pmatrix} = \begin{pmatrix} \cos \omega t & \frac{\sin \omega t}{\omega} \\ -\omega \sin \omega t & \cos \omega t \end{pmatrix} \begin{pmatrix} D_0 \\ \dot{D}_0 \end{pmatrix} \tag{12}$$

Remembering that a distribution multiplied by a smooth function is again a distribution, we have a rewording of "T_t transforms Ω into itself". Of course (12) also follows from (8) by directly evaluating $\exp[tA]$.

Consequence: We now easily see that T_t in $s' \oplus s'$ is determined uniquely by S_t in $s \oplus s$ such that $S_t^* = \exp[tA]$.

B. The method of elementary excitations. Take as initial data $x(j;0) = a(n)\delta_{nj}$, $\dot{x}(j;0) = b(n)\delta_{nj}$, so an "elementary excitation" in n, and look at the response of particle j. Via ordinary fourier analysis and (7), with J as the Bessel function,

$$x(j;t) = [a(n) + b(n) \int_0^t dt]\, J_{2(j-n)}(2t) \quad . \tag{13}$$

The system (7) being linear, one would like to take a linear superposition of (13) with $a(n)$ and $b(n)$ from $s'(\mathbb{Z})$. This is possible indeed by an analiticity argument [5]; even more is true. In addition, the argument works in any dimension.

Clearly both methods, A and B, lead to the same result, though admittedly A is more flexible.

2. EQUILIBRIUM STATES

At temperature $T > 0$, each degree of freedom has (on the average) an energy of $\frac{1}{2}kT$, so the infinite system contains an infinite amount of energy and - as observed before - the hamiltonian H has only a formal meaning. Thus $\exp - \beta H$ does not make any sense, while $dx_1dx_2 \dots dp_1dp_2 \dots$ has no meaning either! But together they do make sense, as a gaussian measure; we characterize this measure by its fourier transform.

Definition: Let Ω be the phase space; so $\Omega = E'$ (*e.g.* $s' \oplus s'$ with $E = s \oplus s$). Let μ be a (radon) measure on Ω describing the statistical state of the system. Then the fourier transform of μ, $F\mu(\xi)$ with $\xi \in E$, is defined by

$$F\mu(\xi) = \int_\Omega e^{i\phi(\xi)(\omega)} d\mu(\omega) \quad . \tag{14}$$

$F\mu$ determines μ uniquely □.

The thermal equilibrium state μ_β at inverse temperature β is given by

$$F\mu_\beta(\xi) = \exp - \frac{1}{2\beta} Q(\xi) \tag{15}$$

with

$$Q(\xi) = (\Phi^{-1/2}\xi_1, \Phi^{-1/2}\xi_1) + (\xi_2, \xi_2) \tag{16}$$

as associated quadratic form; ξ_1 is related to the coordinates and ξ_2 to the momenta (*e.g.* $\xi = \xi_1 \oplus \xi_2 \in s \oplus s$), up to a technical modification when $\nu = 1$ or 2. One says μ_β is gaussian. As usual $(\xi, \eta) = \Sigma \overline{\xi(n)} \eta(n)$. The reader may convince himself that (15) and (16) are true for a finite system, Φ being appropriately modified; then take the limit (non-trivial) and you end up with $\exp - \frac{1}{2\beta} Q(\xi)$. Minlos' theorem - see [5] for an easy proof - tells you that there is a unique μ_β such that (15) holds true. Another approach is outlined in the next section.

As a second application we mention the fact that one may efficiently study the time evolution of a classical state via

$$\lim_{t\to\infty} F(\mu \circ T_{-t})(\xi) = \lim_{t\to\infty} F\mu(S_t\xi) \tag{17}$$

where μ is the $t = 0$ situation and $F\mu$ is known ($S_t\xi$ usually is, cf. section 1). In (17) we have essentially used $T_t = S_t^*$. For a nice example of this technique we refer to [8].

Remark: After having modified the phase space and by using either Peierls' or Mermin's criterion one can show that in one and two dimensions the a priori infinite harmonic lattice in thermal equilibrium (μ_β) is not a harmonic crystal. The salient feature is here that μ_β is the state of the *infinite* system.

3. THE CLASSICAL KMS CONDITION

The idea behind this section is as simple as it is attractive: given the dynamics T_t of the infinite system find the equilibrium state μ_β (thermodynamics) by means of a condition that directly connects dynamics and thermodynamics, in this way circumventing the thermodynamic limit. One then arrives at the so-called KMS condition, originally introduced in quantum statistical mechanics [10] and much later in a classical context [11].

Let $C(\Omega)$ be the collection of all continuous functions on Ω (Ω being equipped with its natural weak*-topology). We introduce a group of automorphisms α_t in $C(\Omega)$ by $\alpha_t(f)(\underline{x}) = f(T_t\underline{x})$, $f \in C(\Omega)$ and $\underline{x} \in \Omega$, for instance (cf. section 1, eq. (5)) $\alpha_t\phi(\underline{\xi})(\underline{x}) = < T_t\underline{x},\underline{\xi} > = < \underline{x},S_t\underline{\xi} > = \phi(S_t\underline{\xi})(\underline{x})$, so that $\alpha_t\phi(\underline{\xi}) = \phi(S_t\underline{\xi})$. In what follows μ is a (radon) probability measure on Ω and we put $\mu(f) = \int f d\mu$. Finally, if f and g are differentiable the Poisson bracket $\{f,g\}$ is defined by

$$\left\{ f,g \right\} = \sum_j \left[\frac{\partial f}{\partial q_j}\frac{\partial g}{\partial p_j} - \frac{\partial f}{\partial p_j}\frac{\partial g}{\partial q_j}\right] \tag{18}$$

whenever the right-hand side is meaningful. We then say that μ satisfies the <u>classical KMS condition</u> if

$$\mu(f\alpha_t g) - \mu(fg) = -\frac{1}{\beta}\int_0^t d\tau\ \mu(\{f,\alpha_\tau g\}),\ \forall f,\ g \in \mathcal{D} \tag{KMS-D}$$

where $\mathcal{D}$ is to be specified later.

<u>Motivation</u>. Let A and B be two quantum observables, $[A,B] = AB-BA$ and $A_t = \alpha_t A$. The quantum KMS condition [10] says - up to some technical refinements - that μ is an equilibrium state at inverse temperature β if

$$\mu\ (A_t B) = \mu\ (BA_{t+i\beta\hbar})\ . \tag{KMS-Q}$$

Thus, subtracting $\mu(BA_t) = \mu(BA_t)$,

$$\mu\left(\frac{[B,A_t]}{i\hbar}\right) = -\beta\ \frac{\mu\ (BA_{t+i\beta\hbar}) - \mu\ (BA_t)}{i\beta\hbar}\ . \tag{19}$$

Take the classical limit $\hbar \to 0$, so formally $[P,Q]/i\hbar \to \{P,Q\}$, put $B = f$, $A = g$, integrate and you get (KMS-D). □

We choose [5] a $\mathcal{D}$ whose elements are smooth and uniformly bounded, that satisfies $\alpha_t\mathcal{D} \subseteq \mathcal{D}$ and is locally dense in $L^2_\mu(\Omega)$, viz.

$$\mathcal{D} = \left\{ \exp\ i\phi(\underline{\xi})\ \middle|\ \underline{\xi} \in E \right\}. \tag{20}$$

<u>APPLICATION OF KMS-D.</u> To get rid of many indices and large parts of alphabets we consider again the harmonic chain (section 1); we take $f = \exp\ i\phi(\underline{\xi})$ and $g = \exp\ i\phi(\underline{\eta})$. We will apply KMS-D in such a way that we get an explicit representation of the fourier transform of μ_β. One part of the calculation is easy, it looks like

$$\mu(fa_t g) - \mu(fg) = \mu(\exp\ i\phi(\underline{\xi}+S_t\underline{\eta})) - \mu(\exp\ i\phi(\underline{\xi}+\underline{\eta})) \equiv \psi(t) - \psi(0). \tag{21}$$

We now write down the relevant Poisson bracket

$$\left\{ f,\alpha_t g \right\} = \left[\sum_{j\in J} - \xi_1(j)\eta_2^t(j) + \xi_2(j)\eta_1^t(j)\right] \exp\ i\phi(\underline{\xi}+S_t\underline{\eta}), \tag{22}$$

the expression between the brackets being equal to

get the following result.

Theorem: By applying the classical KMS-D condition one obtains a unique measure μ_β on the phase space Ω representing the thermal equilibrium state at inverse temperature β. This measure is gaussian and therefore determined completely by a positive quadratic form Q. When E is a nuclear space as introduced in *e.g.* section 1, then
(1) $\nu \leq 2$ implies: $Y \subsetneqq E$ and $\Omega = Y'$, Q is continuous on Y but cannot be extended to E;
(2) $\nu = 3$ implies: $Y = E$ and $\Omega = Y'$, Q is continuous on Y.

4. ERGODICITY

There exists an ergodic hierarchy, starting with simply "ergodic" and ending with "bernoulli". For convenience we first recall the definition of "bernoulli". Given μ on Ω, we say that $\{T_t, -\infty \leq t \leq +\infty\}$ is a bernoulli flow if and only if we can find a $t_0 \neq 0$ (then it is true for every $t \neq 0$!) such that T_{t_0} is a bernoulli automorphism. This animal is defined as follows: The phase space Ω can be divided into countably many subsets of events Ω_n such that $T_{t_0}\Omega_n = \Omega_{n+1}$ and the Ω_n-s are mutually independent (*i.e.* for all $A \in \Omega_n$, $B \in \Omega_m$ with $m \neq n \Rightarrow \mu(A \cap B) = \mu(A)\,\mu(B)$). So T_{t_0} looks like an infinitely repeated roulette wheel experiment. Each time the outcome is from an Ω_n, and because T_{t_0} is an "automorhpism", all the Ω_n may be identified, that is, you "repeat" the experiment. For precise formulations we refer to elsewhere, *e.g.* [5,6] or in this volume.

Clearly the $\omega(k)^2$ of section 1 is nothing but the dynamical matrix D(k) in a situation where $\nu = f = 1$; cf. [4]. D is defined on the brillouin zone and it reflects the periodicity of the lattice. We now have the following theorem [5,6].

Theorem: In studying the ergodic behaviour of the infinite harmonic perfect crystal (lattice) with interaction matrix Φ such that the elements of the dynamical matrix D are infinitely differentiable (this was the case in our example!) we have the following three alternatives: either (1) the point spectrum $\sigma_p(D)$ of the dynamical matrix is not empty and the system is not ergodic, or (2) $\sigma_p(D)$ is empty and the system is ergodic or (3) the spectrum of D is absolutely continuous and the system is bernoulli.

In case of a finite range interaction we have either (1) or (3) and a simple glance at the vibration spectrum suffices to conclude whether or not the system is bernoulli: whenever each dispersion curve $\{\omega_\ell : 1 \leq \ell \leq \nu f\}$ has a non-constant piece, only (3) remains.

Remarks: Let us add that the theorem is also true when we replace D(k) by Φ itself. Then the system is not ergodic if Φ has bound states [12]; this confirms earlier results [13].

A non-trivial example of a vibration spectrum associated with a finite range interaction as discussed above is presented in the appendix.

$$-\left(\underline{\xi},\begin{pmatrix}0 & \mathbb{1}\\ -\mathbb{1} & 0\end{pmatrix}S_t\underline{\eta}\right) = -\left(\hat{\underline{\xi}},\ \begin{pmatrix}\sin\omega t/\omega & \cos\omega t\\ -\cos\omega t & \omega\sin\omega t\end{pmatrix}\hat{\underline{\eta}}\right). \tag{23}$$

The right-hand side of (23) was obtained after a fourier transformation. KMS-D reads

$$\psi(t) - \psi(0) = \frac{1}{\beta}\int_0^t d\tau\ \psi(\tau)\left(\underline{\xi},\begin{pmatrix}0 & \mathbb{1}\\ -\mathbb{1} & 0\end{pmatrix}S_\tau\underline{\eta}\right) \tag{24}$$

or, $\psi(\tau)$ being continuous,

$$\beta\frac{d\psi}{dt} = \psi(t)\left(\underline{\xi},\begin{pmatrix}0 & \mathbb{1}\\ -\mathbb{1} & 0\end{pmatrix}S_t\underline{\eta}\right) \equiv \psi(t)B_t(\underline{\xi},\underline{\eta}) \tag{25}$$

giving

$$\psi(t) = \psi(0)\ \exp\frac{1}{\beta}\int_0^t d\tau\ B_\tau(\underline{\xi},\underline{\eta}). \tag{26}$$

Substituting $\underline{\eta} := -\underline{\xi}$ we can transform (26) into

$$\mu(\exp i\phi((\mathbb{1}-S_t)\underline{\xi})) = \exp -\frac{1}{\beta}\int_0^t d\tau\ B_\tau(\underline{\xi},\underline{\xi}) \tag{27}$$

so that we have to evaluate $\int_0^t d\tau\ B_\tau(\underline{\xi},\underline{\xi})$. Instead of a double argument $(\underline{\xi}\ \underline{\xi})$ we will simply write $(\underline{\xi})$.

It appears that in one and two dimensions we can find a subspace $Y \subsetneq E$ (with $\Omega = Y'$) while in three dimensions $Y = E$, such that

$$\underline{\xi} \in Y \Rightarrow B_t(\underline{\xi}) = -\frac{d}{dt}Q(\underline{\xi},S_t\underline{\xi})\ , \tag{28}$$

as one easily verifies; Q is the quadratic form given in (16). And using $Q(S_t\underline{\xi}) = Q(\underline{\xi})$ we thus obtain

$$\underline{\xi} \in Y \Rightarrow \mu(\exp i\phi((\mathbb{1} - S_t)\underline{\xi})) = \exp -\frac{1}{2\beta}Q((\mathbb{1} - S_t)\underline{\xi}) \tag{29}$$

Note $\underline{\xi} \in Y$ implies $Q(S_t\underline{\xi},\underline{\xi}) = Q(\underline{\xi},S_t\underline{\xi})$, because the elements of Y are real. If we now prove that the closure of $(\mathbb{1} - S_t)Y$ with respect to the inner product given by Q is equal to H, the completion of Y with respect to Q, we have shown the existence of at most one measure μ_β on $\Omega = Y'$ satisfying KMS-D. For (29) has to hold for Y and not only for $(\mathbb{1} - S_t)Y$. Suppose we have

$$\exists\underline{\zeta} \in H \mid \langle\underline{\zeta},(\mathbb{1} - S_t)\underline{\xi}\rangle_Q = 0,\ \forall\underline{\xi} \in Y,\ \forall t \in \mathbb{R}$$

then $S_t\underline{\zeta} = \underline{\zeta}$, and thus

$$t^{-1}(\mathbb{1} - S_t)\underline{\zeta} = 0,\ \forall t \in \mathbb{R} - \{0\} \Rightarrow \begin{pmatrix}0 & -\Phi\\ \mathbb{1} & 0\end{pmatrix}\begin{pmatrix}\zeta_1\\ \zeta_2\end{pmatrix} = 0; \tag{30}$$

$\zeta_1 = 0$ and $\Phi\zeta_2 = 0$, ergo $\zeta_2 = 0$ by some reasonable assumptions we have to make about the vibration spectrum.

Conversely, given the gaussian measure μ_β with associated quadratic form Q defined by (16), one verifies directly that μ_β satisfies the KMS-D condition. The generalization of all these arguments to more complicated situations is evident. In this way we

REFERENCES

1. W.R. Hamilton, Mathematical papers; Cambridge University Press, London (1940).
2. E. Schrödinger, Ann. Phys. 44 (1914) 191.
3. P.C. Hemmer, Dynamic and stochastic types of motion in the linear chain; Det Fysiske Seminar i Trondheim, n° 2 (1959).
4. A.A. Maradudin et al., Theory of lattice dynamics in the harmonic approximation, 2nd edition; Academic Press, New York and London (1971).
5. J.L. van Hemmen, Dynamics and ergodicity of the infinite harmonic crystal; thesis, University of Groningen (1976). University Microfilms n° 77 - 70, 001.
6. O.E. Lanford III, J.L. Lebowitz, in: Springer Lecture Notes in Physics 38 (1975) 38 (1975) 144.
7. Laurent Schwartz, Théorie des distributions; Hermann, Paris (1966).
8. H. Spohn, J.L. Lebowitz, Comm. Math. Phys. 54 (1977) 97.
9. O.E. Lanford III, in: The Boltzmann equation; Springer-Verlag, Wien and New York (1973) 619.
10. R. Haag, N.M. Hugenholtz, M. Winnink, Comm. Math. Phys. 5 (1967) 215.
11. G. Gallavotti, E. Verboven, Nuovo Cimento 28B (1975) 274.
12. J.L. van Hemmen, A generalization of Rayleigh's theorem for the infinite harmonic crystal, J. Stat. Phys. 18 (1978) 53.
13. R.I. Cukier, P. Mazur, Physica 53 (1971) 157.

APPENDIX

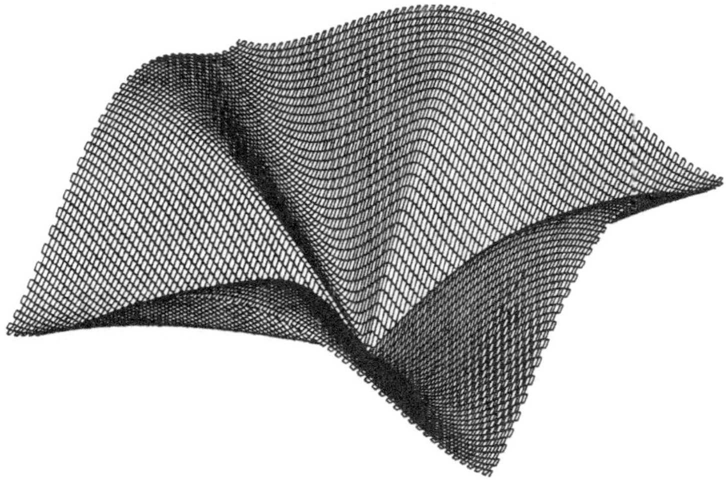

The transverse mode is plotted of the vibration spectrum belonging to a 2-dimensional simple cubic lattice with nearest-neighbour (Hooke's constant α_1) and next nearest-neighbour interaction (Hooke's constant α_2). The mass $m = 1$ and the lattice distance is also equal to one; $\alpha_1 = 0.5$ and $\alpha_2 = 2.0$. We look at the spectrum from (10,5,10). The corresponding characteristic equation is ($-\pi \leq k_i \leq \pi$; $i = 1,2$)

$$\det\begin{pmatrix} [2\alpha_1(1-\cos k_1)+2\alpha_2(1-\cos k_1 \cos k_2)-m\omega(k)^2] & 2\alpha_2 \sin k_1 \sin k_2 \\ 2\alpha_2 \sin k_1 \sin k_2 & [2\alpha_1(1-\cos k_2)+2\alpha_2(1-\cos k_1 \cos k_2)-m\omega(k)^2] \end{pmatrix} = 0 .$$

GEODESIC CORRECTION TO STOCHASTIC PARALLEL DISPLACEMENT OF TENSORS.+

DANIELA DOHRN and FRANCESCO GUERRA
Institute of Physics,++ University of Salerno.
U.E.R. Scientifique de Luminy, University of Aix-Marseille II.
and Centre de Physique Théorique , CNRS Marseille

Abstract

The notion of stochastic parallel displacement of tensors on a Riemannian manifold, as introduced by Itô and Dynkin, is modified by taking into account the correction arising from geodesic deviation. Two applications are given. The first is related to the problem of diffusion of tensors on a manifold and shows that the generator of the diffusion is the Laplace-de Rham operator. The second is related to Nelson stochastic mechanics on a manifold and its connection with quantum mechanics.

+ Expanded version of a talk given by the second named author at the Conference on "Stochastic Behavior in Classical and Quantum Hamiltonian Systems", Como, June 20-24, 1977.

++ Postal address : Istituto di Fisica, Via Vernieri 42, 84100 Salerno, Italy.

1. STOCHASTIC PARALLEL DISPLACEMENT. DIFFUSION OF TENSORS.

Let $g_{ij}(x)$ be the metric tensor on an n-dimensional Riemannian manifold V and Γ^i_{jk} the corresponding Cristoffel symbols. The Brownian motion on V is a diffusion process with the generator

(1) $$\frac{1}{2} g^{ij} \nabla_i \nabla_j ,$$

where ∇_i denotes covariant differentiation. Following Itô [5] , we introduce σ^i_α and m^i by

$$\sum_\alpha \sigma^i_\alpha \sigma^j_\alpha = g^{ij} \qquad \text{and} \qquad m^i = -\frac{1}{2} g^{jk} \Gamma^i_{jk} .$$

The paths of the Brownian motion can be constructed by solving the stochastic differential equation

(2) $$dw^i(t) = m^i(w(t))\,dt + \sum_\alpha \sigma^i_\alpha(w(t))\,d\xi_\alpha(t) ,$$

where ξ_α are n independent copies of the Wiener process with expectations given by

$$E(d\xi_\alpha) = 0 \quad , \quad E(d\xi_\alpha\, d\xi_\beta) = \delta_{\alpha\beta}\, dt.$$

We consider also a more general process q(t) solution of the stochastic differential equation

(3) $$dq^i(t) = f^i(q(t),t)\,dt + dw^i(t) ,$$

where $f^i(x,t)$, $x \in V$, is some given non-random vector field on V .

Following Itô [5] and Dynkin [3] we recall how the process q(t) induces a diffusion in the space of tensors on V . For the sake of simplicity we consider only the case of vectors.

Let $y(s)$, $s_0 \le s \le s_1$, be a smooth curve. According to the classical Levi-Civita definition the family of vectors u(s) , each attached to y(s), is called parallel if

(4) $$\frac{du^i(s)}{ds} + \Gamma^i_{jk}\, u^j \frac{dy^k}{ds} = 0 .$$

By integrating this differential equation under the initial condition $u(s_0)=u_0$ we obtain the parallel dispacement of the vector u_0 along the curve y(s). We now define the stochastic parallel displacement of vectors along trajectories of the process q(t) through the following stochastic generalization of (4)

(5) $$du^i(t) + \Gamma^i_{jk}\, u^j \circ dq^k = 0,$$

where " $\circ$ " is the symmetric product defined by

(6) $$F \circ dG = F\,dG + \frac{1}{2}\, dF\, dG$$

(see [6] for more details).

The stochastic differential equation (5) can be written explicitely in the form

(7) $$du^i + \Gamma^i_{jk}\, u^j dq^k + A^i_j\, u^j dt = 0,$$

where

(8) $$A^i{}_j = \frac{1}{2}\, g^{kl} \left(\partial_l \Gamma^i_{jk} - \Gamma^i_{rk}\, \Gamma^r_{jl} \right)$$

and dq^k is given by (3).

By solving the stochastic differential equations (3) and (7) under the initial conditions $q(0)=q_0$ and $u(0)=u_0$ we get the random family of vectors u(t) , attached to q(t) , which can be considered the stochastic parallel dosplacement of u_0 along the trajectories of the process q(t) starting at q_0 .

As it is shown in [5, 6] and [3] the definition of stochastic parallel displacement given by (7) is equivalent to the following geometric definition.

Let q(t) be the solution of (3) starting at q_0 and u_0 be a vector attached to q_0 . Let

(9) $$\Delta : 0 = t_0 < t_1 < \dots < t_n = t$$

be a partition of the time interval [0, t]. Consider the polygonal curve $y_\Delta(s)$, $0 \leq s \leq t$, made of the geodesic curves connecting $q(t_{i-1})$ with $q(t_i)$, i = 1, 2, ...n, and let $u_\Delta(t)$ be the vector obtained from u_0 by the Levi-Civita parallel displacement along $y_\Delta(s)$. Then Itô [5] and Dynkin [3] show that $u_\Delta(t)$ tends in probability, for each fixed t, to the solution u(t) of the stochastic differential equation (7) , with the initial condition $u(0)=u_0$, as $|\Delta|$ tends to zero, $|\Delta| = \max_i (t_i - t_{i-1})$.

Let us come back to the solutions q(t), u(t) of the equations (3) and (7). Consider the mapping

(10) $$T_{q,t} : u_0 \to u(t)$$

and the vector

(11) $$v(x,t) = E\left(T_{q,t}^{-1}\, v_0(q(t))\right),$$

where $x \equiv q_0$ is the initial position of the process q(t) and $v_0(\cdot)$ is a given vector field on V . Then it is immediately seen [5, 3] that $v(x,t)$ satisfies the diffusion equation

(12) $$\frac{\partial v}{\partial t} = \frac{1}{2} g^{ij} \nabla_i \nabla_j v + b^i \nabla_i v,$$

with the initial condition

(13) $$v(x,0) = v_0(x).$$

We can do the same for more general tensor fields. The process q(t) therefore induces a diffusion in the space of tensors on V characterized by the heat equation (12), where the Laplace-Beltrami operator appears in the generator.

2. GEODESIC CORRECTION TO STOCHASTIC PARALLEL DISPLACEMENT

In this section we show how a suitable modification of the mappirg (10) allows us to define a diffusion of tensors with generator given by the Laplace-Kodaira-de Rham operator Δ [10]. On vector fields Δ acts as

(14) $$(\Delta v)^i = g^{jk} \nabla_j \nabla_k v^i + R^i_j v^j ,$$

where R^i_j is the curvature tensor associated to g (our choice of the sign is opposite to that found in the mathematical litterature).

Since $\Delta \nabla_i = \nabla_i \Delta$, the diffusion generated by Δ has the attractive property that if $v_0^i = \nabla^i \phi_0$ then, calling $v(x,t)$ and $\phi(x,t)$ respectively the diffused vector and scalar field for $b = 0$, we still get at all times $v^i = \nabla^i \phi$. Analogously for higher order tensor fields. Another application to Nelson stochastic mechanics will be given in the next section.

Let us introduce the notion of geodesic correction to stochastic parallel displacement [2] .

Let $y(s)$, $s_0 \leq s \leq s_1$, be a segment of geodesic and u_0 be an initial vector attached to $y_0 = y(s_0)$. Consider the geodesic $\bar{y}(s)$, $s_0 \leq s \leq s_1$, which starts parallel to $y(s)$ from the point $\bar{y}_0 = \bar{y}(s_0) = y_0 + h u_0$, where h is an infinitesimal auxiliary parameter. We define the vector $u(s) = h^{-1}(\bar{y}(s) - y(s))$ attached to $y(s)$, in the limit $h \to 0$, as the result of geodesic displacement of u_0 along $y(s)$. Notice that $u(s)$ differs from the Levi-Civita parallel displacement by terms which are of second order in s , in particular the component of $u(s)$ parallel to $y(s)$ undergoes usual parallel displacement.

Then we follow the geometric construction given in the previous section and based on the partition (9), but now we call $u_\Delta(t)$ the vector obtained from u_0 by successive geodesic displacement along $y_\Delta(s)$. Then it can be immediately shown [2] that, in the limit $|\Delta| \to 0$, $u_\Delta(t)$ tends in probability, for each fixed t, to the solution $u(t)$ of the stochastic differential equation

(15) $$du^i + \Gamma^i_{jk}\, u^j dq^k + A^i_j\, u^j dt = \frac{1}{2} R^i_j\, u^j dt ,$$

with the initial condition $u(0) = u_0$. The term at the r.h.s. of (15) gives the geodesic correction to the equation (7), its origin comes from second order effects who are felt by the stochastic nature of dq.

By proceeding as in the previous section, we arrive at the definition of diffusion of tensors induced by q(t) given by

(16) $$\frac{\partial v}{\partial t} = \frac{1}{2} \Delta v + b^i \nabla_i v$$

where Δ is the Laplace-Kodaira-deRham operator, instead of (12).

The possibility of modifying the mapping (10) in order to get a diffusion generated by Δ is also mentioned by Itô in [5] (compare also [7]).

3. APPLICATION TO NELSON STOCHASTIC MECHANICS ON MANIFOLDS.

We refer to the literature [8, 9, 1, 4, 2] for the general framework of Nelson stochastic mechanics and its extension to manifolds. Here we give a sketch of the main arguments and show in particular how the notion of geodesic correction to stochastic parallel displacement given before allows us to introduce a natural geometric definition of mean forward and backward derivatives of tensor fields, necessary in Nelson treatment of stochastic mechanics. A more complete account can be found in [2] .

Consider a classical dynamical system with Lagrangian

(17) $$\mathcal{L} = \frac{1}{2} M_{ij} \dot{q}^i \dot{q}^j - V(q)$$

and equations of motion

(18) $$a^i + \nabla^i V = 0 ,$$

with

(19) $$a^i = \ddot{q}^i + \Gamma^i_{jk} \dot{q}^j \dot{q}^k .$$

Following Nelson [8,9] we promote the classical configuration variables q(t) to a Markov process q(t) ruled by the stochastic differential equation

(20) $$dq^i = b^i(q(t),t)\, dt + dw^i ,$$

where $b(x,t)$ is a non random field and the Brownian motion dw is such that

(21) $$E\left(dw^i dw^j \,/\, q(t) = x \right) = \hbar\, M^{ij}(x)\, dt ,$$

with

(22) $$M_{ij} M^{jk} = \delta_i^k$$

and E(. | .) the conditional expectation.

In order to find the equation of motion for b , following Nelson we assume that (18) still holds but with a new definition for a

(23) $$a^i = \frac{1}{2} \left(D D_* + D_* D \right) q^i ,$$

where D and D_* are suitably defined mean forward and backward derivatives. The difficulty in giving a proper definition of D and D_* comes from the fact that while the classical displacement dq^i is the prototype for a vector on the contrary the stochastic displacement dq^i does not transform as a vector in the case of non flat manifold, due to second order effects.

Dankel [1] has given a natural definition of Dq^i and $D_* q^i$, so that in particular $Dq^i = b^i$, but was unable to find a natural geometric definition for Db^i. In [2] we show that such natural geometric definition can be based on the notion of geodesic correction to stochastic parallel displacement. It turns out that

$$(Db)(x,t) = \lim_{h \to 0^+} h^{-1} E\left(T^{-1}_{q,t+h}\, b(q(t+h), t+h) - b(x,t) \,\middle|\, q(t) = x \right), \tag{24}$$

where T is the mapping defined in section 2 with geodesic correction.

The Schrödinger equation associated to Nelson stochastic mechanics, defined by (20, (18), (23), (24), can be shown [1, 2] to be

$$i\hbar \frac{\partial \psi}{\partial t} = - \frac{\hbar^2}{2} M^{ij} \nabla_i \nabla_j \psi + V\psi , \tag{25}$$

without further terms involving the curvature tensor.

In conclusion the authors would like to thank Professors G. CASATI and J. FORD for the nice hospitality in Como and Professor K. ITÔ for kindly providing reference [7] .

REFERENCES

[1] T.G. Dankel, Jr., Mechanics on Manifolds and the Incorporation of Spin into Nelson's Stochastic mechanics, Arch. Rational Mech. Anal., 37 (1971)

[2] D. Dohrn and F. Guerra, Nelson Stochastic Mechanics on Riemannian Manifolds, Lettere al Nuovo Cimento, 1978.

[3] E.B. Dynkin, Diffusion of Tensor, Soviet Math. Dokl., 9, 532, (1968)

[4] F. Guerra, On the Connection between Euclidean-Markov Field Theory and Stochastic Quantization, in C^+- Algebras and their Applications to Statistical Mechanics and Quantum Field Theory, D. Kastler, ed., North-Holland, Amsterdam (1976)

[5] K. Itô, The Brownian Motion and Tensor Fields on Riemannian Manifold, Proc. Int. Congress Math. 1962 (Stockholm), 536-539.

[6] K. Itô, Stichastic Parallel Displacement, Spinger Lecture Notes in Math. 451 (1975), 1-7

[7] P. Malliavin, Formules de la Moyenne, Calcul de Perturbations et THéorèmes d'Annulation pour les formes Harmoniques, Journal of Functional Analysis 17, 274-291 (1974)

[8] E. Nelson, Derivation of the Schrödinger Equation from Newtonian Mechanics Phys. Rev. 150 , 1079-1085 (1966).

[9] E. Nelson, Dynamical Theories of Brownian Motion, Princeton 1967

[10] G. de Rham, Variétés différentielles, Actualités Sci. Ind. 1222 (1955), Paris

THE METHOD OF DIRICHLET FORMS *

by

S. ALBEVERIO and R. HØEGH-KROHN
Matematisk Institutt, Universitetet i Oslo,
Z.I.F., Universität Bielefeld, and Centre
de Physique Théorique, C.N.R.S., Marseille

In several contributions to this conference stochastic processes (especially those of diffusion type) have plaid a role. We would like to give a short exposition of some work done recently on the construction of such processes from a point of view which unifies the finite dimensional case (elliptic operators, heat equation, potential theory, quantum mechanics), the infinite dimensional case (variational equations, quantum fields) as well as the non commutative (finite or infinite dimensional) case (positive maps of C^*-algebras, quantum statistical mechanics). The unification is achieved by the "method of Dirichlet forms" (or "energy forms"), which we shall now briefly expose, on the basis of the references [1] - [8], to which we refer for complements and bibliography.

1. THE COMMUTATIVE CASE

1.1. The definition of Dirichlet forms in the finite dimensional case:

Let μ be a σ-finite regular Borel measure on the Euclidean space $\mathbb{R}^d$. Under general assumptions on μ (see Remark 1) the gradient operator ∇ from $C_o^1(\mathbb{R}^d)$ to the space of vector valued square integrable functions $\mathbb{R}^d \otimes L^2(d\mu)$ is closable, i.e. has a smallest closed extension, which we shall denote again by ∇. The sesquilinear positive form

$$E(f,g) \equiv \frac{1}{2}(\nabla f, \nabla g) \equiv \frac{1}{2}\sum_j^d \int_{\mathbb{R}^d} \frac{\partial \bar{f}}{\partial x_j}\frac{\partial g}{\partial x_j}\, d\mu(x)$$

is then closed in $L^2(d\mu)$. This form will be called the "Dirichlet (diffusion) form" (or "energy form") given by μ. The self-adjoint positive operator $H \equiv \frac{1}{2}\nabla^*\nabla$ in $L^2(d\mu)$ is the self-adjoint operator uniquely associated with the Dirichlet form, hence we shall call it the "Dirichlet (diffusion) operator" given by μ. On the dense subset $C_o^2(\mathbb{R}^d)$ we have $H \equiv -\frac{1}{2}\Delta - \beta\cdot\nabla$, where Δ is the

---

* Work supported in part by the Norwegian Research Council for Science and the Humanities.

Laplacian and $\beta \equiv -\frac{1}{2}\nabla^*\Omega$, where Ω is the function identically equal to one in $\mathbb{R}^d \otimes L^2(d\mu)$.

Remark 1 : A sufficient condition on μ in order that ∇ be closable, so that H is well defined, is e.g. $d\mu(x) \equiv \varphi^2(x)dx$, with $\varphi \in L^2_{loc}(\mathbb{R}^d)$, $\nabla\varphi \in L^2_{loc}(\mathbb{R}^d)$. In this case we have $\beta \equiv \varphi^{-1}\nabla\varphi$. For this and other conditions see [5].

If $\varphi^2 > 0$ almost everywhere with respect to Lebesgue measure dx then multiplication by φ gives an unitary map from $L^2(d\mu)$ onto $L^2(dx)$ and we can then realize H as an operator in $L^2(dx)$. It is easy to see, by two partial integrations, that, for any $f \in C_0^2(\mathbb{R}^d)$, $(\varphi^{-1}f, H\varphi^{-1}f)_\mu \equiv (f, (-\frac{1}{2}\Delta + V)f)$, where $(\ ,\ )_\mu$ resp. $(\ ,\ )$ are the scalar products in $L^2(d\mu)$ resp. $L^2(dx)$ and V is formally equal to $\frac{1}{2}(\beta^2 + \nabla\beta) = \frac{1}{2}\Delta\varphi/\varphi$. Hence H is realized in $L^2(dx)$ by an unitary equivalent operator, written $-\frac{1}{2}\Delta + V$, at least in the case where φ is such that V exists as a multiplication operator or a densely defined quadratic form (in which cases $-\frac{1}{2}\Delta + V$ is defined as an operator sum or as a sum of quadratic forms). This is what one usually calls a "Schrödinger operator" and we see thus that every Dirichlet operator can be looked upon as a Schrödinger operator, possibly with very singular interaction. The infimum (zero) of the spectrum of this Schrödinger operator is such as to have a (possibly generalized) eigenfunction φ in L^2_{loc}, different from zero almost everywhere. The converse is also true, so that the study of Schrödinger operators which are lower bounded and have at the infimum of their spectrum (taken to be zero) a (possibly generalized) eigenfunction $\varphi \in L^2_{loc}$ different from zero a.e. is equivalent to the study of Dirichlet operators or of Dirichlet forms.

The advantage of studying Dirichlet forms rather than Schrödinger operators is that the definition of the former poses no problem even in cases where the definition of the latter is given by $-\frac{1}{2}\Delta + V$ only formally (due to high singularities of V). Another advantage of Dirichlet forms and Dirichlet operators is that they are defined also in the infinite dimensional case, where "Schrödinger operators" of the form $-\frac{1}{2}\Delta + V$ have no direct meaning, being a concept relative to a Lebesgue measure, which does not exist on an infinite dimensional Hilbert space. We shall now precisely go over to the consideration of the infinite dimensional case.

1.2. The definitions of Dirichlet forms in the infinite dimensional case:

The procedure to define Dirichlet forms in the infinite dimensional case is entirely parallel to the one used in the finite dimensional case. We shall now give the correspondents of $(\mathbb{R}^d, \mu, \nabla)$. $\mathbb{R}^d$ is replaced by a separable real Hilbert space H. The Borel measure μ is replaced by a finite measure defined on all Borel subsets of finite dimensional subspaces of H and it is no restriction of the generality to suppose that the support of μ is the topological dual Q' of some nuclear space Q densely and continuously contained in H (such that the dualization between Q and Q' restricted to $Q \times H$ coincides with the scalar product in H). With other words, it is no restriction of the generality to suppose that one has a rigged Hilbert space $Q \subset H \subset Q'$ and a measure μ with support in Q'.

For the definition of ∇ (and below of Δ) it is useful to consider the spaces $F C_b^k$, $k = 0, 1, 2, \ldots$ of continuous bounded functions on Q' which are of the form $f(.) \equiv \tilde{f}_L(P_L .)$, where P_L is the continuous extension to Q' of an orthogonal projection in H with finite dimensional range L and $\tilde{f}_L$ is a C_b^k - function on L (bounded continuous derivatives up to order k). Such functions are called cylinder functions based on L. The spaces $F C_b^k$ are dense in $L^2(d\mu)$. ∇ is defined on $F C_b^1$ by $\nabla f = \nabla_L \tilde{f}_L$, where ∇_L is the gradient on $C_b^1(L)$. Then ∇ is a densely defined map from $F C_b^1$ into $H \otimes L^2(d\mu)$.

Sufficient conditions for μ to give a closable ∇ are given in [3]: these conditions are verified e.g. in the applications to the case of quantum fields.

Calling again ∇ the closure, we then define the Dirichlet (diffusion) form given by μ , analogously as in the finite dimensional case, by closing up the form first defined on functions in $F C_b^1$ with base L by

$$E(f,g) \equiv \tfrac{1}{2} (\nabla f, \nabla g) \equiv \tfrac{1}{2} \int \nabla_L \tilde{f}_L \cdot \nabla_L \tilde{g}_L \, d\mu_L ,$$

where μ_L is the projection of μ in L. The positive self-adjoint operator $H \equiv \frac{1}{2} \nabla^* \nabla$ in $L^2(d\mu)$, called "Dirichlet (diffusion) operator", is the self-adjoint operator uniquely associated with the Dirichlet diffusion form given by μ . On the dense subset $F C_b^2$ one has $H = - \frac{1}{2} \Delta - \beta \nabla$, where Δ on functions $f \in F C_b^2$ acts as the Laplacian on $\tilde{f}_L$ and $\beta \equiv - \frac{1}{2} \nabla^* \Omega$, Ω being the function identically one in $H \otimes L^2(d\mu)$.

As an example, in the application to relativistic quantum fields in (s+1) space-time dimensions one takes $H = L^2(\mathbb{R}^s)$, $Q = S(\mathbb{R}^s)$, $Q' = S'(\mathbb{R}^s)$ (real spaces). If we take μ as the Gaussian measure on $S'(\mathbb{R}^s)$ with characteristic functional $e^{-\frac{1}{4}(\varphi, (-\Delta + m^2)^{-1/2} \varphi)}$, $\varphi \in S(\mathbb{R}^s)$ we have that the Dirichlet operator H is the Hamiltonian for free boson scalar quantum fields. If we take, for $s = 1$, μ to be the measure on $S'(\mathbb{R})$ associated with the time zero restriction of the Euclidean measures constructed for polynomial or exponential interactions, see [3] and references therein, we have that the operator H coincides as a quadratic form on a dense domain with the energy operator for the relativistic fields. In the case of the trigonometric interactions (and also in the case of the weak polynomial interactions of Glimm, Jaffe and Spencer and Dobrushin and Minlos (under the assumption that the Borel σ - algebra at infinity is countably generated))we have that μ is the time zero measure associated with the time zero relativistic quantum fields, H coincides as an operator on a dense domain with the energy operator for these fields [3], [4] and these fields are canonical quantum fields in the sense that the time zero fields are multiplication operators $<\xi, \varphi> \equiv \int \xi(x)\, \varphi(x)\, dx$, $\xi \in S'(\mathbb{R})$, $\varphi \in S(\mathbb{R})$ in the physical Hilbert space $L^2(d\mu)$, the conjugate field momentum operators act by translations in $L^2(d\mu)$, one has the equations of motion of the canonical formalism, with H and the infinitesimal generator Λ of Lorentz boosts given on dense domains by

$$\int\int \left(-\frac{1}{2}\frac{\delta^2}{\delta\xi(x)^2} - \beta(\xi(x))\frac{\delta}{\delta\xi(x)}\right) \lambda(x)dx\ d\mu\,(\xi(x)\,),$$

with $\lambda(x) = 1$ resp. $\lambda(x) = x$, the variational derivative $\frac{\delta}{\delta\xi(x)}$ being such that for $f \in F\, C_b^1$, $\int \varphi(x) \frac{\delta f}{\delta\xi(x)} dx \equiv (\varphi\cdot\nabla)\, f\,(\xi)$, where $\varphi\cdot\nabla$ is the derivative in the direction of the basis vector $\varphi \in S(\mathbb{R})$[6], [7]. This justifies the canonical formalism, which was, as well known, at the origins of quantum field theory (Heisenberg, Born, Jordan, Wentzel, ...) and received new momentum as the principle that "the ground state (vacuum) determines everything" by Coester, Haag and Araki. (For recent discussions of this principle see [5] and references therein to the work of Ezawa, Hida, Klauder, Shepp and Streit).

1.3. Dirichlet forms generate Markov processes:

Above we have seen the advantages of considering Dirichlet diffusion forms as a mean to define the Hamiltonians of quantum mechanical systems of finitely or infinitely many degrees of freedom. There are however also other reasons for the interest of Dirichlet (diffusion) forms, the main being that their associated Dirichlet (diffusion) operators are second order elliptic operators (with singular coefficients) which generate Markov (diffusion) processes with invariant measure (if β is linear and $H = \mathbb{R}$ we get the so-called Ornstein-Uhlenbeck velocity process, studied originally particularly by Doob). More generally, both in the case of a finite and infinite dimensional Hilbert space H, we have that the Dirichlet operator given by μ is the infinitesimal generator of a conservative symmetric Markov semigroup $P_t \equiv e^{-tH}$, $t \geq 0$ (conservative means $P_t\, 1 = 1$, symmetric means $P_t^* = P_t$, Markov means $0 \leq f \leq 1 \rightarrow 0 \leq P_t f \leq 1$). This type of results follows from the theory of Beurling, Deny, Fukushima and Silverstein, see e.g. [10], [11], [3]. Having P_t one gets then a Markov process with stationary distribution μ and transition semigroup P_t. This process can be realized in such a way as to have essentially the same properties as the regular processes considered in potential theory, in particular they have the strong Markov property i.e. the Markov property with respect to stopping times. In fact they are Dynkin and Hunt processes, their trajectories can be taken to be continuous (in the $\mathbb{R}^d$ topology resp. in suitable Banach norms in the infinite dimensional case [4]). They solve (weakly) the stochastic differential equation $d\xi_t = \beta(\xi_t)dt + dw_t$, where w_t is the Wiener process (Brownian motion) canonically associated with H. The drift coefficient β is the same quantity appearing in the Dirichlet operator H. In particular it can be singular and grow at infinity (it is only restricted to be in $L^2(d\mu)$ if H is infinite dimensional, and even less in the finite dimensional case [5]), a result which goes beyond the usual assumptions of the theory of stochastic differential equations.

Remark : The above stochastic equation can also be interpreted as the one of stochastic mechanics in the stationary case (for the stochastic mechanics meant here see the lecture of F. Guerra in these Proceedings). It is satisfied by the Euclidean random fields corresponding to quantum field interactions (whenever the Markov property is proven for these [4], [6], [16] [17]) as well as by the Markov process generated by the Dirichlet operator given by the corresponding time zero measure μ. We conjecture that the two solutions can be completely identified.

Let us also remark that there are results stating the equivalence of the ergodicity of the process with the uniqueness and simplicity of the infimum (zero) of the spectrum of H as well as, in the case of the "strict positivity" of μ [$\varphi^2 > 0$ everywhere in the finite dimensional case and, in the infinite dimensional case, strict positivity of the density of the conditional measures obtained by conditioning a Q-quasi invariant measure μ with respect to subspaces of finite codimension] with the quasi invariance and ergodicity of μ with respect to translations by elements in Q. In turn strict positivity follows essentially from "analyticity" of μ (in the sense that Ω is an analytic vector for the infinitesimal generators of translations in $L^2(d\mu)$ i.e. $\pi(q) = \frac{1}{2i}(q\cdot\nabla - (q\cdot\nabla)^*)$, $q\cdot\nabla$ being the derivative in the direction $q \in Q$). Such analyticity is however a consequence of ellipticity (both in finite and infinite dimensions). In particular the measure μ of the quantum fields discussed in Section 1.2 is quasi invariant, analytic, strictly positive and ergodic with respect to translations by elements in $Q = \mathcal{S}(\mathbb{R})$.

Finally a remark which is useful for the extension of the construction of Dirichlet forms in the non commutative case : the Dirichlet forms $E(f,g)$ (of the finite or infinite dimensional cases) have the contraction property that $E(T(f),T(g)) \leq E(f,g)$ where $T(f)$ is suitable contractions e.g. $T(.)$ in the space Lip $(R,0)$ of Lipshitz continuous maps with Lipshitz norm 1 from R into R mapping 0 into 0. This type of results and converse ones are known from the work of Beurling, Deny, Fukushima, Silverstein and others, see e.g. [10], [11], [3] and references therein.

2. THE NON COMMUTATIVE CASE

The study of the non commutative case in a similar spirit has been initiated in [8]. The space H is here replaced by a C^*-algebra A, the measure μ by a faithful lower semicontinuous trace τ on A, such that $A_\tau \equiv \{x \in A \mid \tau(x^* x) < \infty\}$ is dense in A (τ is then semifinite). $L^2(d\mu)$ is then replaced by the closure $L^2(A,\tau)$ of A_τ in the norm given by $\tau(x^* x)$. The correspondent of the order in $L^2(d\mu)$ (used e.g. in the definition of Markov semigroups) is in $L^2(A,\tau)$ the order obtained by identifying elements in $L^2(A,\tau)$ with left multiplications (by the same elements) and using then the order structure of closed operators. A Dirichlet form (as characterized by a contraction property of the type discussed at the end of the previous section) finds its non commutative analogue in a sesquilinear form E with the contraction property that $E(T(x), T(x)) \leq E(x, x)$, for all hermitian elements in a dense domain of the hermitian part of $L^2(A,\tau)$. The correspondent of a Markov semigroup in $L^2(d\mu)$ is then a semigroup Φ_t, $t \geq 0$ of maps of $L^2(A,\tau)$ into itself such that $0 \leq x \leq 1$ implies $0 \leq \Phi_t(x) \leq 1$. This is equivalent with having a positivity preserving semigroup with $\Phi_t(1) \leq 1$ on the W^*-algebra obtained by taking the weak closure of the *-representation of A in $L^2(A,\tau)$ by left multiplication. The Beurling-Deny-Fukushima theorem [10] of the commutative case which gives a one-to-one correspondence between Dirichlet forms and Markov semigroups has then the correspondent in the non commutative case that there is a one-to-one correspondence between symmetric Markov semigroups and Dirichlet forms on $L^2(A,\tau)$, see [8]. An example of Dirichlet forms (non commutative analogue of $\frac{1}{2}\int \nabla f^2 \, d\mu$) in the case where A is a subalgebra of the algebra of all bounded operators on a Hilbert space K is the following:

$$E(x, x) \equiv \operatorname{tr}(x^2 M) + \sum_{i=1}^{\infty} \operatorname{tr}([x, m_i]^* [x, m_i]), \quad \text{where } x \in L^2(A,\tau),\ M \geq 0$$

is self-adjoint (possibly unbounded) in K, m_i are bounded operators with $\operatorname{tr}(m_i^* m_i) < \infty$, tr being the trace and $[\ ,\ ]$ the commutator. From the point of view of the construction of "processes" (cf. L. Accardi's talk in these Proceedings) as well as from the point of view of quantum statistical physics the proper non commutative correspondent of "positive maps" are the "completely positive maps" (rather than the positive maps themselves). For this reason it is useful to consider "completely Markov semigroups", which are Markov semigroups Φ_t, $t \geq 0$ in $L^2(A,\tau)$ which have the additional property of being such that $\Phi_t \otimes 1_n$ is also a Markov semigroup, for all n, where 1_n is the identity in the algebra M_n of nxn complex matrices. It is also useful to introduce the concept of "completely Dirichlet forms" as the Dirichlet forms E on $L^2(A,\tau)$ which have the additional property that $\sum_{i,j} E(x_{ij}, x_{ij})$ is a Dirichlet form an $L^2(A,\tau) \otimes M_n$, where $x_{ij} \in L^2(A,\tau)$ are the components of $x \in L^2(A,\tau) \otimes M_n$.

Then one has a theorem (Theor. 3.2 in [8]) which says that a symmetric semigroup Φ_t is completely Markov if and only if the correspondent Dirichlet form E is completely Dirichlet. This is the case if and only if there exists a net of weights w_α, ρ_α on A resp. $A \otimes \bar{A}$ [$\bar{A}$ is the conjugate algebra to A, so that scalar multiplication is $(\lambda, a) \to \bar{\lambda} a$, $a \in A$] such that

$$E(x, x) = \lim_{\alpha \to \infty} [\, w_\alpha(x^2) + \rho_\alpha((x \otimes 1 - 1 \otimes x)^2)\,].$$

An example of a completely Dirichlet form is the one given above $E(x,x) = tr(x^2 M) + \sum_{i=1}^{\infty} tr([x, m_i]^* [x, m_i])$. Recalling that completely Markov semigroups are in particular strongly continuous semigroups of completely positive maps of C^*-algebras, we see that by constructing completely Dirichlet forms we provide in particular examples of strongly continuous semigroups of completely positive maps of C^*-algebras which go beyond the ones constructed by other techniques (cf. e.g. [12] and references therein). In this connection it should be recalled that recently semigroups of completely positive maps have attracted much attention in connection e.g. with non equilibrium quantum statistical mechanics, see e.g. [13] - [15], [17], [18] and references therein.

Finally lut us remark that for completely Markov semigroups on von Neumann algebras one has connections between spectral properties and ergodic properties, correspondent to those in the commutative case (Frobenius, Krein-Rutman, von Neumann ...). In fact for any completely Markov semigroup Φ_t on a von Neumann algebra M, looked upon, by the G.N.S. construction, as acting on the Hilbert space H obtained by applying M on a cyclic and separating vector Ω_ξ invariant for Φ_t (giving a vector state ξ invariant for Φ_t) we have e.g. the result [9] that if Φ_t is ergodic then the spectrum of the infinitesimal generator is in the closed upper half-plane, and on the real axis the possible eigenvalues α form a subgroup Γ of the additive group of the real line (Γ is either dense or discrete in $\mathbb{R}$). The spectrum of the semigroup is invariant under Γ, and $\{e^{2\pi i\alpha}, \alpha \in \Gamma\}$ are simple eigenvalues. To the eigenvectors corresponding to these eigenvalues there correspond unitary operators u_α in M and $\alpha \to u_\alpha$ is a multiplier representation of Γ (unitary if Γ is cyclic). The restriction of Φ_t to the von Neumann algebra generated by the u_α, $\alpha \in \Gamma$ is an ergodic automorphism and the restriction of ξ to this von Neumann algebra is a trace.

REFERENCES :

[1] S. ALBEVERIO, R. HØEGH-KROHN,
J. Math. Phys. 15, 1745-1747 (1974)

[2] S. ALBEVERIO, R. HØEGH-KROHN,
pp. 11-59 in F. Guerra, D.W. Robinson, R. Stora Eds., Les méthodes mathématiques de la théorie quantique des champs, Coll. Intern. CNRS, no 248, Marseille 1975, CNRS 1976

[3] S. ALBEVERIO, R. HØEGH-KROHN,
Zeitschr. f. Wahrscheinlichkeitsth. verw. Geb. 40, 1-57 (1977)

[4] S. ALBEVERIO, R. HØEGH-KROHN,
Ann. Inst. H. Poincaré B13, (1977)

[5] S. ALBEVERIO, R. HØEGH-KROHN, L. STREIT,
J. Math. Phys. 18, 907-917 (1977)

[6] S. ALBEVERIO, R. HØEGH-KROHN,
Proceedings of the Intern. Conf. on the Mathematical Problems in Theoretical Physics, Rome, June 1977 (to appear at Springer-Verlag).

[7] S. ALBEVERIO, R. HØEGH-KROHN,
Canonical relativistic quantum fields,
to appear in Ann. Inst. H. Poincaré A.

[8] S. ALBEVERIO, R. HØEGH-KROHN,
Dirichlet forms and Markov semigroups on C*-algebras,
Comm. math. Phys. 56, 173-187 (1977)

[9] S. ALBEVERIO, R. HØEGH-KROHN,
Frobenius theory for positive maps of von Neumann algebras,
Oslo University Preprint, May 1977 (to appear in Comm. math. Phys.)

[10] M. FUKUSHIMA,
pp. 46-79 in "Proceedings of the second Japan-USSR Symposium on Probability Theory", Edts. G. Maruyama, Yu. V. Prohorov, Springer, Berlin, 1973

[11] M.L. SILVERSTEIN,
Symmetric Markov processes,
Springer, Berlin, 1974

M.L. SILVERSTEIN,
Boundary theory for symmetric Markov processes,
Springer, Berlin, 1976

[12] V. GORINI, A. FRIGERIO, M. VERRI, A. KOSSAKOWSKI, E.C.G. SUDARSHAN,
Properties of quantum markovian master equations,
University of Texas Preprint, November 1976

[13] G.G. EMCH,
Non-equilibrium quantum statistical mechanics,
to appear in the Proc. of the XV Intern. Universitätswochen für Kernphysik, Schladming, Febr. 1976 (Phys. Austr. Suppl.)

[14] E.B. DAVIES,
Quantum theory of open systems,
Academic Press, London, 1976

[15] PH. MARTIN,
Modèles en mécanique statistique des processus irréversibles,
Cours du 3ème cycle de la Physique en Suisse Romande (Hiver 1976)
EPF - Lausanne

[16] R.L. DOBRUSHIN, R.A. MINLOS,
pp. 23-49 in B. JANCEWISZ, ed., "Functional and probabilistic methods in quantum field theory,
Vol. I, Acta Univ. Wratisl., n° 368, XII-th Winter School of Theor. Phys. Karpacz 1975, Wroclaw 1976

[17] S. ALBEVERIO, R. HØEGH-KROHN,
in preparation

[18] L. ACCARDI,
Adv. in Math. 20, 329-366 (1976)
and these Proceedings

[19] G.G. EMCH, S. ALBEVERIO, J.P. ECKMANN
Rep. Math. Phys. 13, 73-85 (1978)

[20] H. SPOHN,
Rep. Math. Phys. 10, 283-296 (1976)

REGULAR AND IRREGULAR SPECTRA OF MOLECULES

by

Ian C. Percival

Department of Applied Mathematics

Queen Mary College

Mile End Road

London E1 4NS

Abstract

KAM theory and numerical methods are applied to the motion of triatomic molecules, in particular OCS. The theory of regular and irregular spectra is reviewed.

1. Molecules

Compare molecules with the solar system. Both are made up of bodies interacting through inverse square forces. In the case of molecules the bodies are a variety of positively charged nuclei and identical negatively charged electrons. A typical ratio of the electron mass to a nuclear mass is similar to a typical ratio of a planetary mass to the mass of the Sun. However unlike the Sun's planetary system the forces between electrons are comparable to the forces between electrons and nuclei.

Because the characteristic frequencies of typical electron motion are much higher than those of typical nuclear motion and also because of quantum mechanics the motion of the nuclei may be considered to take place in the average field of the electrons so that the hamiltonian for the motion of the nuclei of an n-atomic molecule has the form

$$H(\underset{\sim}{r}_1, \ldots, \underset{\sim}{r}_n, \underset{\sim}{p}_1, \ldots, \underset{\sim}{p}_n) = V(\underset{\sim}{r}_1, \ldots, \underset{\sim}{r}_n) + T(\underset{\sim}{p}_1, \ldots, \underset{\sim}{p}_n). \tag{1.1}$$

The functional form of the kinetic energy T is a simple quadratic but the functional form of the potential energy V is much more complicated; it is strongly repulsive at short distances, weakly attractive at longer distances of the order of 10^{-10}m and tending to zero at larger distances. The interaction between two atoms is crudely represented by a potential of the form

$$U(r) = \frac{A}{r^{12}} - \frac{B}{r^6}. \tag{1.2}$$

Actual molecular potentials are much more complicated.

There is a potential minimum at one (or more) stable equilibrium points, which corresponds to the standard configuration of the molecule, but the atoms vibrate about this equilibrium and the whole molecule also rotates. The vibration and rotation of the nuclei and their attendant swarms of electrons may cause the emission or absorption of radiation, typically at infrared frequencies for vibration and at radio frequencies for rotation.

The vibration and rotation of molecules are particularly interesting from the theoretical viewpoint because they are on a borderline between those problems which can be solved by using classical mechanics only and those which require quantum mechanics.

The vibrational and rotational radiation shows distinct and sharp spectral lines, clear evidence of quantization, but nevertheless the motion is sufficiently close to being classical that solutions of the classical equations of motion can be used to obtain many interesting properties, using semiclassical mechanics. A review is given in (1).

For polyatomic molecules, with more than two atoms, the hamiltonian (1.1) is non integrable, and the phenomena which are the subject of this meeting are predicted to occur, in particular a transition between regular and irregular (or stochastic) behaviour. A particular molecule carbonyl sulfide or OCS will be discussed; it is used as a standard in molecular spectroscopy.

For real triatomic molecules the potentials V of equation (1.1) are not very well known. Calculations of the mean motion of the electrons on the basis of quantum theory constitute an industry which consumes a significant fraction of universities' computer time, but nevertheless the errors are still of the order of a few percent at best. Observational spectroscopy can provide frequencies accurate to one part in 10^6 or better, and these are used to obtain potentials in the neighbourhood of the equilibrium configuration by inverting the usual dynamical problem of obtaining the spectrum from the potential. The potential is usually presented as a polynomial, typically with terms up to quartic in some co-ordinate system. Non-linear effects are evident. The observations are due for rapid improvement and extension to higher energies, well away from equilibrium by double resonance experiments using lasers.

Even if the hamiltonian (1.1) were known the solution of the quantum mechanical equations of motion for the higher vibrational and rotational states, which are becoming accessible to laser experiments, would be a formidable task, involving the diagonalization of $10^4 \times 10^4$ matrices or larger. The corresponding classical equations are much easier to solve, at least in the regular region of phase space.

Classical and semiclassical mechanics can be used to obtain approximate energy levels from given potentials where quantum theory is too difficult.

We are concerned with the behaviour of dynamical systems over long periods of time, when measured in units of a typical unperturbed frequency. Since no system is perfectly isolated, the assumption that its motion can be described by a time-dependent hamiltonian is only valid so long as external perturbations are negligible, or for as long as the system has evolved. This period, when measured in units of a characteristic frequency has the following <u>very</u> approximate values for different systems

Solar system	10^{10}	
Particle in storage ring	10^{10}	
Atom in molecule	10^{6}	
Star in galaxy	300	(1.3)

For the first computers cannot integrate the trajectories directly for the requisite time. Theorems like the KAM theorem, which applies for all time are essential. For molecules integration is possible for sufficiently stable trajectories, but the expense would usually be prohibitive. For a star in a simplified model galactic field numerical integration presents relatively few problems.

2. Quantization rules.

Let a system of N degrees of freedom have coordinates q_k and canonically conjugate momenta p_k. Let the N action integrals I_k be defined by the integrals

$$I_k = \frac{1}{2\pi}\oint p_k dq_k \qquad (k = 1,\ldots,N) \tag{2.1}$$

where each integral is taken around one cycle of the motion of the coordinate q_k. The Sommerfeld-Wilson quantization rule then states that for each stationary state of the system

$$I_k = n_k\hbar \quad (k = 1,\dots,N) \ , \tag{2.2}$$

where $2\pi\hbar$ is Planck's constant and n_k is an integer.

The condition depends on the choice of coordinates; if it is valid for one choice it cannot be valid for all. Schwarzchild and Epstein (2) proposed that the coordinates be chosen so that the action function S_E, which is a solution of the classical Hamilton-Jacobi equation, can be expressed as a sum over N action functions, each dependent on one coordinate alone. This proposal is limited to separable systems, and the theory is not invariant under the canonical transformations of classical mechanics.

Jammer has written a good review (3) of this early work but does not mention the paper of Einstein (4), who was naturally dissatisfied with the lack of invariance of the Schwarzschild-Epstein theory.

Einstein based his theory on the invariant differential sum

$$\sum_{k=1}^{N} p_k \, dq_k \ . \tag{2.3}$$

Einstein's quantization rule is given by invariant line integrals of the form

$$I = \frac{1}{2\pi} \oint \sum_k p_k \, dq_k = n\hbar \tag{2.4}$$

along closed curves in coordinate space that have <u>no need</u> to be classical trajectories. In order to obtain N independent quantum conditions. Einstein was led to consider the invariance of the integrals under continuous transformations, and the topology (connectivity) of the momenta p_k as functions of the coordinate q_k.

For the simple case of a particle of one degree of freedom oscillating in a potential well, there are two momenta of opposite sign for almost all classically accessible positions, and the momentum may be considered as a continuous two-valued function of these positions, or alternatively as a single-valued function on two sheets, by analogy with the Riemann theory of complex variables.

For more degrees of freedom we quote Einstein in a fairly free translation:

"We now come to an absolutely essential point -. We follow the unrestricted motion of a single system for an unlimited period of time, and think of the corresponding trajectory traced out in q-space. Two possibilities arise:

1. There exists a part of q-space such that the trajectory comes arbitrarily close to every point in this N-dimensional subspace in the course of time.

2. The trajectory can be kept within a continuum of less than N dimensions. A special case of this is the exact closed trajectory.

Case 1 is general. Cases 2 are generated from 1 by specialization. As an example of Case 1 think of a particle under the action of a central force, the motion of which is described by 2 coordinates which determine the position of the particle within the plane of its orbit (for example polar coordinates r and ϕ). Case 2 occurs, for example, when the central force is attractive and exactly proportional to $1/r^2$ and when the relativistic corrections to the Kepler motion are neglected. In that case the orbit is closed and the points on it form a continuum of only one dimension. When considered in three-dimensional space, the motion under the central force is always of Type 2, because the orbit can be put into a continuum of two dimensions. From the three-dimensional viewpoint, one has to regard the central motion as a special case of a motion which is defined by a complicated law of force (for example the motion studied by Epstein in the theory of the Stark effect.)

The following is concerned with the general Case 1. Look at an element $d\tau$ in the q-space. The trajectories that we have considered pass through this element any number of times. Each such passage gives rise to a momentum vector p_k. A priori 2 fundamentally different types of trajectories are possible.

Type 1a): the vectors p_k repeat themselves so that only a finite number of them belong to each $d\tau$. In this case for the trajectories considered the p_k represent one- or many-valued functions of the q_k.

Type 1b): An infinite number of p_k vectors pass through the place considered. In this case p_k cannot be expressed as a function of the q_k.

One notices that the formulation of the (Einstein) quantum conditions is not

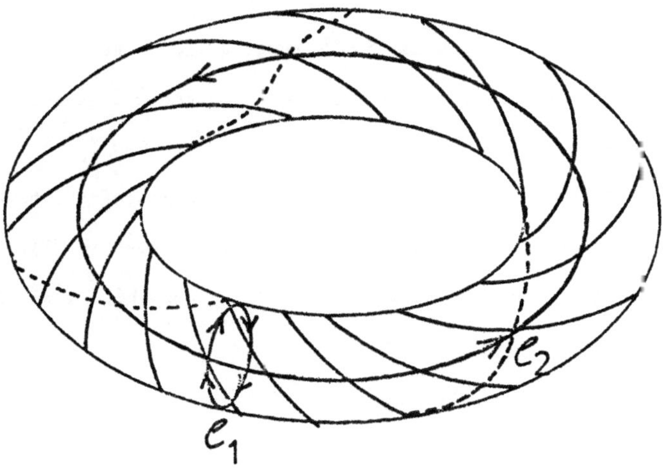

Figure 1. Phase space trajectory and invariant toroid for two degrees of freedom. $_1$ and $_2$ are curves for the definition of action integrals I_1 and I_2. The toroidal helix is the trajectory: normally it is not closed.

possible for the case (b). On the other hand classical statistical mechanics holds only for the case (b); so only in this case is the microanonical ensemble equivalent to the time ensemble."

Einstein's quantization rule is canonically invariant, and for separable systems it is equivalent to the condition of Epstein and Schwarzschild. In phase space Einstein's rule may simply be expressed in terms of invariant tori. For two degrees of freedom the integrals (2.4) are taken around independent curves $\mathcal{C}_k$ on a torus as illustrated in Figure 1. Because each torus is a lagrangian variety the values of the integrals are invariant under continuous deformation on the torus.

The last paragraph of Einstein is remarkable. He was clearly aware of the difficulty of quantization in the presence of ergodic behaviour of trajectories, a difficulty that is still with us.

Since Einstein's time, Brillouin, Keller and Maslov (5) have put his quantization rule on a firm theoretical foundation, as reviewed in (1). In doing so it has had to be modified to take into account the possibility of half-integer quantum numbers, giving

$$I_\kappa = \hbar(n_\kappa + \alpha_\kappa/4) \qquad (2.5)$$

where n_κ is an integer quantum number and α_κ is a "Maslov index" which takes the value of 0 or 2 depending on the topology of the embedding of the torus in the phase space. In semiclassical mechanics only tori with I_κ satisfying (2.5) correspond to quantal states. The energies of these states, which give rise to the observed states are given by

$$E(I_1,\ldots,I_N), \qquad (2.6)$$

where $E(\underset{\sim}{I})$ is the functional dependence of the energy on the action integrals. This quantization is known as Einstein-Brillouin-Keller or "EBK" quantization.

3. Practical EBK quantization

Three numerical methods have been used. Any method that can be used for stars or planets can also be used for molecules, but the form of the problem is different. Accurate values of the energy function (2.6) are required for values of the action integrals given by (2.5), but not elsewhere. Tabulation of the function, and methods adapted to such tabulation, have been preferred to power series expansion.

M1) Eastes and Marcus, Noid and Marcus and Sorbie and Handy (6) use step-wise integration of trajectories to obtain the invariant tori and surfaces of section to derive the action integrals. This allows arbitrary distortion of the tori, but precise estimation of the action integrals is difficult, particularly when the number of degrees of freedom increases beyond two.

M2) The Hamilton-Jacobi method. This does not allow arbitrary distortion of tori but action integrals can easily be obtained. It was developed by Chapman, Garrett and Miller and has been applied by Handy, Colwell and Miller (7) to vibrational states of H_2O and SO_2.

M3) The iteration-variation method of Percival and Pomphrey (8). This makes use of a parametric representation of the torus and allows arbitrary distortion, although large distortions are more difficult to handle than for M1. The estimation of action integrals is easy and it follows from a variational principle (9) that the errors in the energy function are of second order in the errors in the torus. It has been applied (10) to the energy levels of carbonyl sulfide (OCS).

Both methods M3 and M2 use Fourier expansions and iteration, and the convergence of each is tested numerically.

The results for H_2O, SO_2 and OCS obtained by using these methods are presented in Tables 1 and 2. The traditional unit of energy is $hc^{-1}cm^{-1}$ where h is the Plank constant and c is the velocity of light. By tradition it is written as simply cm^{-1}.

For purely harmonic oscillations the semiclassical theory is exact. The

Table 1: Semiclassical energy and quantal energy levels of SO_2 and H_2O after Handy, Colwell and Miller (7) (non-linear molecules).

SO_2

LEVEL $(v_1 v_2 v_3)$	ν^{SC} (cm^{-1})	ν^{QM} (cm^{-1})	$\Delta\nu^{anharmonic}$ (cm^{-1})
0 0 0	0	0	-
0 1 0	516.24	516.21	8.79
0 2 0	1026.74	1026.61	23.39
1 0 0	1155.99	1156.03	14.97
0 0 1	1359.95	1359.93	18.07

H_2O

LEVEL $(v_1 v_2 v_3)$	ν^{SC} (cm^{-1})	ν^{QM} (cm^{-1})	$\Delta\nu^{anharmonic}$ (cm^{-1})
0 0 0	0	0	-
0 1 0	1597.44	1597.35	50.65
0 2 0	3159.69	3159.55	136.45
1 0 0	3715.79	3717.36	114.64
0 0 1	3821.47	3820.77	121.23

Table 2: OCS - sample semiclassical energy levels of Percival and Pomphrey, compared with quantal levels of Foord, Smith and Whiffen (16), for $\ell = 0$. For complete table see (10).

LEVEL $(v_1 v_2 v_3)$	ν^{SC} (cm^{-1})	$\nu^{QM} - \nu^{SC}$ (cm^{-1})	$\Delta\nu^{anharmonic}$ (cm^{-1})
$0\ 0^o\ 0$	0	0	-
$1\ 0^o\ 0$	859.3	- 0.3	16.7
$0\ 2^o\ 0$	1046.8	0.2	0.2
$2\ 0^o\ 0$	1710.6	0.0	40.8
$1\ 2^o\ 0$	1891.8	0.3	30.8
$0\ 0^o\ 1$	2061.9	0.1	30.4
$0\ 4^o\ 0$	2105.2	0.9	-11.7
$3\ 0^o\ 0$	2554.8	0.3	72.1
$2\ 2^o\ 0$	*	*	67.9
$1\ 0^o\ 1$	2918.4	- 0.3	50.1
$1\ 4^o\ 0$	2935.2	1.2	33.8
$0\ 2^o\ 1$	3095.1	0.1	44.5
$0\ 6^o\ 0$	3174.0	2.7	-34.9
$4\ 0^o\ 0$	3391.9	3.3	107.7
$3\ 2^o\ 0$	*	*	110.6
$2\ 4^o\ 0$	*	*	83.3
$2\ 0^o\ 1$	3767.8	0.2	75.8
$1\ 2^o\ 1$	*	*	77.7
$1\ 6^o\ 0$	*	*	17.2
$0\ 0^o\ 2$	4104.1	-2.3	83.5
$0\ 4^o\ 1$	4141.2	0.6	45.1
$3\ 0^o\ 1$	4609.1	1.2	109.4
$2\ 2^o\ 1$	*	*	117.4
$1\ 0^o\ 2$	4958.4	-4.0	106.2
$1\ 4^o\ 1$	*	*	91.1

ability of the semiclassical theory to deal with anharmonicities is determined by comparing the difference semiclassical and quantal results and $\Delta\nu^{anh}$, which is a measure of the anharmonicity for that state. The numerical errors in the estimates of $E(\underset{\sim}{I})$ are negligible.

The results show that the semiclassical theory is effective, though in the present form of the theory the accuracy is not as high as good spectroscopic measurements with errors of the order of $0.01 cm^{-1}$ or better. In the OCS calculations 48 semiclassical levels were obtained using 151 seconds of CDC 7600 computing time. For details consult the original papers.

4. Regular and irregular spectra.

The results of the previous section give us confidence that classical and semiclassical methods can be used to predict properties of vibrational and rotational spectra of molecules.

The KAM theorem says that if a bounded system is sufficiently close to being separable (or, more generally, integrable) then its phase space is almost always dominated by invariant toroids, given certain subsidiary conditions. However, the situation is not as clear as it might appear to be. There is a residual region, to which the theorem does not apply; although this residual region occupies a smaller volume of phase space than the region of invariant toroids, it does so in a very complicated fashion. Any invariant toroid has arbitrarily close to it trajectories from the residual region.

For systems of two degrees of freedom, the invariant toroids of two dimensions lie in a three-dimensional energy shell, so that a residual trajectory between two invariant toroids remains trapped between them. For more degrees of freedom there is no such trapping.

The KAM theorem shows that the invariant toroids required by EBK quantization are common for most nearly separable systems, but says nothing about the nature of the residual regions nor about systems that are far from being separable.

The many numerical integrations of classical trajectories that have now been carried out are consistent with the view that the residual regions contain unstable trajectories but that invariant toroids are common for many systems that are far from being separable. For examples of the numerical integrations of trajectories of model hamiltonians see Henon and Heiles, Contopoulos, and Walker and Ford (11).

Numerical experiments do not constitute proofs, but on the basis of these experiments we are able to supplement the proofs of the KAM theory by plausible hypotheses.

The theory and hypotheses are combined in the following list of properties of regions of phase space containing only bounded trajectories of a nonseparable system of N degrees freedom with an analytic Hamiltonian function. For simplicity, we suppose that there are no constants of motion, such as momentum, which are derived from the symmetry of the Hamiltonian function, except for the energy E.

C.1. The solutions of the classical equations of motion have a very complicated structure.

C2. There are no constants of the motion defined throughout an energy shell besides the energy E (strictly, no isolating integrals of the motion).

C3. The motion is not ergodic: time averages are not equal to averages over an energy shell.

C4. Almost all of the phase space of the bounded trajectories may be divided into regions of two types.

a. Regular regions, made up entirely of invarinat toroids of N dimensions (Einstein case 1a).

b. Irregular regions, made up almost entirely of unstable trajectories (Einstein 1b).

C5. At those energies for which the system is nearly separable, the regular regions normally occupy most of the phase space, but irregular regions remain.

C6. At those energies for which the system is far from being separable, irregular regions may occupy most of the phse space, but regular regions remain.

C7. The transition between C5 and C6 may or may not be rapid.

C8. There are irregular regions arbitrarily close to any point of a regular region. The trajectories of the regular regions are therefore not strictly stable.

C9. For systems of two degrees of freedom, predominantly regular regions surround stable periodic trajectories, irregular regions surround unstable periodic trajectories.

C10. The larger irregular regions are associated with low-order resonances. The trajectories of these regions are very unstable: those trajectories that result from a very small initial perturbation of a given trajectory soon wander through large part of an irregular region of dimension greater than N.

C11. The smaller irregular regions are associated with high-order resonances. The instabilities of the trajectories of these regions may take a long time to appear, and very precise calculation is required to find them.

It is the property C8 that causes the most complication. It is because of this property that we do not refer to the regions as "stable" and "unstable". If higher-order resonances are neglected, say by suppressing the high frequencies of the motion, the situation is much simpler. Properties C1, C2 and C8 are replaced by the properties:

Simplified properties of nonseparable classical systems

CS1. The structure of the solutions is greatly simplified.

CS2. There may be constants of the motion besides the energy E.

Where the system is nearly separable, the irregular regions are small or even absent, except for some special cases. If they are absent then the simplified system has N physically significant constants of the motion (isolating integrals of the motion).

CS8. The regular regions each fill a finite volume of phase space, and there is a finite volume around almost every point of every regular region containing no unstable trajectories.

This is similar to the simplified picture presented by Hénon and Heiles.

The correspondence principle has been applied to the properties of regular and irregular regions of phase space (12).

In the asymptotic limit as $\hbar \to 0$ the quantal systems have all the complications of the classical systems given by C1 to C11, together with the additional problems of quantization. This limit is not of such physical importance as the properties of systems with high quantum numbers, where $\hbar$ is finite, but small when compared with the most important action variables of the classical motion.

In the limit $\hbar \to 0$ all regular and irregular regions must be considered, and there are an infinite number of them, but for finite $\hbar$ those regions that are much smaller in volume than $(2\pi\hbar)^N$ can almost all be neglected, and only a finite number of regions of either type remain.

A similar simplification can be made by putting a limit on the frequency. The higher frequencies of the quantized motion are given by relations of the type

$$\omega_{\gamma'\gamma} = \frac{E_{\gamma'} - E_{\gamma}}{\hbar} \tag{4.1}$$

and are not adequately represented by high multiples of classical frequencies. Such high multiples may be neglected without introducing significant additional error into the process of quantization. There remain a finite number of possible classical resonances, and these can produce no more than a finite number of regular and irregular regions of phase space.

Therefore the physical problem of quantization, with finite but small $\hbar$, should commence with the simplified classical model of a nonseparable system, in which the complex properties (C1, C2, C8) are replaced by the simpler properties (CS1, CS2, CS8).

The KAM theorem, which proves the existence of invariant toroids occupying finite volumes of phase space, shows that EBK quantization is valid in those parts of phase space. This fact, together with the hypotheses suggested by numerical experiment, implies the following properties of quantized nonseparable systems with N degrees of freedom. It is assumed throughout that $\hbar$ is small but finite.

A high quantum level of the discrete energy spectrum of a bound quantal system belongs to either (R), a regular energy spectrum, or (I), an irregular energy spectrum.

The regular energy spectrum and its associated states have the following properties:

R1. A quantal state may be labeled by the vector quantum number

$$\underset{\sim}{n} = (n_1, n_2, \ldots, n_N) .$$

R2. A state with quantum number n corresponds to those phase-space trajectories of the corresponding classical system that lie in an N-dimensional invariant toroid with action constants I_κ given by quantum conditions

$$I_\kappa = (n_\kappa + \alpha_\kappa/4)\hbar .$$

R3. The quantal state must resonate at frequencies close to those of the corresponding classical motion. Given two quantal states with one n_κ differing by unity and the others the same, the Planck relation for their energy difference is

$$\Delta E_\kappa = \hbar\omega_\kappa \tag{4.2}$$

where ω_κ is a fundamental frequency on the corresponding toroid.

R4. A "neighbouring state" to a state $\underset{\sim}{n}^o$ with energy E^o is a state with vector quantum number $\underset{\sim}{n}$ close to $\underset{\sim}{n}^o$, with energy difference no more than a small multiple of the maximum $|\Delta E_\kappa|$.

R5. We now use the correspondence principle for weak perturbations, whose most well-known form is the correspondence principle for intensity of radiation. Under weak external perturbations the state $\underset{\sim}{n}^o$ is much more strongly coupled to neighbouring states than to other states, with the coupling tending to decrease rapidly with $|\underset{\sim}{n} - \underset{\sim}{n}^o|$.

Not all states close in energy to a given state are neighbouring states, not even all those belonging to the same regular spectrum. However, experiments that are able to select a few high-n states of a regular spectrum have a high probability of selecting neighbouring states, so it is possible to observe the regularity. If transitions with energy difference ΔE^1 and ΔE^2 are observed, approximate multiples and integer combinations of them should also be observable. Bound integrable systems have a regular spectrum.

The correspondence principle predicts properties of an irregular spectrum in striking contrast to those of a regular spectrum:

11. There is no unambiguous assignment of a vector quantum number to a state ψ_o.

12. The discrete bound-state quantal spectrum must tend to a continuous classical spectrum in the classical limit. The frequencies

$$\left[E(\psi) - E(\psi_o) \right] / \hbar = \omega \qquad (4.3)$$

for fixed stationary state ψ_o and varying ψ form a discrete distribution that tends to the continuous distribution in ω. The distribution of levels of the irregular spectrum could take on the appearance of a random distribution.

13. By applying the corresponding principle for weak perturbations, there are no neighbouring states in the sense of (R4) and (R5). Except for selection rules and accidents, a state of an irregular spectrum is coupled by a weak perturbation with intensities of similar magnitude to all those states of a similar energy that correspond to the same irregular region of classical phase space. The number of such states is very large, of order

$$n_j^{N-1} \qquad (4.4)$$

where n_j is a typical quantum number.

14. The energies of the irregular spectrum are more sensitive to a slowly changing or fixed perturbation than those of the regular spectrum.

In the Born-Oppenheimer approximation, the vibrational energy spectrum of a polyatomic molecule is observed to be regular near equilibrium, except for Fermi resonances, which are a residual form of irregular spectrum. Near the dissociation limit the spectrum should be mainly irregular, with a large number of weak optical lines in place of a small number of strong ones. A regular progression should be observed to terminate abruptly at a maximum energy below the dissociation limit. An irregular spectrum could easily be confused with a continuum under poor conditions of observation.

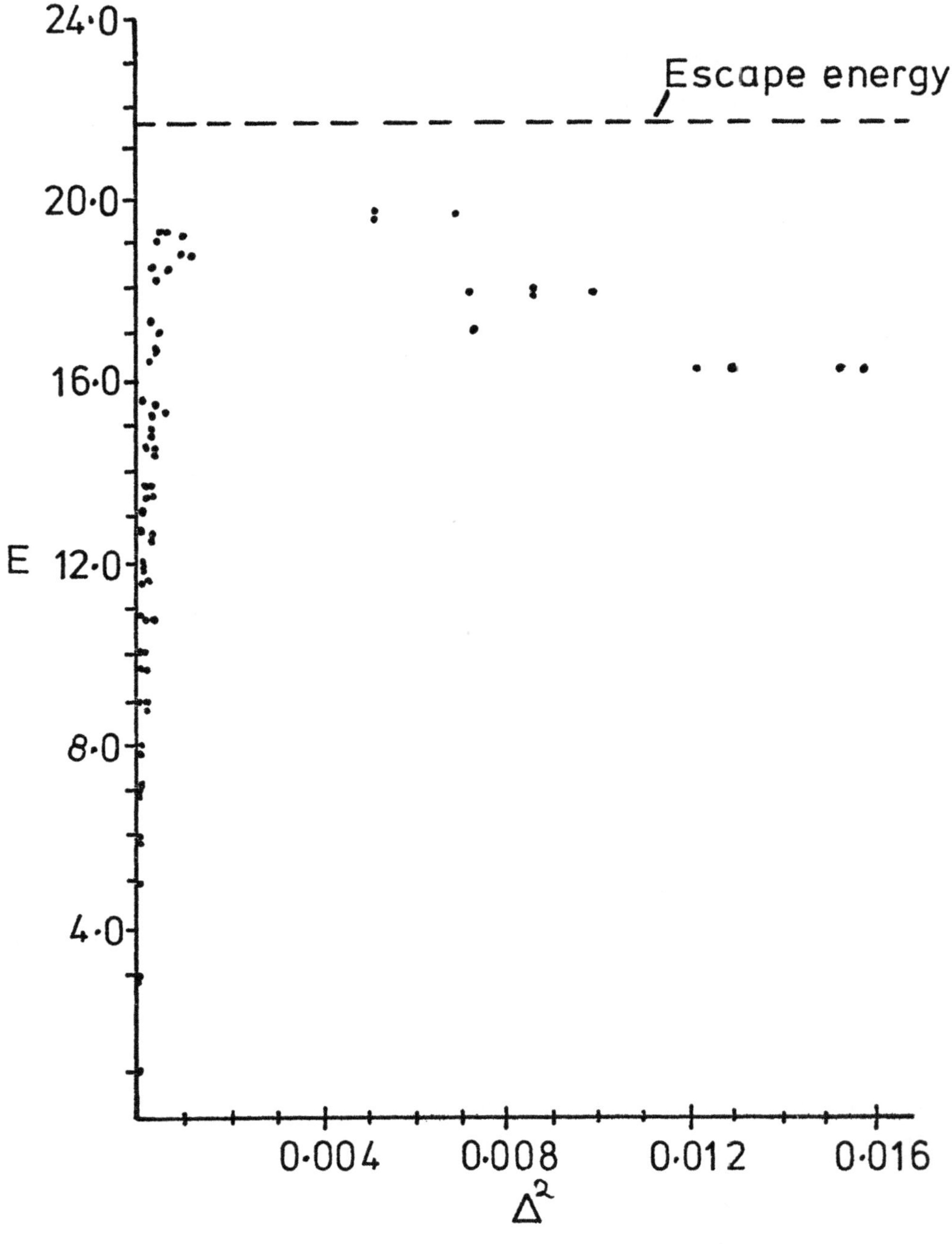

Figure 2. Second differences of energy levels with respect to the perturbation parameter α, after Pomphrey.

The instability of the irregular spectrum probably extends to instability under variations in $\hbar$. If that is the case the semiclassical limit does not exist for the irregular spectrum, except in a statistical sense.

No unambiguous observations of individual states of irregular spectra are yet known to the author.

5. Numerical Calculations on Regularity

Pomphrey (13) has studied the energy spectrum of the Henon-Heiles potential. It was important <u>not</u> to use semiclassical methods, although for part of the spectrum they are clearly more efficient, because of the risk of implicitly assuming what is to be proved. Pomphrey used the Rayleigh-Ritz method and an harmonic-oscillator basis, the eigenfunctions of the unperturbed Hamiltonian

$$H^o(\omega) = \tfrac{1}{2}(\underset{\sim}{p}^2 + \omega^2 \underset{\sim}{r}^2)$$

where ω is a free parameter. The Hénon-Heiles Hamiltonian takes the form

$$H = H^o(\omega) + \frac{\alpha r^3}{3} \sin 3\theta + \tfrac{1}{2}(\omega^2 - 1)r^2$$

Errors are introduced by the inevitable truncation of the basis set, but they were shown to be less than unity by extensive checks.

Eigenvalues were obtained for values of the perturbation parameter α from 0.086 to 0.090. The behaviour of the spectrum under perturbation was examined by obtaining the magnitudes of the second differences Δ_i^2 of the energy levels with respect to the increments $\Delta\alpha = 0.001$ in α. They are

$$\Delta_i^2 = \left| E_i(\alpha + \Delta\alpha) - 2E_i(\alpha) + E_i(\alpha - \Delta\alpha) \right| , \tag{5.1}$$

The values of Δ_i^2 for the 71 values of E_i are given in Figure 2.

For energies less than $E = 16.0 \approx 0.74D$, all second differences are very small; D is the depth of the Henon-Heiles well. For energies greater than $E = 16.0$, however, eigenvalues are found with corresponding Δ_i^2 values orders of magnitude larger. These eigenvalues are evidently very sensitive to small changes $\Delta\alpha$ in the perturbation. High-order terms in the perturbation expansion become important for these eigenvalues.

Thus two types of eigenvalue are distinguished by their behaviour under a slowly changing perturbation.

From the results of Hénon and Heiles the relative area α_I of the surfaces of section covered by unstable trajectories is given approximately as a function of the energy E by

$$\alpha_I(E) = 0 \qquad E < 0.68D$$

$$= 3.125(E/D) - 2.125 \qquad E > 0.68D$$

Pomphrey (13) compared the classical integral

$$I(E) = \int^{E} \alpha_I(E)\,dE ,$$

which is the integrated area covered by unstable trajectories up to an energy E, with the quantal sum

$$S(E) = \frac{1}{D} \sum_i{}' \; n_I(E_i) < \Delta E_i > \qquad (E_i < E)$$

In this sum, set $n_I(E_i) = 1$ if the eigenvalue E_i is very sensitive to the slight changes in the perturbation, and $n_I(E_i) = 0$ otherwise. $<\Delta E_i>$ is a mean separation between an eigenvalue E_i and its two neighbouring eigenvalues E_{i-1} and E_{i+1}.

$$< E_i > = (E_{i+1} - E_{i-1})/2$$

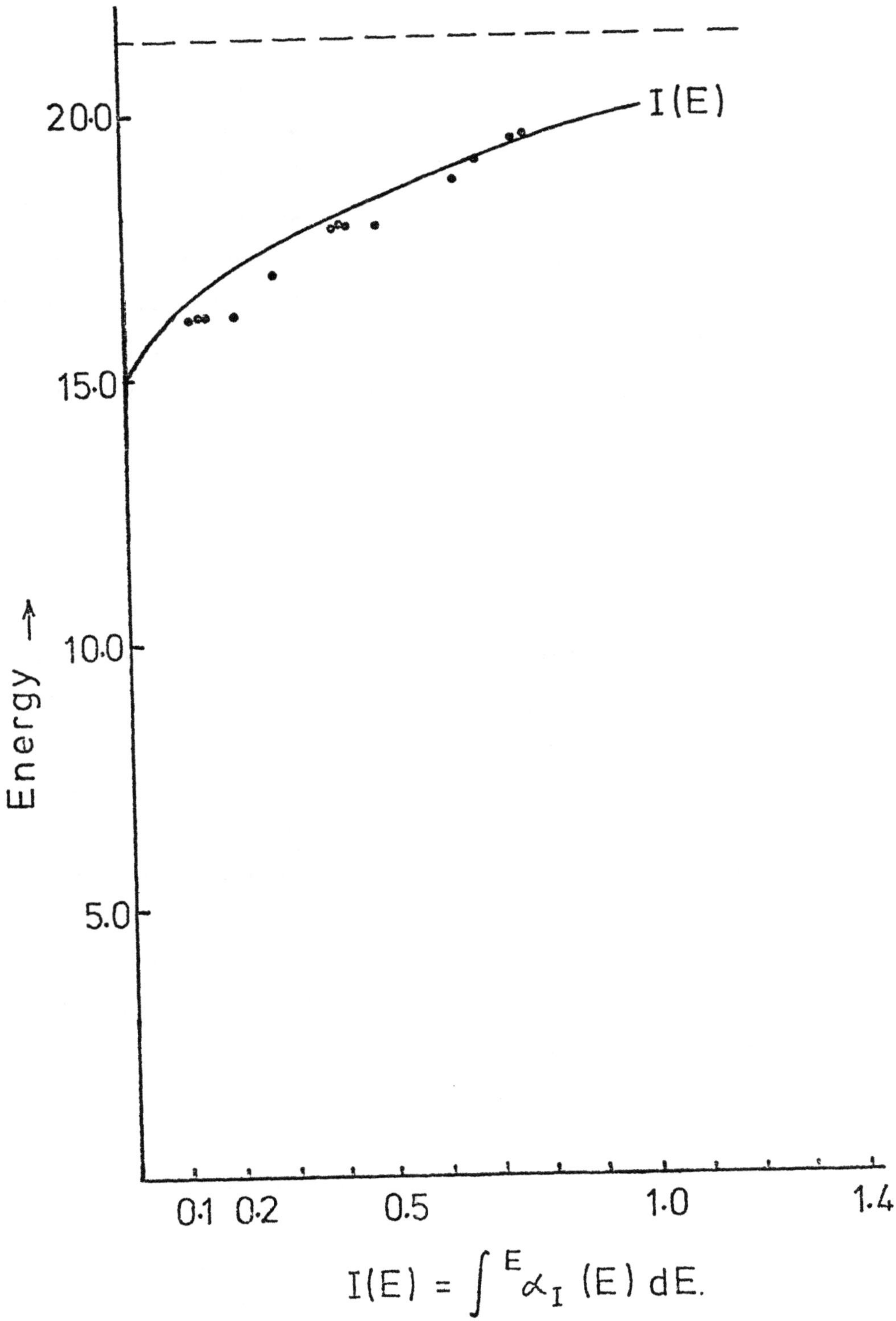

Figure 3. Comparison of classical and quantal irregular regions.

S(E) represents the number of states of the irregular spectrum below energy E. Figure 3 shows a plot of I(E) as a continuous curve and S(E) as a series of points. We see that qualitatively the points (quantal spectrum results) follow the shape of the curve (classical trajectory results). The two sets of results cannot hope to agree quantitatively since the sum S(E) corresponds to a volume of phase space of dimension four while the integral is over three dimensions. However, the energy at which the eigenvalues first become sensitive to the changing perturbation (≈ 0.74D) agrees well with the critical energy of Hénon and Heiles (≈ 0.68D).

Thus eigenvalues of the Hénon and Heiles nonintegrable Hamiltonian with energy less than a critical energy $E^c \approx 0.74D$ were found to be insensitive to a slight change in the perturbation. These eigenvalues belong to a regular spectrum. Above the critical energy, which is known to within a narrow band of energy, eigenvalues are found that are very sensitive to a small change in the perturbation. These eigenvalues belong to an irregular spectrum.

For this particular system the existence of two types of spectrum is confirmed insofar as the energy levels of the irregular spectrum are more sensitive to a fixed perturbation than those of the regular spectrum, in second order.

Early classical trajectory calculations on model molecular potentials were carried out by Thiele and Wilson and by Bunker (14) and distinguished between regular and irregular trajectories. They were carried out with a view to investigating the validity of conflicting theories of unimolecular reactions. More recently Nordholm and Rice have carried out quantal calculations on two-dimensional model potentials, and MacDonald and Marcus (15) have used the classical trajectory method to study the flow of energy in a system similar to the molecules CD_3Cl and CD_3H.

References

(1) I.C. Percival, Adv. Chem. Phys. 36, 1 (1977)

(2) K. Schwarzschild, Berliner Berichte, 548 (1916); P.S. Epstein, Ann. der Physik, 50, 439 (1916).

(3) M. Jammer, Conceptual Developments of Quantum Mechanics, McGraw-Hill, New York, 1966 Chapters 2 and 3.

(4) A. Einstein, Verhand. Deut. Phys. Ges., 19, 82 (1917)

(5) M.L. Brillouin, J. de Physique (ser. 6) 7, 353 (1926); J.B. Keller, Ann. Phys. (N.Y.) 4, 180 (1958); V. Maslov, Theorie des Perturbations, Dunod, Paris, 1972.

(6) W. Eastes and R.A. Marcus, J. Chem. Phys., 61, 4301 (1974); D.W. Noid and R.A. Marcus, J. Chem. Phys., 62, 2119 (1975); K.S. Sorbie and N.C. Handy, Molec. Phys., 32, 1327 (1976).

(7) S. Chapman, B.C. Garrett and W.H. Miller, J. Chem. Phys., 64 (1976); N.C. Handy, S.M. Colwell and W.H. Miller, Faraday Disc. Chem. Soc., No. 62, 29 (1977).

(8) I.C. Percival and N. Pomphrey, Molec. Phys., 31, 97 (1976), J. Phys. B: Atom. Molec. Phys., 9, 3131 (1976).

(9) V. Trkal, Proc. Camb. Phil. Soc., 21, 80 (1922); J.H. Van Vleck, Phys. Rev., 22, 547 (1923); I.C. Percival, J. Phys. A: Math. Nucl. Gen., 7, 794 (1974).

(10) I.C. Percival and N. Pomphrey, Molec. Phys., 35, 649 (1978).

(11) M. Hénon and C. Heiles, Astron. J., 69, 73 (1964); G. Contopoulos, Astrophys. J., 138, 1297 (1963); G. Contopoulos, Astron. J., 75, 96 (1970); J. Ford, Advan. Chem. Phys., 24, 155 (1973); J. Ford and G.H. Lunsford, Phys. Rev. A., 1, 59 (1970).

(12) I.C. Percival, J. Phys. B: Atom. Molec. Phys., 6, L229 (1973).

(13) N. Pomphrey, J. Phys. B: Atom. Molec. Phys., 7, 1909 (1974).

(14) E. Thiele and D.J. Wilson, J. Chem. Phys., 35, 1256 (1961); D.L. Bunker, J. Chem. Phys., 37, 393 (1962).

(15) K.S.J. Nordholm and S.A. Rice, J. Chem.Phys., 61, 203 (1974);
―――――――――――――――――――― 61, 768 (1974);
J.D. McDonald and R.A. Marcus, J. Chem. Phys., 65, 2180 (1976).

(16) A. Foord, J.G. Smith and D.H. Whiffen, Molec. Phys., 29, 1685 (1975).

SEMICLASSICAL STUDIES OF BOUND STATES AND MOLECULAR DYNAMICS

R. A. Marcus,[1], D. W. Noid[1] and M. L. Koszykowski[1]

Department of Chemistry
Noyes Laboratory
University of Illinois
Urbana, Illinois 61801

Abstract

Semiclassical studies on molecular bound states, molecular collisions, and time-dependent dynamical systems are described. Several methods have been evolved for determining the action variables of nonseparable bound state systems, and thereby for calculating their quantum mechanical eigenvalues. Nonresonant and resonant systems are included, and surfaces of section in higher (>2) dimensional as well as in two-dimensional systems are obtained. Further, from the Fourier transform of an autocorrelation function the molecular spectrum can be determined directly from a trajectory. The effect of an oscillating electric field on a molecule is discussed semiclassically with a view to obtaining a relation between the classical mechanical and quantum mechanical treatments.

1. Introductory Remarks

In the present paper we would like to describe some recent results obtained in the semiclassical theory of bound states of molecules, molecular collisions, and time-dependent phenomena. The theory permits one to use classical mechanical trajectories and quasi-periodic concepts to calculate quantum mechanical properties.

In the treatment of molecular collisions [1], the semiclassical wavefunction $\psi(q)$ of the coordinates q is expressed in terms of the phase integral S along a trajectory

$$\Psi(q) \sim |\det \partial^2 S/\partial q \partial P| \exp(iS/\hbar) \qquad (1.1)$$

$$S = \int_{q_0}^{q} p \cdot dq + p_0 \cdot q_0 - (M\pi/2) \qquad (1.2)$$

Here, p is the N-dimensional momentum conjugate to the N-dimensional coordinate q, with initial values of (p_0,q_0). $M\pi/2$ is the accumulated loss of phase due to the trajectory having been reflected from caustics or having passed through foci. The determinant in (1.1) is an N x N one. P denotes a set of constants of the motion--usually the action variables for the N-1 bounded degrees of freedom and the total energy E. (There is one unbounded coordinate, a radial distance coordinate, in molecular collisions.) $\hbar$ is Planck's constant divided by 2π.

When there is more than one topologically independent trajectory that proceeds to q, ψ consists of a sum of terms of the above form. The wavefunction over all space is obtained by appropriate choice of the q_0's. The action variables contained in P are the classical analogs of the quantum numbers n, and usually equal $(n+\frac{1}{2})h$, depen-

---

[1]Present addresses: California Institute of Technology, Pasadena, CA, Oak Ridge National Laboratory, Oak Ridge, TN, Sandia Livermore, Livermore, CA, respectively.

ding on the degree of freedom.

The semiclassical theory has been extensively reviewed for elastic collisions [2]--i.e., collisions which involve no interconversion of translational and internal motion. In most elastic collision studies it was possible to separate variables in the classical motion and so obtain p as a function of q. Since 1970 the major development [1] has been in the treatment of inelastic collisions [3]. Here, in these no longer separable systems the classical mechanical equations of motion are now integrated numerically using computers to obtain p(q). Once the wavefunction for the collision has thereby been determined one can calculate any desired properties associated with the collision, such as transition probabilities from one internal state of a molecule to another of its states. Some of these transitions are classical dynamically-forbidden but not energetically-forbidden, and then one uses trajectories in which p is a complex-valued function of q. One then obtains transition probabilities which are exponentially small.

We focus our attention in the present paper, however, on some results for bound state rather than collision problems, and finally we consider briefly a time-dependent problem such as occurs for a molecule in an intense coherent or incoherent light pulse.

2. Bound State Problems

In 1917 EINSTEIN [4] pointed out that to quantize a system one should find the topologically independent canonical invariants-- the $\oint p\cdot dq$ equal to $(n+\delta)h$, where δ is a known constant, typically 1/2 for nondegenerate vibrations. I.e.

$$\oint_{C_i} p\cdot dq = (n_i+\tfrac{1}{2})h \qquad (i = 1,\ldots,N), \tag{2.1}$$

where C_i are the topologically independent paths and N is the number of coordinates.

KELLER showed how to obtain in this way the quantum mechanical eigenvalues of the systems in which a particle moved in various shaped regions of constant potential energy inside and infinitely high at the boundary [5]. Until recently no method was available for determining these $\oint p\cdot dq$ integrals for nonseparable two or more dimensional systems when the potential energy function is smoothly varying. An example of a trajectory for such a system having a Henon-Heiles Hamiltonian (2.2) is given in Fig.1 for the case of incommensurate ω_x and ω_y.

$$H = \tfrac{1}{2}(p_x{}^2+p_y{}^2+\omega_x{}^2x^2+\omega_y y^2) + \lambda x(y^2+\eta x^2) \tag{2.2}$$

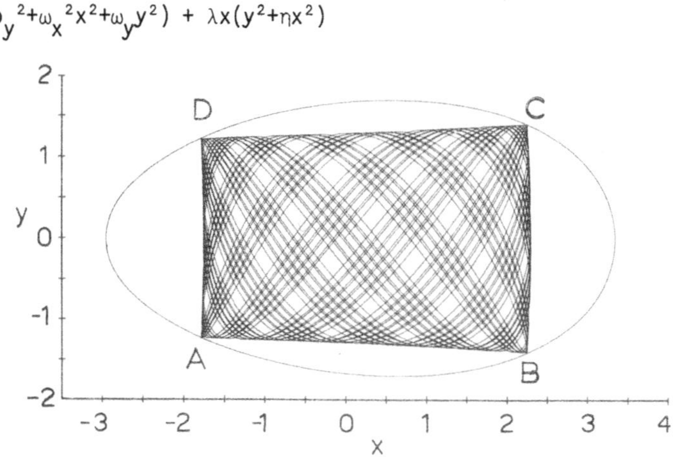

Fig.1 Trajectory for the Hamiltonian (2.2), with ω_x and ω_y incommensurate.

By integrating along the caustic in Fig.1 from A to B and back and also from A to C and back, EASTES in this laboratory was able to evaluate for the first time the two independent phase integrals [6]. Using an iteration procedure it was possible to find trajectories at energies such that (2.1) were simultaneously satisfied. Thereby, the quantum mechanical eigenvalues were determined.

The topologically independent paths C_i are also depicted for the Fig 1 system in Fig.2. NOID [7] made use of Poincaré surfaces of section to evaluate the phase integral. E.g., for the surface of section at y = 0 it is the area under the curve of p_x versus x. Results for the eigenvalues obtained in this way are given in Table I. Other methods [8] for obtaining the eigenvalues have been developed since. The methods tend to complement each other each other. (Table I is at end of paper.)

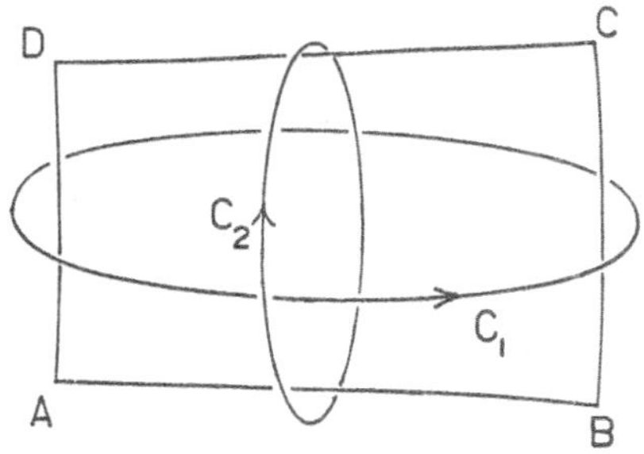

Fig.2 Topologically independent paths C_1 and C_2 for calculating action integrals for the trajectory in Fig.1

One sees from Fig.1 that when ω_x and ω_y are incommensurate the anharmonicity in (2.2) causes only a minor distortion of the original box-like shape and, correspondingly, only minor energy transfer between the degrees of freedom x and y. The situation is quite different when ω_x and ω_y are commensurate. When $\omega_x = \omega_y$ the trajectories are either of the librating or precessing type, an example of the latter being given in Fig.3. Curvilinear (e.g., radius=constant) and rectilinear (y = 0) surfaces of section were used to obtain the phase space data to evaluate the topologically independent phase integrals and again evaluate the quantum eigenvalues thereby [9].

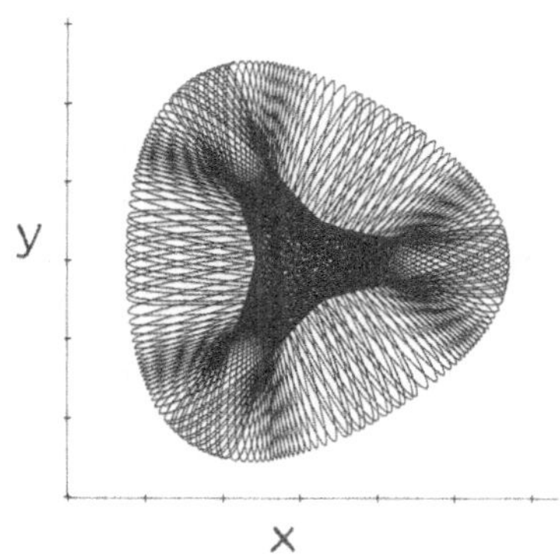

Fig.3 Trajectory for the Hamiltonian (2.2) with $\omega_x = \omega_y$

Another example showing extensive energy sharing between the x and y coordinates is the Fermi resonance case ($\omega_x = 2\omega_y$). Here a secular term in the anharmonic contribution can be shown to arise (unlike the $\omega_y = 2\omega_x$ case) and to cause the extensive distortion, as in Fig.4, again unlike the $\omega_y = 2\omega_x$ case. Because of the shape of the region swept out in Fig.4 one surface of section was selected along a parabolic coordinate curve, and the other at $y = 0$, to evaluate the two independent phase integrals, and, thereby, to evaluate the quantum eigenvalues [10].

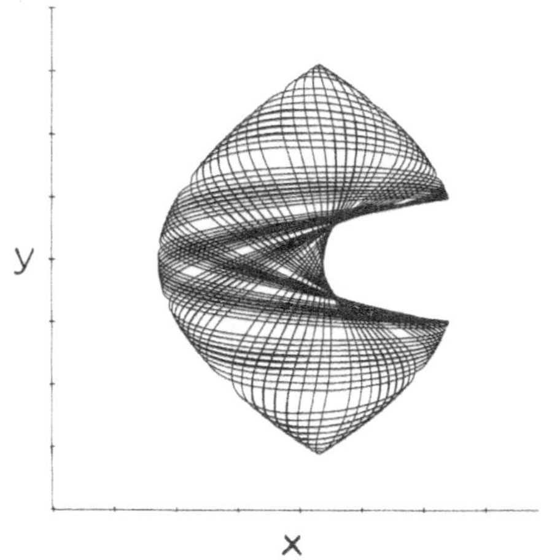

Fig.4 Trajectory for the Hamiltonian (2.2) for the Fermi-resonance case, $\omega_x = 2\omega_y$

We have extended this surface of section method to higher dimensions [11]. Whereas before we noted, say, the value of p_x and of x each time the trajectory crossed the $y = 0$ axis in a positive direction, now we note each time the value of xyz along the trajectory passes through zero, then collect the $x = 0$ cases and, for plotting p_z vs z, y occurring in the small interval $(-\varepsilon,\varepsilon)$, with $p_x > 0$, $p_y > 0$. In practice the phase integrals were evaluated in action-angle coordinates rather than conventional ones to reduce the number of points needed for the evaluation. The entire procedure was also applied to four-dimensional systems as well. The quantum mechanical energy eigenvalues were again evaluated by iteration, using (2.1).

A direct method for evaluating the quantum mechanical spectrum (differences of energy eigenvalues) from classical trajectories has also been developed here [12]: one computes an autocorrelation function such as $\langle x(0)x(t)\rangle$ from the trajectory, and takes the Fourier transform. A suitable form for the spectral function $I(\omega)$ which avoids the averaging over initial phases in $\langle\ \rangle$ was shown to be

$$I(\omega) = \frac{1}{2\pi} \lim_{T\to\infty} \frac{1}{T} \langle | \int_0^T x(t)\exp(-i\omega t)dt|^2\rangle , \tag{2.3}$$

where the remaining averaging now appears only in some statistical mechanical ensembles (e.g., a canonical one). An example of a spectrum for a finite trajectory time T is given in Fig.5 for the function $x(t)+y(t)$. The spectrum is seen to consist of sharp lines, as indeed it should in the quasi-periodic regime. Results comparing these semiclassical spectral lines with the quantum mechanical ones are given in [12] for this two-dimensional system. Analogous results were obtained for three-dimensional systems also [12].

In the "ergodic"[2] region the spectrum obtained by us was "broadened" under low resolution (cf [12], but with increasing length of trajectory time the spectrum

[2] Throughout this paper the term ergodic has been used somewhat loosely to mean a system reaching most of the energetically accessible phase space. It is still assumed that the time-average average equals the average over that space.

appears, increasingly, to consist of sharp lines.)

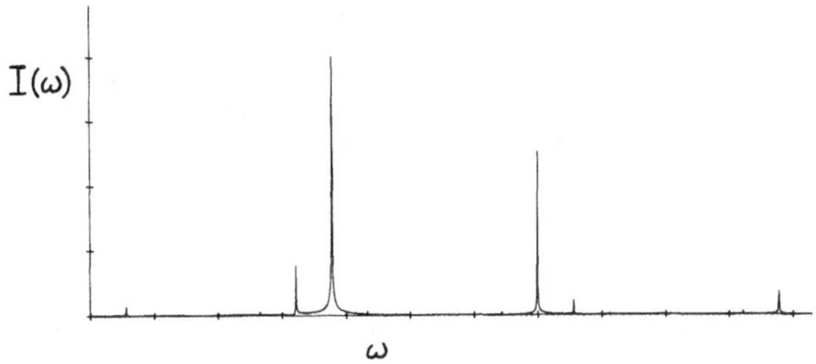

Fig.5 Power spectrum I(ω) of the function x(t) + y(t) for the case of two dimensions and the Hamiltonian (2.2), in the quasi-periodic regime

An example is given in Fig.6 for the system with the Hamiltonian (2.2) and at an energy for which one has exponential separation in time of neighboring trajectories. Three spectra are given, corresponding to three initial conditions at the same energy [11]. The trajectories for the spectra in Fig.6 were for long times, but nevertheless back integrated to two places (and conserved energy to seven places). Spectra for increasingly short times back integrated of course to more figures, but became increasingly less "structured." We have not yet run a spectrum which is a phase space average of these spectra. Unlike the quasi-periodic case the spectral intensities are approximately independent of the initial conditions at the given energy. The spectra are seen to consist of numerous "lines", perhaps evidence for CHIRIKOV's idea of overlapping resonances. (Resonances typically introduce additional lines.)

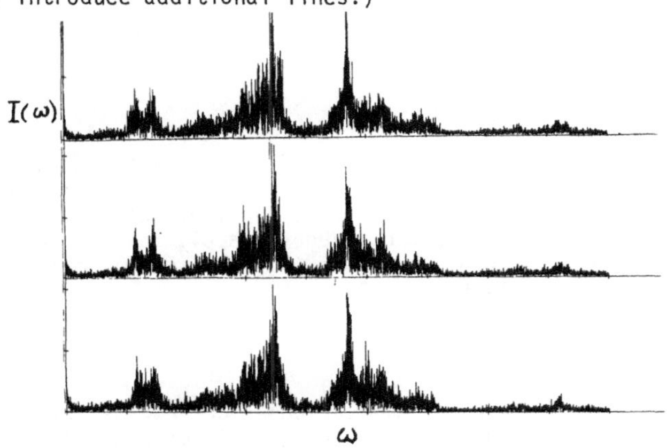

Fig.6 Legend as in Fig.5 but for the ergodic regime: the three spectra are for trajectories of a given energy but different initial condition

In the ergodic regime we find that trajectories at different initial conditions with the same total energy yield spectra which become increasingly similar the more the trajectories are able to average over all of phase space: If the trajectory has not been allowed sufficient time to pass through a large portion of phase space, the spectrum is still broad, and a longer trajectory would yield a somewhat different spectrum. Presumably the length of time needed for the spectrum in the "ergodic" regime to be independent of initial conditions at given energy is the recurrence time. Application of the spectral method to determine energy distribution of high energy molecules among various normal modes was made in a recent classical trajectory calculation [13].

3. Time-Dependent Problems

We consider here the motion of the atoms in a molecule under the influence of an oscillating electric field, a light pulse for example. The pulse may be coherent or incoherent. In either case the Hamiltonian for the motion of the molecule will depend on the coordinate q, the momenta p and, now, explicitly on the time t. One can show that the appropriate semiclassical wavefunction now for the time-dependent wave equation is again given by (1.1), but with S given by

$$S = \int_{q_0}^{q} p \cdot dq + p_0 q_0 - \int_0^t H\, dt, \tag{3.1}$$

and the P's now all being initial action variables. The average semiclassical value of, say, x(t), is given by

$$\langle x(t) \rangle_{sc} = \int |\psi(q,t)|^2 \, x \, dq, \tag{3.2}$$

where ψ is the semiclassical wavefunction,

$$\psi_{sc}(q,t) = |\det \partial^2 S/\partial q \partial P| \exp(iS/\hbar)$$

$$\equiv |\det \partial q_0/\partial q| \exp(iS/\hbar), \tag{3.3}$$

where P denotes the initial action variables for this bound state system.

One can then show that the $\langle x(t) \rangle_{sc}$ in (3.2) is equal to its classical value,

$$\langle x(t) \rangle_c = \int_0^1 x(t)\, dq_0 \tag{3.4}$$

where the integration is over initial phases (initial angle variables) of the classical motion. I.e.,

$$\langle x(t) \rangle_{sc} = \langle x(t) \rangle_c \tag{3.5}$$

The conditions used to obtain (3.4) from (3.2) were (i) to neglect tunneling, if any, along the trajectory, i.e., to neglect the case where p in (3.1) is complex-valued over some portion of the path, and to consider the case where $\det|\partial q/\partial q_0|$ had no zeros. Remarks similar to those about x apply to any other functions of the q and p, such as that portion of the energy H_{mol} of the molecule not explicitly involving the oscillating electric field.

$$\langle H_{mol}(t) \rangle_{sc} = \langle H_{mol}(t) \rangle_c \tag{3.6}$$

The result (3.5) helps explain the similar behavior reported by Dr. Lamb at this conference, where he compared classical and quantum mechanical results for an oscillator in an oscillating electric field. As he suggested, one should compute $\langle x(t) \rangle_c$ by averaging over initial phases w_0, rather than begin with the oscillator classically at rest.

The derivation of Eq. (3.6) also provides some justification for the recent classical trajectory treatment of the behavior of a polyatomic molecule $CH_3C\ell$ under the influence of an intense oscillating electric field [14].

Further studies [15] utilizing these concepts are currently being performed in an attempt to characterize the effects of ergodicity on a classical calculation of infrared multiphoton disassociation.

4. Ergodic Wavefunctions

We have made both a classical trajectory study as well as a quantum mechanical variational calculation using the Hamiltonian (2.2) with $\omega_x = 2\omega_y$. At low energies the motion is quasi-periodic and trajectories which resemble curved rectangles are somewhat similar to those in Fig.4. The wavefunctions corresponding to a state in this energy region occupy a similarly shaped region of configuration space, as seen in Fig.7. At higher energies the classical motion becomes stochastic and the wavefunction occupies a correspondingly larger region of configuration space in Fig.8 [11]. Further aspects of this behavior are currently being examined [10].

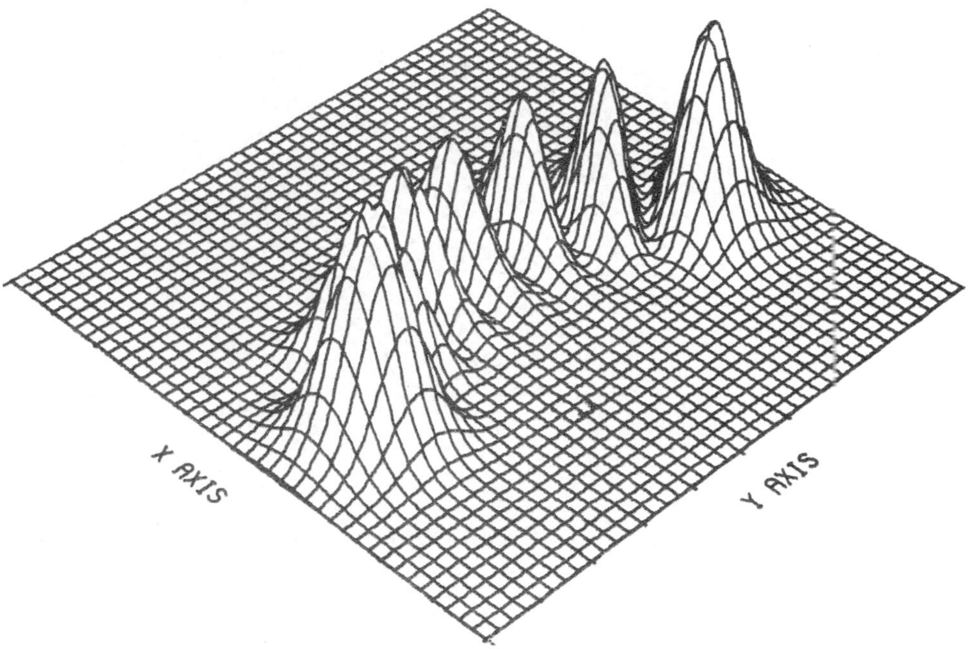

Fig.7 A wave function for the Hamiltonian (2.2) at an energy where the classical motion is quasi-periodic, for the Fermi-resonance case

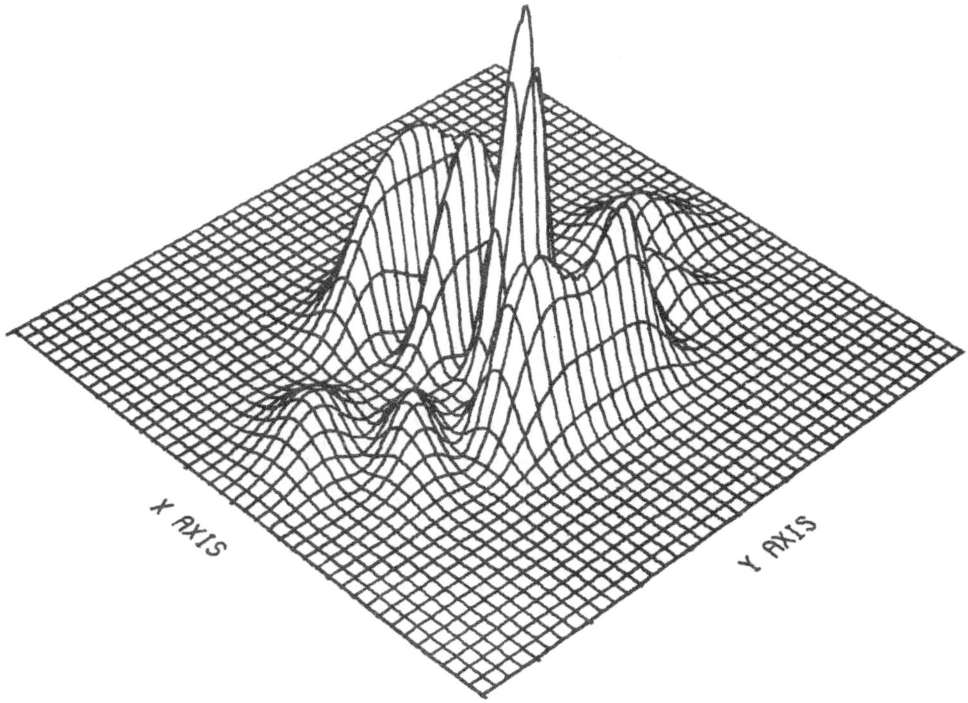

Fig.8 Legend as in Fig.7, but for the case that the classical motion is "ergodic"

Additional studies [16] using a numerical integration of the time-dependent Schrodinger equation for a system in the presence of an external laser field have been done at various values of the laser power and frequency. The wave functions have been observed to change from a localized type to one spread out with many ripples, after the system has absorbed sufficient energy from the field.

5. Concluding Remarks

In the present paper we have surveyed some recent results of semiclassical theory of anharmonic systems. The results draw on some of the classical concepts being treated at this conference.

TABLE I

Comparison of the semiclassical and quantum energy levels for various systems

System ω_x^2, ω_y^2 λ, η		Quantum numbers (n_x,n_y)	Eigenvalues Quantum	Eigenvalues Semicl.	Uncoupled $\lambda = 0$
0.29375,	2.12581,	0,0	0.9916	0.9920	1.0000
		1,0	1.5159	1.5164	1.5420
-0.1116,	0.08414	2,0	2.0308	2.0313	2.0840
		0,1	2.4188	2.4198	2.4580
0.36,	1.96	0,0	0.9939	0.9942	1.0000
-0.1,	0.1	1,0	1.5809	1.5812	1.6000
		2,0	2.1612	2.1616	2.2000
0.49,	1.69	0,0	0.9955	0.9954	1.0000
		1,0	1.6870	1.6870	1.7000
-0.1,	0.1	0,1	2.2781	2.2785	2.3000
		2,0	2.3750	2.3751	2.4000
		1,1	2.9583	2.9588	3.0000
		0,2	3.5479	3.5480	3.6000
0.81,	1.21	0,0	0.9980	0.9978	1.0000
		1,0	1.8944	1.8944	1.9000
-0.08,	0.1	0,1	2.0890	2.0889	2.1000
		2,0	2.7899	2.7900	2.8000

Acknowledgment

This research was supported by the National Science Foundation. Some of the wavefunctions were computed at Oak Ridge National Laboratory. We are pleased to acknowledge, too, discussions with Professor E. Atlee Jackson.

References

1. R. A. Marcus, J. Chem. Phys. 59, 5135 (1973) and references cited therein; W. H. Miller, Adv. Chem. Phys. 25, 69 (1974) and references cited therein.

2. E.g., K. W. Ford and J. A. Wheeler, Ann. Phys. (N.Y.) 7, 259, 287 (1959); R. B. Bernstein, Adv. Chem. Phys. 10, 75 (1966).

3. A recent review appears in M. S. Child, in Modern Theoretical Chemistry, W. H. Miller, ed. (Plenum, New York, 1977).

4. A. Einstein, Verh. Deut. Phys. Ges. 19, 82 (1917).

5. J. B. Keller, Ann. Phys. (N.Y.) 4, 180 (1958).

6. W. Eastes and R. A. Marcus, J. Chem. Phys. 61, 4301 (1974); cf discussion of semiclassical bound states, R. A. Marcus, Faraday Disc. Chem. Soc. 55, 9 (1973).

7. D. W. Noid and R. A. Marcus, J. Chem. Phys. 62, 2119 (1975).

8. I. C. Percival and N. Pomphrey, Mol. Phys., 31, 917 (1976); S. Chapman, B. Garrett, and W. H. Miller, J. Chem. Phys. 64, 502 (1976); K. S. Sorbie and N. C. Handy, Mol. Phys. 32, 1327 (1976); K. S. Sorbie, ibid. 32, 1577 (1976); N. C. Handy, S. M. Colwell and W. H. Miller, Faraday Disc. Chem. Soc. 62, 29 (1977); I. C. Percival, Adv. Chem. Phys. 36, 1 (1977); K. S. Sorbie and N. C. Handy, Mol. Phys. 33, 1319 (1977); S. M. Colwell and N. C. Handy, ibid. 35, 1183 (1978); J. B. Delos and R. T. Swimm, Chem. Phys. Lett. 47, 76 (1977).

9. (a) D. W. Noid, Ph.D. Thesis, University of Illinois, Urbana (April, 1976).
(b) D. W. Noid and R. A. Marcus, J. Chem. Phys. 67, 559 (1977).

10. D. W. Noid, M. L. Koszykowski and R. A. Marcus, J. Chem. Phys. (to be submitted).

11. (a) M. L. Koszykowski, Ph.D. Thesis, University of Illinois, Urbana (April, 1978).
(b) M. L. Koszykowski, D. W. Noid and R. A. Marcus, J. Chem. Phys. (to be submitted).

12. D. W. Noid, M. L. Koszykowski and R. A. Marcus, J. Chem. Phys. 67, 404 (1977). The ergodic spectrum there is for the case of $\omega_x = \omega_y$.

13. J. D. McDonald, R. A. Marcus, J. Chem. Phys. 65, 2180 (1976).

14. D. W. Noid, M. L. Koszykowski, R. A. Marcus and J. D. McDonald, Chem. Phys. Lett. 51, 540 (1977).

15. J. R. Stine, M. L. Koszykowski, D. W. Noid and R. A. Marcus, J. Chem. Phys. (to be submitted).

16. A movie of this calculation has been made at Oak Ridge National Laboratory, D. W. Noid, G. S. Sewell and M. L. Koszykowski, J. Chem. Phys. (to be submitted).

THE ROLE OF PERIODIC ORBITS IN SEMICLASSICAL QUANTIZATION

Michael Tabor†
Department of Physical Chemistry
Hebrew University of Jerusalem
Jerusalem
Israel

1. Introduction

PERCIVAL [1,2,3] has suggested that, in the semiclassical limit, the quantal bound state spectrum can be classified as either 'Regular' or "Iregular'. This idea is intimately tied up with the underlying dynamics and the KAM theorem. The Regular spectrum arises when the classical motion is integrable: The phase space is filled with invariant tori and the so-called EBK quantization condition (which depends on the existence of the tori) can be applied. Sufficient departures from integrability cause most of the tori to be destroyed. Then the phase space is dominated by stochastic trajectories and there is no known (semiclassical) quantization condition. The very different properties of the two classes of spectrum reflect the different properties of the underlying dynamics. Most of the practical methods for obtaining energy levels semiclassically are based on the EBK quantization condition and hence only apply to the Regular spectrum. These methods are reviewed by PERCIVAL [2,3].

A completely different approach is due to GUTZWILLER [4,5] who, on the basis of an analysis of the semiclassical propagator, obtained a quantization condition involving closed periodic orbits. These results appear to be in complete contradiction with the EBK condition. This problem, however, can be resolved when one distinguishes between the Regular and Irregular spectrum. In the case of the Regular spectrum we were able to show [6] that EBK quantization and periodic orbit theory are entirely equivalent. Before this is described we briefly summarize the relevant classical mechanics.

2. Integrable Hamiltonians and Periodic Orbits

Consider an integrable Hamiltonian of N-degrees of freedom. If it is integrable then we can write it as a function of N action variables, i.e.

$$E = H(\underset{\sim}{I}) \qquad , \underset{\sim}{I} = I_1, \ldots\ldots, I_N \quad . \tag{1}$$

Hamilton's equations give us the N-component frequency vector $\underset{\sim}{\omega}$:

$$\underset{\sim}{\omega} = \underset{\sim}{\omega}(\underset{\sim}{I}) = \nabla_{\underset{\sim}{I}} H(\underset{\sim}{I}) \qquad , \underset{\sim}{\omega}(\underset{\sim}{I}) = \omega_1(\underset{\sim}{I}), \ldots\ldots, \omega_N(\underset{\sim}{I}) \tag{2}$$

Notice that the frequencies depend on the torus, i.e., the corresponding set of action variables $\underset{\sim}{I}$. This is the generic case and we have the nondegeneracy condition

$$\det\left|\frac{\partial^2 H}{\partial I_i \partial I_j}\right| > 0 \tag{3}$$

In the case of two degrees of freedom (i.e. $E = H(I_1,I_2)$) we can plot contours of the 'action surface', for each value of the energy, in the I_1,I_2 plane.

†Present address: California Institute of Technology, Pasadena, California.

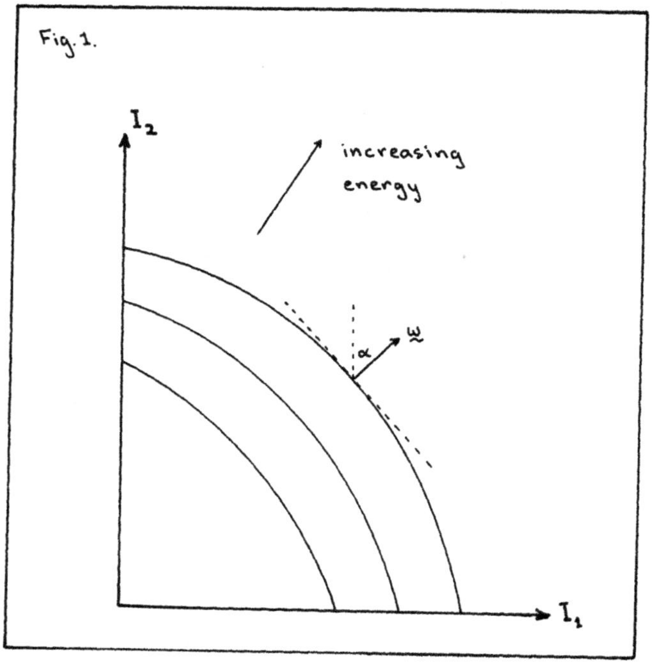

Fig.1 Contours of "action surface" for Hamiltonian $E = H = H(I_1,I_2)$. Normal to contour is frequency vector $\underset{\sim}{\omega}$ with direction $\tan\alpha = \omega_1/\omega_2$.

Every point on a given contour corresponds to a torus in phase space. From (2) we can see that the normal at each point gives the frequency vector associated with that torus. If the normal has a rational direction, i.e.

$$\tan\alpha = \frac{\omega_1}{\omega_2} = \frac{M_1}{M_2} \tag{4}$$

where M_1 and M_2 are integers, then the corresponding torus is covered by infinite one parameter family of stable closed orbits. The topology of the closed orbit family is completely specified by the integers M_1,M_2. The tori of closed orbits, although dense, form a subset of zero measure in the phase space. In Fig.2 we sketch some simple topologies of closed orbits that can occur for a central force potential with repulsive core.

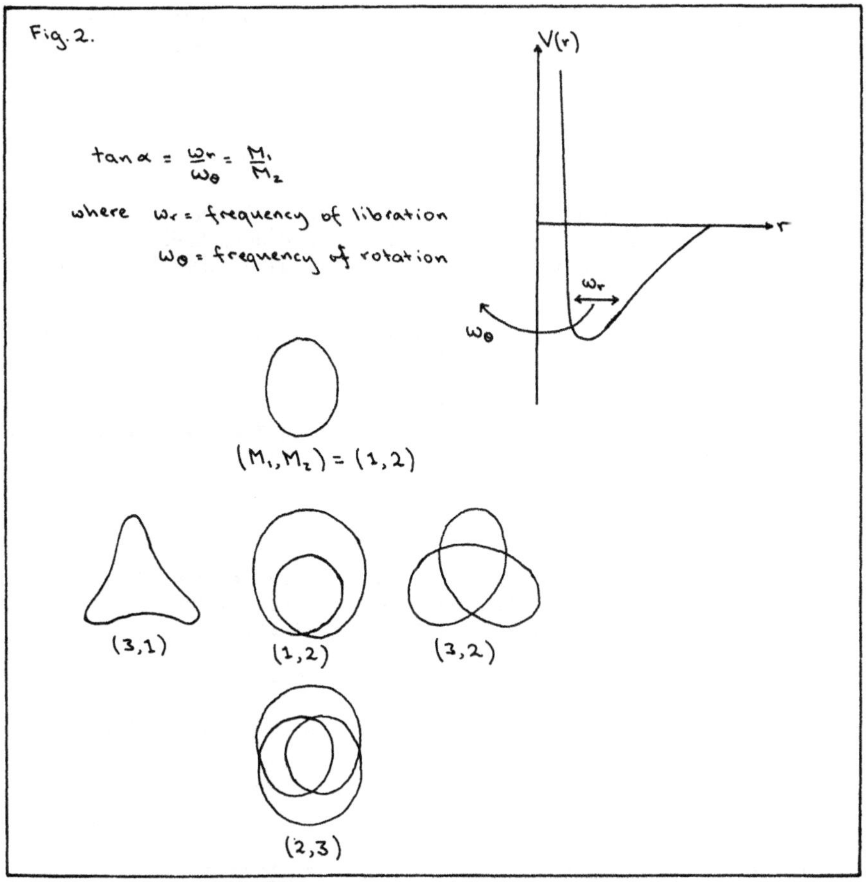

Fig.2 Some simple orbit topologies (M_1,M_2) for central force potential V(r) with repulsive core

3. EBK Quantization and Closed Orbits

EBK quantization tells us that each action component I_i can be quantized as

$$I_i = (m_i + \frac{\alpha_i}{4})\hbar \qquad , \; m_i = 0,1,2, \; . \; . \; . \; . \tag{5}$$

where m_i is the quantum number. For details of this formulation and the significance of the 'Maslov Index' α_i (α_i = 2 for libration) see reference [1]. Thus for N-degrees of freedom we can write

$$E_{\underset{\sim}{m}} = H(\underset{\sim}{I} = (\underset{\sim}{m} = \frac{\underset{\sim}{\alpha}}{4})\hbar) \tag{6}$$

where $\underset{\sim}{m} = (m_1, \; . \; . \; . \; . \; ., m_N)$ is the vector of quantum numbers and $\underset{\sim}{\alpha} = (\alpha_1, \; . \; . \; . \; .,$ $\alpha_N)$ the vector of Maslov indices. Using (6) we can write the semiclassical density of states as

$$n(E) = \sum_{\underset{\sim}{m}} \delta(E - H((\underset{\sim}{m} + \frac{\underset{\sim}{\alpha}}{4})\hbar)) \tag{7}$$

We transform the sum over the discrete index $\underset{\sim}{m}$ into an integration over the continuous classical variable $\underset{\sim}{I}$ (where $\underset{\sim}{I} = (\underset{\sim}{m} + \frac{\underset{\sim}{\alpha}}{4})\hbar$) by means of the Poisson Sum Formula. This transformation is exact and yields

$$n(E) = \frac{1}{\hbar^N} \sum_{\underset{\sim}{M}} \exp\{-i\tfrac{1}{2}\pi\underset{\sim}{\alpha}.\underset{\sim}{M}\} \int d^n I \delta(E - H(\underset{\sim}{I})) \exp\{\frac{2\pi i}{\hbar}\underset{\sim}{M}.\underset{\sim}{I}\} \quad (8)$$

Here $d^n I = dI_1 dI_2 \ldots\ldots dI_N$ and $\underset{\sim}{M} = (M_1,\ldots\ldots,M_N)$ where M_i is an arbitrary integer. The full details of the subsequent analysis of (8) are described in [6]. Here we just summarize the basic points. (i) The M = 0 term gives the 'Thomas-Fermi" density of states

$$n_{TF}(E) = \frac{1}{\hbar^N}\int d^n I \delta(E - H(\underset{\sim}{I})) = \frac{1}{(2\pi\hbar)^N}\int d\underset{\sim}{p}\int d\underset{\sim}{q}\ \delta(E - H(\underset{\sim}{p},\underset{\sim}{q})) \quad (9)$$

which is usually derived on a simple statistical basis assuming each quantal state occupies a cell of volume $(2\pi\hbar)^N$ in phase space. (ii) The $M \neq 0$ terms are oscillatory integrals of the form

$$\int d^n I\ \delta(E - H(\underset{\sim}{I}))\exp\{\frac{2\pi i}{\hbar}\ \underset{\sim}{M}.\underset{\sim}{I}\} \quad (10)$$

In the limit $\hbar \to 0$ these integrals can be evaluated by the method of Stationary Phase. It can be shown that, for these integrals, the stationary points arise when the components of the frequency vector $\underset{\sim}{\omega}$ are rationally related, i.e.

$$\underset{\sim}{\omega}(\underset{\sim}{I}) \propto \underset{\sim}{M} \quad , \quad \omega_1 : \omega_2 : \ldots : \omega_N = M_1 : M_2 : \ldots : M_N . \quad (11)$$

That is, the only contributions to the integrals (10) come from the tori of closed orbits.

Our final result is

$$n(E) = n_{TF}(E) + \frac{2}{\hbar^{\frac{1}{2}(n+1)}}\ \mathrm{Re}\sum_{\underset{\sim}{M}(\neq 0)} \frac{1}{|\underset{\sim}{M}|^{\frac{1}{2}(n-1)}|\underset{\sim}{\omega}(\underset{\sim}{I}_M)|\sqrt{|K(\underset{\sim}{I}_M)|}} \times \exp\{i[\frac{W}{\hbar}(\underset{\sim}{M}) - \tfrac{1}{2}\pi\underset{\sim}{M}.\alpha - \tfrac{1}{4}\pi\beta(\underset{\sim}{M})]\} \quad , \quad (12)$$

where

(i) $W(\underset{\sim}{M})$ is the total action about the closed orbit of topology specified by the vector $\underset{\sim}{M}$,
(ii) $K(\underset{\sim}{I}_M)$ is the Gaussian Curvature, at the point $\underset{\sim}{I}_M$, of the action surface $E = H(\underset{\sim}{I})$ (see (3)), and
(iii) $\beta(\underset{\sim}{M})$ is the signature of the matrix of $K(\underset{\sim}{I}_M)$.

Series similar to (12) have been obtained by BALIAN and BLOCH (7) for the density of eigenmodes in a hollow cavity and by GUTZWILLER [4] for the density of states of spherically symmetric potential.

Our result (12) shows that the density of states can be expressed as a 'topological sum' over all closed orbits. As more and more terms in the sum are included, corresponding to the contribution from orbits of increasing topological complexity, the smooth 'Thomas-Fermi' background is gradually eaten away and replaced by sharp peaks in the density of states. These spikes correspond to individual quantum states. We thus see that it is the sum over all closed orbits that is responsible for quantization and not just individual closed orbits. This result has been subjected to extensive numerical tests and shown to give excellent results [6]. Other refinements such as level clustering and uniform approximations are also discussed in [6].

4. Path Summation and Closed Orbits

In the last section we started at the quantal end of the problem and by means of EBK quantization were able to find a semiclassical representation of the density of states in terms of the closed orbits. We again derive this result, but this time the starting point is the expression for n(E) in terms of the trace of the time dependent Green function, i.e.

$$n(E) = \text{Trace}\ \delta\ (E-\hat{H}) = \text{Re}\frac{1}{\pi\hbar}\int dt\ \exp\{-\frac{iEt}{\hbar}\}\int d^N q K(\underset{\sim}{q},\underset{\sim}{q};t) \tag{13}$$

where K is defined as

$$K(\underset{\sim}{q}_B,\underset{\sim}{q}_A;t) = \langle\underset{\sim}{q}_B|\exp\{-\frac{i\hat{H}t}{\hbar}\}|\underset{\sim}{q}_A\rangle \quad . \tag{14}$$

In the semiclassical approximation K depends on the classical paths (labelled by index r) from $\underset{\sim}{q}_A$ to $\underset{\sim}{q}_B$ in the time t. The general form is (see e.g. Berry and Mount [8])

$$K_{sc}(\underset{\sim}{q}_B,\underset{\sim}{q}_A;t) = \frac{1}{(2\pi\hbar i)^{N/2}}\sum_r |D_r|^{\frac{1}{2}}\exp\{i[\frac{W_r}{\hbar} - \alpha_r\frac{\pi}{2}]\} \tag{15}$$

where W_r is the action along the rth path, α_r the Maslov index for that path and D_r the Nth order determinant defined as $D_r = (d\underset{\sim}{q}_B/d\underset{\sim}{p}_{A_r})^{-1}$.

In reference [9] we show that it is possible, for integrable Hamiltonians, to rewrite K_{sc} in terms of action-angle variables. It then turns out that the tracing operation (now an integration over the angle variables) becomes trivial and after some complicated analysis a result identical to (12) is obtained. At no stage are orbital stability parameters involved. This analysis reveals the vital flaw in GUTZWILLER's analysis [5] which is that when taking the trace of the Green function [15] it is assumed that the closed orbits are isolated. For integrable Hamiltonians (and hence the Regular spectrum) this is not correct; the closed orbits form infinite families lying on tori. This defect is also present in later elaborations of GUTZWILLER's results by MILLER [10], DASHEN et al. [11,12] and RAJARAMAN [13].

5. Conclusion

We have shown that for the Regular spectrum, corresponding to integrable classical motion, EBK quantization can also be expressed as a "topological sum" over all (stable) closed orbits. Quantization conditions involving isolated closed orbits are not consistent with EBK quantization. (It is possible, however, that they might provide an approximation to certain portions of the regular spectrum - see VOROS [14,15]).

The role of periodic orbits in describing the Irregular spectrum is still an open question. GUTZWILLER [5,16] has suggested a quantization involving the unstable periodic orbits. However, the resulting states only have a finite lifetime (a sort of 'quasi-mode' - see ARNOL'D [17] and BERRY [18]) and this must cast some doubt on the validity of the theory.

Finally it is perhaps worth mentioning another aspect of the Regular and Irregular spectrum, namely their statistical properties. For the Regular spectrum it is possible to show that, for the generic case, the nearest neibhbour level spacing takes on a simple random distribution. This was a difficult result to derive [19] and in the non-generic case, such as systems of harmonic oscillators, the spectrum shows unexpected correlations. The statistical properties of the Irregular spectrum are still unknown.

References

1. Percival, I. C. 1973 J. Phys. B. 6, L229-32.
2. Percival, I. C. 1977 Adv. Chem. Phys. 36, 1.
3. Percival, I. C. these proceedings.
4. Gutzwiller, M. C. 1970 J. Math. Phys. 11, 1791-1806.
5. Gutzwiller, M. C. 1971 J. Math. Phys. 12, 343-358.
6. Berry, M. V. and Tabor, M. 1976 Proc. Roy. Soc. A349, 101-123.
7. Balian, R. and Bloch, C. 1974 Ann. Phys. 85, 514-45.
8. Berry, M. V. and Mount, K. E. 1972 Rep. Prog. Phys. 35, 315-397.
9. Berry, M. V. and Tabor, M. 1977 J. Phys. A. 10, 371-379.
10. Miller, W. H. 1974 J. Chem. Phys. 63, 996-999.
11. Dashen, R. F., Hasslacher, B. and Neveu, A. 1974 Phys. Rev. D 10, 4114-29.
12. ______________________________ 1975 Phys. Rev. D 11, 3424-50.
13. Rajaraman, R. 1974 Phys. Rev. 21, 227-313.
14. Voros, A. 1976 Ann. Inst. Henri Poincaré A 24, 31-90.
15. Voros, A. 1977 Dissertation, Universite de Paris-Sud, Centre d'Orsay.
16. Gutzwiller, M. C. these proceedings.
17. Arnol'd, V. I. 1972 Funct. anal. applns. 6, 94-101.
18. Berry, M. V., J. Phys. A. 11, L193-L194.
19. Berry, M. V. and Tabor, M. 1977 Proc. Roy. Soc. A356, 375-394.

SEMICLASSICAL EIGENVALUES FOR ROTATING TRIATOMIC MOLECULES*

S. M. Colwell, N. C. Handy
University Chemical Laboratory
Lensfield Road
Cambridge, England

and

W. H. Miller‡
Department of Chemistry
University of California
Berkeley, California 94720

Abstract

The rigid asymmetric rotor is a system with only one non-conserved degree of freedom, and the effective potential function for this motion has a symmetric double well. The uniform semiclassical eigenvalue condition for such potentials, which takes account of tunneling through the barriers, is applied and shown to yield rotational eigenvalues in good agreement with the correct quantum mechanical values, much better than the "primitive" WKB eigenvalues.

I. Introduction

There has been a great deal of interest and progress in ways of determining semiclassical energy levels for small polyatomic (i.e., triatomic) molecules.[1-9] A variety of methods have been used--numerical integration of classical trajectories to determine the necessary "invariant toroids",[1-3] an algebraic solution of the Hamilton-Jacobi equation in action-angle variables,[4-5] a variational expression[6] for the invariant toroids, and others.[7-9] Most applications to date have been to model problems[1-4,6-9] of two coupled oscillators, although there has been one treatment[5] for realistic models of non-rotating triatomic molecules, which are essentially three coupled oscillators.

To explore the feasibility of describing rotational as well as vibrational states of triatomic molecules, we have considered first the semiclassical treatment of pure rotational eigenvalues of asymmetric rotors (e.g., non-vibrating triatomic molecules). This paper summarizes this recent work of ours.[10] Early semiclassical theory for the asymmetric rotor was the "primitive" WKB treatment by King.[11] Comparison of these "primitive" WKB results with the correct quantum mechanical energy levels is somewhat discouraging, the agreement being not nearly so good as for vibrational energy levels.

To do better, it is necessary to take account of tunneling in the rotational motion. The results we have obtained show that if this is done in the appropriate (uniform) manner, the semiclassical energy levels are in much better agreement with the correct quantum mechanical values.

In one sense, though, our results are discouraging since it is necessary to include the effects of tunneling to achieve acceptable accuracy. This is easy enough to do for systems with one degree of freedom--such as the rigid asymmetric rotor--but one does not know how to do this for systems with more than one degree of freedom. All of the methods[1-9] developed recently for multidimensional systems are of the "primitive" semiclassical variety and include no classically forbidden effects such as tunneling.

II. The Rigid Asymmetric Rotor; Primitive Semiclassical Eigenvalues

The rigid asymmetric rotor is a system of three degrees of freedom, the canonical varibles of which may be chosen as the action-angle varibles (j,q_j), (m,q_m), and (k,q_k). j, m, and k are the total angular momentum, its projection onto a space-fixed axis, and its projection onto a body-fixed axis, respectively, and q_j, q_m and q_k are the corresponding angle variables.

j and m are constants of the motion, so that the Hamiltonian for the system is not a function of q_j and q_m, and because of the isotropy of space it is also independent of m. If A, B, and C $(A \geq B \geq C)$ are the three rotation constants of the rotor, then the classical Hamiltonian in these canonical variables is[12]

$$H(j,q_j,m,q_m,k,q_k) = Ak^2 + B(j^2-k^2)\cos^2 q_k + C(j^2-k^2)\sin^2 q_k$$

$$\equiv H_j(k,q_k) \quad . \tag{2.1}$$

Since j is conserved, this is the Hamiltonian for a system with only one (non-trivial) degree of freedom. It is customary to define a reduced Hamiltonian $\tilde{H}_j$ by

$$H_j(k,q_k) = \frac{1}{2}(A + C)j^2 + \frac{1}{2}(A - C)\tilde{H}_j(k,q_k) \quad , \tag{2.2}$$

and from Eq. (2.1) one finds

$$\tilde{H}_j(k,q_k) = j^2(-\sin^2 q_k + \kappa\cos^2 q_k) + k^2(1 + \sin^2 q_k - \kappa\cos^2 q_k) \quad , \tag{2.3}$$

where κ is the asymmetry parameter,

$$\kappa = \frac{2B-A-C}{A-C} \quad . \tag{2.4}$$

The "primitive" WKB eigenvalues[11] of the reduced Hamiltonian of Eq. (2.3) are given by the standard Bohr-Sommerfeld eigenvalue condition

$$2n\pi = \oint dq_k \; k(q_k,E) \quad , \tag{2.5}$$

where $k(q_k,E)$ is determined by energy conservation,

$$\tilde{H}_j(k,q_k) = E$$

and is

$$k(q_k,E) = \left[\frac{E + j^2\sin^2 q_k - \kappa j^2\cos^2 q_k}{1 + \sin^2 q_k - \kappa\cos^2 q_k}\right]^{\frac{1}{2}} \quad . \tag{2.6}$$

It is useful to consider $k(q_k,E)$ to be of the form

$$k(q_k,E) \sim [E - V(q_k)]^{\frac{1}{2}} \quad ,$$

where the "potential" $V(q_k)$ is identified as

$$V(q_k) = j^2[\kappa\cos^2 q_k - \sin^2 q_k]$$

$$= j^2[\kappa - (1+\kappa)\sin^2 q_k] \quad , \tag{2.7}$$

and is shown in Figure 1. Eq. (2.5) thus falls into two cases:

(a) $E > \kappa j^2$. In this "above barrier" case there are no classical turning points, and Eq. (2.5) is given explicitly by

$$2n\pi = \int_0^{2\pi} dq_k\, k(q_k,E)$$

or

$$n\pi = \phi(E) \quad , \tag{2.8a}$$

where

$$\phi(E) = \int_0^{\pi} dq_k\, k(q_k,E) \quad . \tag{2.8b}$$

The classical motion is that of a hindered rotor.

(b) $E < \kappa j^2$. In this "below barrier" case there are four classical turning points, so that Eq. (2.5) becomes (since $\oint dq_k = 2\int_{q_0}^{\pi-q_0} dq_k$)

$$(n + \tfrac{1}{2})\pi = \phi(E) \quad , \tag{2.9a}$$

where here ϕ is the phase integral across one of the wells,

$$\phi(E) = \int_{q_0}^{\pi-q_0} dq_k\, k(q_k,E) \quad ; \tag{2.9b}$$

as usual, "n" has been replaced by "n + ½" because of the two classical turning points. In this case the classical motion is that of a torsional vibration.

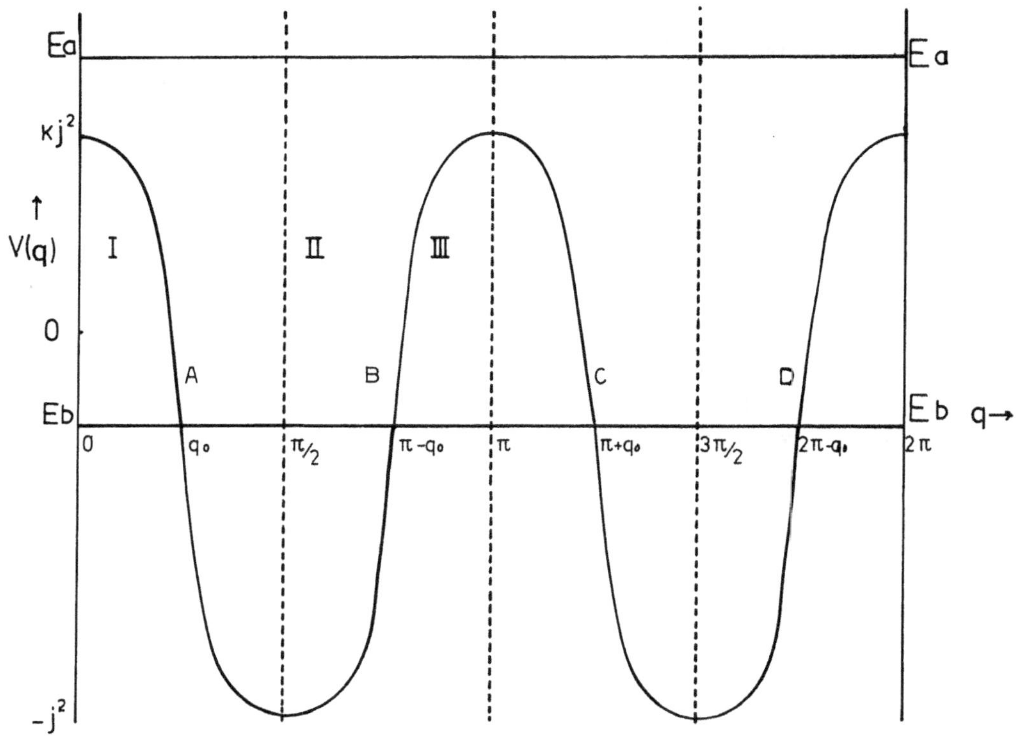

A sketch of the potential function defined by Eq. (2.7). Energies E corresponding to the above barrier case (a) and below barrier case (b) are shown.

In both cases (a) and (b) it is easy to conclude that Eq. (2.5) implies that there are two eigenvalues for each root of Eq. (2.8) and Eq. (2.9); i.e., the "primitive" semiclassical eigenvalues occur in doubly degenerate pairs.

In the symmetric rotor limits, $\kappa \to \pm 1$, Eqs. (2.8) and (2.9) give

$$\text{(a)} \quad \kappa = -1 \quad : \quad E_n = -(j+\tfrac{1}{2})^2 + 2n^2 \qquad (2.10a)$$

$$\text{(b)} \quad \kappa = +1 \quad : \quad E_n = (j+\tfrac{1}{2})^2 - 2(n-j)^2 \quad , \qquad (2.10b)$$

where the Langer modification, $j^2 \to (j+\frac{1}{2})^2$, has also been made. Apart from a constant, Eq.(2.10) is the correct quantum mechanical result for the prolate and oblate symmetric rotors.

Although correct in both symmetric rotor limits, the primitive semiclassical eigenvalues are quite poor for $|\kappa| \neq 1$. As noted above, the eigenvalues always occur in degenerate pairs, while the correct quantum mechanical eigenvalues have this degeneracy split. The primitive semiclassical eigenvalues are poorest for $|\kappa| << 1$ and for energy levels closest to the top of the barrier.

III. Uniform Semiclassical Eigenvalues

To do better than the primitive WKB eigenvalues it is necessary to take account of tunneling through the barriers. The uniform semiclassical quantum condition for a symmetric double well potential, which takes correct account of tunneling through the barrier in case (b) and reflection by the barrier in case (a), is[13,14]

$$n\pi = \phi + \Delta\phi \pm \tan^{-1}(e^{\theta}) \quad , \qquad (3.1)$$

where ϕ is the action integral across the well and is given by Eqs. (2.8b) and (2.9b); θ is the action integral through the barrier

$$\text{(a)} \quad E > \kappa j^2 \quad : \quad \theta = -\int_{-\tilde{q}_0}^{\tilde{q}_0} dq_k \left[\frac{E-j^2\sinh^2 q_k - \kappa j^2 \cosh^2 q_k}{1-\sinh^2 q_k - \kappa \cosh^2 q_k}\right]^{\frac{1}{2}} \qquad (3.2a)$$

$$\text{(b)} \quad E < \kappa j^2 \quad : \quad \theta = \int_{-q_0}^{q_0} dq_k \left[\frac{-E-j^2 \sin^2 q_k + \kappa j^2 \cos^2 q_k}{1+\sin^2 q_k - \kappa \cos^2 q_k}\right]^{\frac{1}{2}} \quad . \qquad (3.2b)$$

and $\Delta\phi$ is a "phase shift"

$$\Delta\phi = \arg \Gamma(\tfrac{1}{2} + i\,\tfrac{\theta}{\pi}) - \tfrac{\theta}{\pi} \log\left(\tfrac{|\theta|}{e\pi}\right) \quad . \qquad (3.3)$$

If $\Delta\phi$ is neglected and one takes the limits $\theta \to -\infty$ and $+\infty$ in case (a) and (b), respectively, then the "primitive" quantum condition of Section II is recovered. Eq. (3.1) is uniformly valid for E above, below, or close to the top of the barrier.

IV. Example

Table I shows the comparison of the "primitive" WKB eigenvalues (WKB), the uniform semiclassical eigenvalues (USC), and the correct quantum mechanical eigenvalues (QM), for the case $j = 3$. Such a small value of j is a severe test of semiclassical theory, which will in general be better the larger j.

Since the WKB, USC, and QM eigenvalues all obey the symmetry relation

$$E(\kappa) = -E(-\kappa) \tag{4.1}$$

--i.e., if $E(\kappa)$ is an eigenvalue, then so is $-E(-\kappa)$--Table I gives results only for $\kappa \geq 0$. Also, for each value of κ the eigenvalues are given relative to the lowest eigenvalue.

One sees that the uniform semiclassical eigenvalues are indeed superior to the primitive WKB ones and in general are in quite good agreement with the quantum mechanical values.

In conclusion, a semiclassical description of the energy levels for a rigid asymmetric rotor is seen to do reasonably well provided classically forbidden dynamics (i.e., tunneling) is taken into account. Since one expects this also to be the case if vibrational motion is included along with rotation, it is clear that one needs a multidimensional analog of the uniform quantum condition, Eq. (3.1) in order to treat vibrating and rotating states of triatomic molecules.

References

* Supported by the National Science Foundation under Grant CHE76-22621.

‡ Camille and Henry Dreyfus Teacher-Scholar.

1. R. A. Marcus, Faraday Disc. Chem. Soc. **55**, 34-44 (1973).
2. W. Eastes and R. A. Marcus, J. Chem. Phys. **61**, 4301-06 (1974).
3. D. W. Noid and R. A. Marcus, J. Chem. Phys. **62**, 2119-24 (1975).
4. S. Chapman, B. C. Garrett and W. H. Miller, J. Chem. Phys. **64**, 502-509 (1976).
5. N. C. Handy, S. M. Colwell and W. H. Miller, Faraday Disc. Chem. Soc. **62**, 29 (1977).
6. I. C. Percival and N. Pomphrey, Mol. Phys. **31**, 97-114 (1976).
7. K. S. Sorbie, Mol. Phys. **32**, 1577-1590 (1976).
8. K. S. Sorbie and N. C. Handy, Mol. Phys. **32**, 1319-1347 (1976).
9. K. S. Sorbie and N. C. Handy, Mol. Phys. **33**, 1319-1334 (1977).
10. S. M. Colwell, N. C. Handy, W. H. Miller, J. Chem. Phys. **68**, 745 (1978).
11. G. W. King, J. Chem. Phys. **15**, 820-830 (1947).
12. S. D. Augustin and W. H. Miller, J. Chem. Phys. **61**, 3155-3163 (1974).
13. W. H. Miller, J. Chem. Phys. **48**, 1641 (1968).
14. M. S. Child, J. Mol. Spectr. **53**, 280-301 (1974).

Table I. Semiclassical and Quantum Mechanical Eigenvalues[a] for the Rigid Asymmetric Rotor. j = 3

κ = 0.0			κ = 0.2		
W.K.B.(b)	U.S.C.(c)	Q.M.(d)	W.K.B.	U.S.C.	Q.M.
15.35828	15.70170	15.79796	15.40923	15.78310	15.88007
15.35828	15.55660	15.64495	15.40923	15.50173	15.58215
8.65988	9.73065	9.79796	9.74714	10.79125	10.85172
8.65988	7.85085	7.89898	7.53543	8.18734	8.22586
6.69840	5.97105	6.0	7.53543	6.93824	6.97165
0.0	0.14515	0.15301	0.0	0.06614	0.06957
0.0	0.0	0.0	0.0	0.0	0.0

κ = 0.4			κ = 0.6		
W.K.B.	U.S.C.	Q.M.	W.K.B.	U.S.C.	Q.M.
15.57068	16.02520	16.13157	15.87602	16.44269	16.56448
15.57068	15.52438	15.60049	15.87602	15.61419	15.68367
10.98145	11.92390	12.02664	13.86476	13.19821	13.29060
8.25870	8.57767	8.61332	8.89978	9.01608	9.04530
8.25870	7.86068	7.89507	8.89978	8.69435	8.72613
0.0	0.02494	0.02615	0.0	0.00663	0.00693
0.0	0.0	0.0	0.0	0.0	0.0

κ = 0.8			κ = 1.0		
W.K.B.	U.S.C.	Q.M.	W.K.B.	U.S.C.	Q.M.
16.43053	17.04918	17.18841	18.0	18.0	18.0
15.09351	15.79997	15.81976	16.0	16.0	16.0
15.09351	14.53787	14.62054	16.0	16.0	16.0
10.00493	9.48963	9.51027	10.0	10.0	10.0
10.00493	9.41019	9.43213	10.0	10.0	10.0
0.0	0.00075	0.00078	0.0	0.0	0.0
0.0	0.0	0.0	0.0	0.0	0.0

(a) Eigenvalues are given for each κ relative to the lowest eigenvalue.

(b) Primitive WKB eigenvalues from Eqs. (2.8) and (2.9).

(c) Uniform semiclassical eigenvalues from Eq. (3.1).

(d) Quantum mechanical eigenvalues from H. C. Allen and P. C. Cross, Molecular Vib-Rotors, Wiley, New York, 1903.

SEMICLASSICAL CALCULATION OF VIBRATIONAL ENERGY LEVELS FOR NONSEPARABLE POTENTIALS

Randall T. Swimm and John B. Delos

Physics Department
College of William and Mary
Williamsburg, Virginia 23185

ABSTRACT

The energy levels for a nonseparable system of two coupled oscillators are calculated by transforming the classical Hamiltonian to the Birkhoff-Gustavson normal form, and then quantizing. The agreement with exact quantum levels is excellent.

I. INTRODUCTION

The semiclassical method is a widely recognized procedure for obtaining approximate energy levels for the Schrodinger equation. Recently, several authors have presented methods for applying the semiclassical method to the case of a multi-dimensional, nonseparable potential[1-4]. In this paper, we develop and apply a method for calculating vibrational energy levels by means of a transformation of the classical Hamiltonian to an approximate normal form, followed by semiclassical quantization.

The classical Hamiltonian for a set of coupled oscillators can be written in the form

$$H(p_k,q_k) = \sum_{k=1}^{N} \frac{1}{2m} p_k{}^2 + \frac{m\omega_k{}^2}{2} q_k{}^2 + V(q_1 \cdots q_N) \tag{1}$$

where it is assumed that $V(q_1 \cdots q_N)$ can be expanded in a power series in the variables $(q_1 \cdots q_N)$ and that this series contains no linear or quadratic terms (any such terms can be incorporated into the second degree part of H by the standard normal mode analysis). Many years ago, Birkhoff[5] proved that (provided the zero-order frequencies ω_n are incommensurable) any such Hamiltonian can, by a sequence of canonical transformations,

$$\left.\begin{matrix} p_k \\ q_k \\ H \end{matrix}\right\} \rightarrow \left\{\begin{matrix} P_k \\ Q_k \\ \Gamma \end{matrix}\right. \tag{2}$$

be put into a "normal form." In this normal form, Γ is a function only of one dimensional harmonic oscillator Hamiltonians; i.e., P_n and Q_n enter only in the combination

$$\frac{1}{2}\ (P_k{}^2 + Q_k{}^2) = h_k \tag{3}$$

and

$$\Gamma = \Gamma(h_1,\ldots,h_N) \tag{4}$$

At the time, this theorem was interesting because of what it seemed to show about the formal properties of dynamical systems: for a Hamiltonian of the form (4), the functions h_n provide N isolating integrals,[6] and the equations of motion are easily solved; thus, whenever the sequence of canonical transformations (2) converges, this theorem guarantees the existence of quasiperiodic motion.

Unfortunately, it was later shown that the sequence of transformations diverges "in general" (for the meaning of this, see Moser[7]). Nevertheless, numerical studies of Gustavson[8] later showed that a truncated sequence of normalizing transformations gives an accurate representation of the quasiperiodic trajectories that appear in the Hénon-Heiles system.[9] Gustavson[8] also provided a definition of a kind of normal form for systems with zero-order commensurabilities, which we will discuss later.

The Birkhoff-Gustavson normal form is particularly useful for the semiclassical quantization of non-separable systems: for a quantum mechanical Hamiltonian of the form (3) the energy levels are just $(n + \frac{1}{2})\hbar$, and so the energies associated with the full Hamiltonian (4) are obtained by replacing

$$\frac{1}{2}(P_k{}^2 + Q_k{}^2) \rightarrow (n_k + \frac{1}{2})\hbar \tag{5}$$

giving

$$E(n_1,\ldots,n_N) = \Gamma[(n_1 + \frac{1}{2})\hbar,\ (n_2 + \frac{1}{2})\hbar,\ldots,(n_N + \frac{1}{2})\hbar] \tag{6}$$

The above quantization prescription (5) is equivalent to the Bohr-Sommerfeld prescription, as can be shown by transforming the harmonic oscillator terms to action variables.[10] It is, therefore, reasonable to assume that the quantization of the normal form provides a semiclassical approximation to the eigenvalues in the regular part of the spectrum. (It is generally believed that there also exists an "irregular" part of the spectrum, in which the quantum states correspond to ergodic classical orbits; such orbits are not described by the classical normal form (4) and so we cannot expect that the quantized normal form (6) would describe the corresponding eigenvalues.)

In the following section, we consider a Hamiltonian for a pair of oscillators with a non-separable coupling; we transform this Hamiltonian to normal form, and quantize according to the prescription (6). The resulting eigenvalues are shown to be in excellent agreement with exact quantum eigenvalues that have been computed for us by Noid.[11]

II. TRANSFORMATION TO NORMAL FORM

Consider the Hamiltonian

$$H = \frac{1}{2}(p_1^2 + p_2^2 + \omega_1^2 q_1^2 + \omega_2^2 q_2^2) + \lambda q_2(q_1^2 + \eta q_2^2) \tag{7}$$

$$= H^{(2)} \qquad\qquad + H^{(3)} \tag{8}$$

where $H^{(2)}$ and $H^{(3)}$ are respectively the second and third degree parts of H. We assume that the frequencies ω_1, ω_2 are incommensurable, and we follow the method of Gustavson[8] to transform this to normal form.

We first normalize the second degree parts by the substitution

$$p_k \rightarrow \omega_k^{\frac{1}{2}} p_k \tag{9a}$$

$$q_k \rightarrow \omega_k^{-\frac{1}{2}} q_k \tag{9b}$$

giving

$$H^{(2)} = \sum_{k=1}^{2} \frac{1}{2}(p_k^2 + q_k^2)\, \omega_k \tag{10a}$$

$$H^{(3)} = \frac{\lambda}{\omega_1\omega_2^{\frac{1}{2}}} q_1^2 q_2 + \frac{\lambda\eta}{\omega_2^{3/2}} q_2^3 \tag{10b}$$

This Hamiltonian (10) can now be normalized by a sequence of canonical transformations; each transformation is defined by means of a type 2 generating function[10] $W^{(s)}(P,q)$ such that

$$Q = q + \partial W^{(s)}/\partial P \tag{11a}$$

$$p = P + \partial W^{(s)}/\partial q \tag{11b}$$

Beginning with s = 3, the transformations sequentially normalize the terms of degree s in the Hamiltonian; it is easy to show that this can be achieved by taking $W^{(s)}$ to be a homogeneous polynomial of degree s. The normalization of terms of degree s does not affect the terms of lower degree, but in general it modifies all terms of higher degree.

For the purposes of this section, we may write $H = H(p,q)$ and $\Gamma = \Gamma(P,Q)$, where Γ is the new Hamiltonian after the transformation. Then the equations of canonical transformation require

$$H(p,q) = H(P + \partial W^{(s)}/\partial q, q) = \Gamma(P, q + \partial W^{(s)}/\partial P) = \Gamma(P,Q) \tag{12}$$

Gustavson has shown that by separating H and Γ into degrees as in equation (8), expanding functions of $(P + \partial W^{(s)}/\partial q)$ in a Taylor series about P and functions of $(q + \partial W^{(s)}/\partial P)$ in a Taylor series about q, and finally collecting and equating all terms of equal degree, eq. (12) gives

$$H^{(i)}(P,q) = \Gamma^{(i)}(P,q) \qquad (i<s) \tag{13a}$$

$$D\, W^{(s)}(P,q) = \Gamma^{(s)}(P,q) - H^{(s)}(P,q) \qquad (i=s) \tag{13b}$$

$$\Gamma^{(i)}(P,q) = H^{(i)}(P,q) + \sum_j \frac{1}{j!}\left[\left(\frac{\partial W^{(s)}}{\partial q}\right)^j \frac{\partial^{|j|} H^{(\ell)}}{\partial P^j} - \left(\frac{\partial W^{(s)}}{\partial P}\right)^j \frac{\partial^{|j|}\Gamma^{(\ell)}}{\partial q^j}\right]$$

$$\ell-|j|+|j|(s-1)=i \qquad (i>s)$$
$$1\leq|j|<\ell<i$$
$$\ell\geq 2 \quad s\geq 3 \tag{13c}$$

where $j! = j_1!j_2!$. In these equations

$$H^{(i)}(P,q) = H^{(i)}(p,q)\big|_{p\,=\,P}$$

and

$$\Gamma^{(i)}(P,q) = \Gamma^{(i)}(P,Q)\big|_{Q\,=\,q}$$

as arises from the Taylor expansion. Thus $H^{(i)}(P,q)$ effectively means $H^{(i)}(p,q)$ with p replaced by P, and $\Gamma^{(i)}(P,q)$ means $\Gamma^{(i)}(P,Q)$ with Q replaced by q. In eq. (13b), the operator D is

$$D = \sum_{k=1}^{2} -\omega_k\left(q_k \frac{\partial}{\partial P_k} - P_k \frac{\partial}{\partial q_k}\right) \tag{14}$$

Equations (13a) show that the transformation of degree s leaves lower degree terms unaffected; (13b) is the equation to be solved for $W^{(s)}$, and when this is done (13c) gives the higher degree terms in the new Hamiltonian. (In this last equation, we use an abbreviated notation for terms in a multidimensional Taylor expansion.)

To solve eq. (13b) for $W^{(s)}$, we temporarily transform to variables in which D is diagonal: define

$$P_k = 2^{-\frac{1}{2}}(\eta_k + i\;\xi_k) \tag{15a}$$

$$q_k = i\;2^{-\frac{1}{2}}(\eta_k - i\;\xi_k) \tag{15b}$$

It follows that functions of the form $\eta_1^{\ell_1}\eta_2^{\ell_2}\xi_2^{m_1}\xi_2^{m_2} = \Phi_{\ell_1\ell_2 m_1 m_2}$ are eigenfunctions of D with eigenvalue $i\sum_k \omega_k(m_k-\ell_k)$; hence,

$$D^{-1}\,\Phi_{\ell_1\ell_2 m_1 m_2} = \left[i\sum_k \omega_k(m_k-\ell_k)\right]^{-1} \ell_1\ell_2 m_1 m_2 \tag{16}$$

Applying this operation to (13b) after the latter is transformed to variables η,ξ, we see that

$$W^{(s)} = D^{-1}\left[\Gamma^{(s)} - H^{(s)}\right] \tag{17}$$

Clearly, in order that $W^{(s)}$ be finite, $\Gamma^{(s)}$ must be chosen so as to exactly cancel any terms in $H^{(s)}$ that would give a vanishing demonimator in (16); that provides the definition of $\Gamma^{(s)}$. So long as the frequencies ω_k are incommensurable, the only terms that have to appear in $\Gamma^{(s)}$ are those for which $m_k = \ell_k$ for all k; those terms are

$$(i\,\eta_1\xi_1)^{m_1}\,(i\,\eta_2\xi_2)^{m_2} = \left[\tfrac{1}{2}(P_1{}^2 + q_1{}^2)\right]^{m_1}\left[\tfrac{1}{2}(P_2{}^2 + q_2{}^2)\right]^{m_2}. \qquad (18)$$

$\Gamma^{(s)}(P,Q)$ is then obtained by replacing q by Q, and we see that $\Gamma^{(s)}$ is a function only of h_1 and h_2. The desired normal form (4) is thus achieved up to (and including) terms of degree s.

Now that $\Gamma^{(s)}$ is defined, $W^{(s)}$ is obtained from (17) through the action of D^{-1} on the terms that remain after $\Gamma^{(s)}$ has cancelled troublesome terms in $H^{(s)}$. Then (13c) gives the higher degree terms in the new Hamiltonian. The process can then be repeated to normalize terms of degree s+1.

III. CALCULATIONS AND RESULTS

Although the method is straightforward in principle, even for a relatively simple Hamiltonian the algebraic manipulations involved can rapidly get out of hand. We found it most convenient to carry out the sequence of canonical transformations using MACSYMA, a computer language that permits analytical calculations. The basic calculation reported below required less than 4 minutes CPU time on a PDP 10.[12]

For the Hamiltonian (7), with parameters

$$\omega_1 = 1.3 \quad \lambda = -0.1$$
$$\omega_2 = 0.7 \quad \eta = 0.1$$

the quantized normal form is (up to harmonic oscillator terms of degree 8)

$$\begin{aligned}\Gamma((n_1+\tfrac{1}{2})\hbar,\ (n_2+\tfrac{1}{2})\hbar) = {} & 1.3(n_1+\tfrac{1}{2}) + 0.7(n_2+\tfrac{1}{2})\\
& - .0015\ 6185\ (n_2+\tfrac{1}{2})^2\\
& - .0102\ 3323\ (n_1+\tfrac{1}{2})(n_2+\tfrac{1}{2})\\
& - .0058\ 0199\ (n_1+\tfrac{1}{2})^2\\
& - .0000\ 1092\ (n_2+\tfrac{1}{2})^3\\
& - .0001\ 7719\ (n_1+\tfrac{1}{2})(n_2+\tfrac{1}{2})^2\\
& - .0002\ 7327\ (n_1+\tfrac{1}{2})^2(n_2+\tfrac{1}{2})\\
& - .0000\ 8625\ (n_1+\tfrac{1}{2})^3\\
& - .0000\ 0013\ (n_2+\tfrac{1}{2})^4\\
& - .0000\ 0529\ (n_1+\tfrac{1}{2})(n_2+\tfrac{1}{2})^3\\
& - .0000\ 1183\ (n_1+\tfrac{1}{2})^2(n_2+\tfrac{1}{2})^2\end{aligned}$$

$$- .0000\ 1071\ (n_1 + \tfrac{1}{2})^3 (n_2 + \tfrac{1}{2})$$
$$- .0000\ 0229\ (n_1 + \tfrac{1}{2})^4 \qquad (19)$$

Recall that this quantized normal form (degree 4 in the quantum numbers) is obtained from the Birkhoff-Gustavson normal form by letting

$$\tfrac{1}{2}(P_k^{\ 2} + Q_k^{\ 2}) \to (n_k + \tfrac{1}{2})\hbar$$

($\hbar$ = 1 here).

Table I compares the first 42 energy levels predicted by our method against quantum mechanical values calculated with a 20 x 20 (400 function) harmonic oscillator basis set.[11] The percent error in the correction to the unperturbed energy is listed. As shown, the correction to the unperturbed energy is accurate to within 2% for all levels listed here.

The transformation to normal form therefore provides an extremely powerful method for calculating semiclassical eigenvalues of nonseparable systems with no zero-order commensurabilities.

IV. COMMENSURABLE CASE

The demoninator of equation (16) may equal zero if the frequencies are commensurable. For example, if

$$\omega_1 = 2\ \omega_2$$

then

$$D\ \eta_1\xi_2^{\ 2} = \left[\omega_1(-1) + \omega_2(2)\right]\ \eta_1\xi_2^{\ 2} = 0$$

Therefore, when such terms appear in $H^{(s)}$, they must also be included in $\Gamma^{(s)}$ in order that they be cancelled before D^{-1} is allowed to operate in equation (17). Consequently, the final result for Γ will not be expressible as a function of 1-dimensional harmonic oscillator Hamiltonians. Nevertheless, it is possible to transform the normal form partially into action-angle variables. For one commensurability condition, it is found that the Hamiltonian can be written with all but one variable as an action variable. In the present case (two degrees of freedom), that means that the Hamiltonian can be written as

$$H = H(J;J',\omega')$$

where J is an action variable (therefore, a constant), but the canonical momentum J' is not, since its conjugate coordinate appears in the Hamiltonian. In this form, H may be considered to be a 1-dimensional Hamiltonian, although not a harmonic oscillator Hamiltonian. Since it is always possible to quantize a 1-dimensional bound state problem, semiclassical energy levels can be found in this case, too, although they may have to be found one at a time by numerical methods, instead of all at once by analytical methods, as was done in the incommensurable case. We are currently carrying out calculations for the commensurable case.

V. CONCLUSION

The Birkhoff-Gustavson normal form provides a very powerful way to calculate semiclassical energy levels for nonseparable quantum mechanical systems. In contrast with other methods, the normal form provides an analytical expression for the energy levels as a function of the quantum numbers; all the levels are therefore obtained in a single calculation instead of one at a time. Although the infinite sequence of transformations leading to the normal form is known to diverge, numerical calculations presented here and elsewhere[8] show that a finite sequence of transformations gives accurate approximations to the classical orbits and the quantum energy levels. It is assumed, however, that the quantized normal form describes only the "regular" part of the spectrum.

Present calculations have been limited to systems with no zero-order commensurability, but we are now extending the calculations to consider the commensurable case.

ACKNOWLEDGMENT

We thank E. A. Remler for many productive conversations. We thank Don Noid for generously performing the quantum mechanical calculations for the case we treated. We thank Carl Andersen for his assistance during the writing of the MACSYMA programs used for our calculations. The Mathlab group at the Laboratory for Computer Science, MIT, which is responsible for writing MACSYMA, is supported, in part, by the United States Energy Research and Development Administration under Contract Number E(11-1)-3070 and by the National Aeronautics and Space Administration under Grant NSG 1323. This research was supported in part by grants from the Research Corporation and from the National Science Foundation.

TABLE I. Energy Levels[a)]

Quantum Numbers n_1	n_2	$E_{present}$	$E^{b)}_{QM}$	$E^{c)}_{uncoupled}$	$\frac{E_{present}-E_{QM}}{E_{uncoupled}-E_{QM}}$
0	0	.9955	.9955	1.0	0
0	1	1.6870	1.6870	1.7	0
1	0	2.2782	2.2781	2.3	.5%
0	2	2.3750	2.3750	2.4	0
1	1	2.9584	2.9584	3.0	0
0	3	3.0595	3.0596	3.1	-.2%
2	0	3.5480	3.5479	3.6	.2%
1	2	3.6348	3.6347	3.7	.2%
0	4	3.7404	3.7404	3.8	0
2	1	4.2164	4.2162	4.3	.2%
1	3	4.3071	4.3069	4.4	.2%
0	5	4.4176	4.4176	4.5	0
3	0	4.8045	4.8043	4.9	.2%
2	2	4.8803	4.8799	5.0	.3%
1	4	4.9753	4.9749	5.1	.3%
0	6	5.0910	5.0909	5.2	.1%
3	1	5.4601	5.4597	5.6	.3%
2	3	5.5397	5.5390	5.7	.4%
1	5	5.6393	5.6385	5.8	.5%
0	7	5.7605	5.7601	5.9	.3%
4	0	6.0468	6.0463	6.2	.3%
3	2	6.1108	6.1099	6.3	.5%
2	4	6.1944	6.1931	6.4	.6%
1	6	6.2989	6.2975	6.5	.7%
0	8	6.4260	6.4253	6.6	.4%
4	1	6.6889	6.6878	6.9	.5%
3	3	6.7564	6.7546	7.0	.7%
2	5	6.8443	6.8419	7.1	.9%
1	7	6.9540	6.9515	7.2	1.0%
0	9	7.0873	7.0861	7.3	.6%
5	0	7.2742	7.2730	7.5	.5%
4	2	7.3255	7.3234	7.6	.8%
3	4	7.3967	7.3935	7.7	1.0%
2	6	7.4891	7.4850	7.8	1.3%
1	8	7.6044	7.6003	7.9	1.4%

0	10	7.7445	7.7428	8.0	.7%
5	1	7.9019	7.8996	8.2	.8%
4	3	7.9564	7.9524	8.3	1.2%
3	5	8.0315	8.0259	8.4	1.5%
2	7	8.1288	8.1221	8.5	1.8%
1	9	8.2500	8.2439	8.6	1.7%
0	11	8.3972	8.3968	8.7	.1%

a) For case with $\omega_1 = 1.3$, $\omega_2 = .7$, $\lambda = -.1$, $\eta = .1$.

b) Calculated by D. Noid.

c) i.e., $\lambda = 0$ case.

REFERENCES

1. M. C. Gutzwiller, J. Math, Phys. 8, 1979 (1967); 10, 1004 (1969); 11, 1791 (1970); 12, 343 (1971); 18, 806 (1977).
2. R. A. Marcus, Discussions Faraday Soc. 55, 34 (1973);
   W. Eastes and R. A. Marcus, J. Chem. Phys. 61, 4301 (1974);
   D. W. Noid and R. A. Marcus, J. Chem. Phys. 62, 2119 (1975); 67,559 (1977);
   D. W. Noid, M. L. Koszykowski and R. A. Marcus, J. Chem. Phys. 67, 404 (1977).
3. W. H. Miller, J. Chem. Phys. 63, 996 (1975); 64, 2880 (1976);
   S. Chapman, B. C. Garrett and W. H. Miller, J. Chem. Phys. 64, 502 (1976).
4. I. C. Percival, J. Phys. B6, L229 (1973); J. Phys. A7, 794 (1974);
   N. Pomphrey, J. Phys. B7, 1909 (1974);
   I. C. Percival and N. Pomphrey, Mol. Phys. 31, 87 (1976).
   See also M. V. Berry and M. Tabor, Proc. Roy. Soc. (London) A349, 101 (1976); J. Phys. A10, 371 (1977); K. S. Sorbie and N. C. Handy, Mol. Phys. 32, 1327 (1976); K. S. Sorbie, Mol. Phys. 32, 1577 (1976).
5. G. D. Birkhoff, Dynamical Systems, A.M.S. Colloquium Publications, New York, Vol. IX (1966).
6. A. Wintner, Analytical Foundations of Celestial Mechanics (Princeton University Press, Princeton, 1947) p. 96.
7. J. Moser, Lectures on Hamiltonian Systems, Memoirs Am. Math. Soc. 81, 1 (1968).
8. F. G. Gustavson, Astron. J. 71, 670 (1966).
9. H. Hénon and C. Heiles, Astron. J. 69, 73 (1964).
10. H. Goldstein, Classical Mechanics (Addison-Wesley Publishing Co., Inc., Reading, Massachusetts, 1965).
11. D. Noid, private communication.
12. Actually we did the calculation twice: once in single precision (~8 figures) and once in 1½ precision (~12 figures). The higher precision was used as a check of the results, and to obtain the very small coefficient of $(n_2 + \frac{1}{2})^4$ in

eq. (19). This calculation took 24 minutes of CPU time, plus 14 minutes of GC time ("garbage collection" time - an overhead time unique to LISP based languages). The single precision calculation gave essentially identical results, and required only four minutes of CPU time, plus 1 minute of GC time

CLASSICAL QUANTIZATION CONDITIONS FOR A DYNAMICAL SYSTEM WITH STOCHASTIC BEHAVIOR?

by

Martin C. Gutzwiller

IBM Research Center

Yorktown Heights, New York 10598

Abstract

The connection between classical and quantum mechanics is poorly understood when the classical trajectories show stochastic behavior. The author's method, based on Green's function, is reviewed. The importance of the periodic orbits for finding the (approximate) energy eigenvalues is emphasized. The anisotropic Kepler problem is discussed as an ergodic system where the (approximate) quantization can be carried out on the basis of classical mechanics.

0. Introduction

Many Hamiltonian systems in classical mechanics have been investigated in recent years in order to understand their qualitative behavior. Very often such systems have a well defined analog in quantum mechanics, particularly when there are only a few degrees of freedom. The following question can be asked: Is the knowledge of the classical trajectories of any use in order to approximate the eigenvalues of the energy for the corresponding quantum system?

A positive answer was proposed in the days of the "old quantum mechanics" on the assumption that certain convenient coordinate systems in phase space could be found through canonical transformations. In 1917, however, Einstein[1] asked whether such coordinate systems exist at all, independently of the task of constructing them. He came up with a very clear answer in two parts. A) If the trajectories lie on invariant tori of dimension not larger than the number of degrees of freedom, then the canonical transformations exist although they are really not needed to formulate the quantization conditions. The latter require that the action integrals $\oint pdq$ over any closed path on the torus which cannot be contracted to zero, have to equal a multiple of Planck's quantum B) If the trajectories have stochastic behavior, no set of simple quantization conditions can be formulated.

While statement A has been amply confirmed and exploited in the intervening 60 years, there has been virtually no effort to deal with statement B. The discovery of Schrodinger's equation in 1925 may have relieved the physicists from worrying about the case B. But our ignorance remains and we are unable to establish a connection between classical and quantal mechanics in the majority of practical cases. The author[2] has worked on this outstanding problem for some time, and the present report is meant to give a short survey over the steps by which he hopes to tackle the difficulties that Einstein was the first to recognize.

It is required to find an algorithm which uses only information from the classical mechanical system and leads to approximate values for the energy levels of the corresponding quantal system, even when the classical system has stochastic behavior. Such an algorithm has been devised by the author on the basis of certain heuristic arguments which will be confirmed eventually by more rigorous reasoning, it is hoped. The main points will be stated in the following pages in order to bring out quite clearly the path which the author has chosen to follow. For the purpose of organizing the whole scheme, it has been broken up, somewhat artificially, into seven steps. Each of these contains a certain amount of personal judgement and can be contested on various grounds.

1. Wavefunctions for stochastic systems?

It will be difficult to construct wave functions for a classical system with stochastic behavior. The propagator $K(q''q't)$, however, has a well defined semi-classical approximation $\tilde{K}(q''q't)$ even with stochastic trajectories. K is the solution of the time dependent Schrodinger equation with the inhomogeneous term $\delta(q''-q')\ \delta(t)$ and the initial condition $K = 0$ for $t<0$, q'' and q' are the final and the initial positions of the system, they are vectors with as many components as the system has degrees of freedom. The propagator K is the probability amplitude for the system to go from the position q' to the position q'' in the time t .

2. The test of the energy spectrum.

The quality of the quasiclassical approximation has to be judged by comparing the exact (quantum mechanical) spectrum with the approximate spectrum based on the classical mechanics of the system. Also, we are primarily interested in bound states. Therefore, one has to deal with Green's function

$$G(q''q'E) = \frac{1}{2\pi ih}\int_0^\infty dt\ K(q''q't)\ \exp\ (iEt/h) = \sum_j \frac{\psi_j(q'')\ \psi_j^*(q')}{E-E_j} \qquad (1)$$

where the summation goes over all eigenstates $\psi_j(q)$ of the system. E_j is the corresponding eigenvalue of the energy.

The classical approximation $\tilde{G}(q''q'E)$ will contain the classical behavior out to infinite times where the random character of the trajectories will be important. Nevertheless, it is hoped that G has poles as a function of E even if they are not located any longer on the real axis or if the poles are broadened into well defined branch cuts due to the stochastic behavior.

3. Path integrals as a guide.

Feynman's expression of K as a sum over all conceivable paths from q' to q'' seems ideally suited to understand the limit as Planck's quantum becomes small with respect to the characteristic scales of the system[3]. If this idea is carried out systematically, one ends up with the approximation $\tilde{K}$ as a summation over all classical trajectories from q' to q'' in the given time t . The details are not important at this point because we will immediately proceed to perform the Fourier transform (1) on $\tilde{K}$ in order to obtain $\tilde{G}$. The integral is carried out by the stationary phase method and yields a sum over all classical trajectories from q' to q'' at the given energy E .

Each term in the sum consists of an amplitude and a phase factor. The amplitude is easily interpreted in terms of the density of neighboring trajectories. The phase is simply the action integral $\int$pdq along the trajectory divided by h , and there is a phase loss of $\pi/2$ at each conjugate point along the way. There are, in general, infinitely many trajectories between given endpoints q' and q'' for a fixed negative energy (bound states). That is the reason why $\tilde{G}$ can be expected to have singularities for certain special values of E .

4. The approximate resonance function.

The task of summing over all classical trajectories between two given endpoints can be simplified if one renounces finding the wave functions in addition to the energy levels. If we set q' = q'' = q in (1) and integrate over q , we find the resonance function

$$g(E) = \int d^3q\, G(qqE) = \sum_j \frac{1}{E-E_j} , \qquad (2)$$

whose poles as a function of E give the eigenvalue of the energy.

This simplification can be carried out on $\tilde{G}$(q''q'E) in exactly the same way. In doing the integration over q , however, it easily recognized that the contribution of a particular trajectory contributes in leading order of h , if not only the initial and final positions, q' and q'' , but also the initial and final momenta, p' and p'' , agree (are equal). Moreover, it is important to distinguish between an ordinary periodic orbit (where q' = q'' and p' = p'') and a primitive periodic orbit. The latter can not be considered as a multiple pass through a simpler (shorter) periodic orbit. To each periodic orbit (po) belongs a primitive periodic orbit (ppo).

For simplicity's sake the formula for the approximate resonance function $\tilde{g}(E)$ will be written down only for unstable periodic orbits in two degrees of freedom, the simplest non-trivial Hamiltonian system with stochastic behavior. Each po has an action integral $S(E) = \oint pdq$ and a real stability index β, as well as a number of n of conjugate points. One also associates with each po a period T_o which is calculated along the corresponding ppo. With these notations one finds that

$$\tilde{g}(E) = -\frac{i}{h} \sum_{po} \frac{(\pm)T_o}{2\sinh\beta/2} \exp\left[\frac{i}{h} S(E)\right] , \tag{3}$$

where the upper sign holds for $n = 0$, 3 and 4, 7 and 8, ... and the lower sign holds for $n = 1$ and 2, 5 and 6, ...

If a particular po consists of m passes through the associated ppo, one can write $S(E) = m.S_o(E)$, $\beta = m\beta_o$, and $n = m.n_o$. Let us moreover assume that n_o is a multiple of 2, i.e., $n_o = 2\bar{n}$. Then we have

$$\tilde{g}(E) = \sum_{ppo} \frac{T_o}{ih} \sum_{m=1}^{\infty} \left[2\sinh(m\beta_o/2)\right]^{-1} \exp\left\{im\left[\frac{1}{h}S_o(E)-\bar{n}\pi\right]\right\} . \tag{4}$$

Each ppo is seen to contribute a series which involves only β_o, $S_o(E)$, and $\bar{n}$. The occurrence of 2 sinh $(m\beta_o/2)$ in the denominator can be related to the transverse degree of freedom[4] i.e., the one perpendicular to the particular po. Each sum over m is convergent, but it is not obvious whether or not the sum over ppo is convergent in any particular system, at least for certain energies E.

5. Periodic orbits and energy levels.

Formula (4) and its generalizations to stable orbits and more than two degrees of freedom are remarkable because they imply that the periodic orbits of a classical mechanical system determine by themselves the approximate energy spectrum. It has been argued occasionally that one primitive periodic orbit determines a series of approximate energy levels, because if we restrict the sum over ppo in (4) to one term, we still have the sum over m and its singularities. It should be emphasized that such an argument is misleading. It may, at best, serve to justify preliminary results when only one ppo of the system is known, but even then, it remains questionable.

The structure of periodic orbits in a domain of phase space with invariant tori is relatively simple in two degrees of freedom. One does well to distinguish between the case where a second integral of motion (e.g. angular momentum) exists besides the energy, and the case where such an additional integral is absent. In the former case[5], each po is characterized by two integers k and ℓ, the number of times the orbit goes around each of the topologically different paths on the torus before the trajectory closes smoothly. If the two integers have no common factor, the periodic orbit is primitive. Resonance occurs when po's with adjacent integers (like k and k+1, or ℓ and $\ell+1$) have their phases in the analog of (3) differing by multiples, κ and λ, of 2π. This condition determines the two integrals of motion like energy and angular momentum, and the quantum numbers are just the multiples, κ and λ, of 2π by which the phases of adjacent po's differ.

6. The number of periodic orbits.

It is not difficult to define for each po an integer L which can be considered as its length. If we use a surface of section and the Poincare map between successive intersections of the trajectories with this surface then each po is a fix-point of some power L of this map. It is then important to know how many po's have the given length L. In the case of a system with two integrals of motion (and two degrees of freedom), the answer follows from the preceding section: The number P(L) of po's with length L is proportional to L.

The structure of po's when the invariant tori form a Cantor set in phase space (i.e., when there is no second integral of motion), is not known at present, and no estimate of P(L) exists as far as the author is aware. On the other hand, some estimates have been made for completely stochastic systems. In particular it is easy to find P(L) in certain special systems where a complete enumeration of all po's is possible such as in the anisotropic Kepler problem below. It seems that for large L the function P(L) always increases exponentially, $P(L) \sim \exp(aL)$.

The summation in (3) is, therefore, much larger than in the case with two integrals of motion. There may be singularities for real values of E in spite of the "damping" due to the instability index β. In a sense, there is competition between the damping of the periodic orbits in a stochastic Hamiltonian system and their proliferating number.

7. An ideal example of stochastic behavior.

Hamiltonian systems with stochastic behavior are obviously difficult to understand in general terms. Therefore, it is imperative to have an explicit example which is thoroughly understood and where the ideas can be tested. The geodesics on a surface negative curvature satisfy such a need. They have been discussed since Hadamard's first paper on the subject in 1898,[6] but they are not satisfactory as an example of physical interest. Various special cases of the three-body problem have always eluded complete description of all phase space at a given energy. This leaves only the anisotropic Kepler problem which was first considered by the author in 1970, with more details given in 1973 and 1977.[7]

The Hamiltonian for the anisotropic Kepler problem is given in terms of normalized momenta u,v and normalized position coordinates x,y by

$$H = \frac{u^2}{2\mu} + \frac{v^2}{2\nu} - \frac{1}{\sqrt{x^2+y^2}} , \tag{5}$$

where $\mu\nu = 1$ and $\mu/\nu > 9/8$. Typically, $\mu \sim 5\nu$, which corresponds to the effective mass anisotropy in a silicon crystal. If u(t), v(t), x(t), y(t) is a solution of the equations of motion with the Hamiltonian H equal to the constant value $E < 0$, then $u'(t') = u(t)/\sqrt{-2E}$, $v'(t') = v(t)/\sqrt{-2E}$, $x'(t') = -2Ex(t)$, $y'(t') = - 2E\, y(t)$, and $t' = t/(-2E)^{3/2}$ is a solution of the equations of motion with the Hamiltonian H equal to $-1/2$. The enumeration of all the trajectories is, therefore, independent of the constant value of H , and we can set it equal to $-1/2$. The action integral for a particular po depends very simply on E since we have $S(E) = S(-1/2)/(-2E)^{1/2}$ just as in the ordinary Kepler problem. The stability indices and the number of conjugate points is independent of E .

A convenient surface of section is the x-axis. Therefore, we consider the consecutive pairs (u,x) for a particular trajectory every time $y = 0$. We have the inequality

$$|x| \left(1 + \frac{u^2}{\mu}\right) \leq 2 , \tag{6}$$

as a consequence of $y = 0$. This domain in the (u,x) plane can be mapped into a (U,X) plane by an area conserving, one-to-one and continuous, transformation

$$U = \sqrt{\mu}\ \mathrm{arc\ tg} \frac{u}{\sqrt{\mu}} , X = x \left(1 + \frac{u^2}{\mu}\right) , \tag{7}$$

where U and X are restricted to the rectangle $|U| \leq \sqrt{\mu\pi}/2$, $|X| \leq 2$. Any point in this rectangle is a possible initial condition for a trajectory which we will call (U_o, X_o) at $t = 0$.

The sequence of intersections with the x-axis is characterized by a sequence ...., (U_{-1}, X_{-1}), (U_o, X_o), (U_1, X_1), ..., of points in the rectangle. A binary sequence .... a_{-1} , a_o , a_1 , ... , is now defined through $a_j = \text{sign } X_j = \text{sign } x_j$. This binary sequence is mapped into two real numbers

$$\xi = \sum_{j=1}^{\infty} a_j \left(\frac{1}{2}\right)^j , \eta = \sum_{j=0}^{\infty} a_{-j} \left(\frac{1}{2}\right)^{j+1} , \tag{8}$$

which are all located in the square $-1 \leq \xi , \eta \leq +1$. Thus we have a mapping Q from the rectangle into the square.

Both the "rectangle"(7) and the "square"(8) have a natural topology since they are imbedded in a Euclidean space. With these topologies we now have the

Theorem: The mapping Q is one-to-one and continuous.

It is important to notice that our topology for the binary sequences is finer than the topology which is ordinarily chosen for binary sequences (Bernoulli sequences). Thus, the two sequences ++−−−−−−.... and +−++++++.... for $j > 0$ are considered identical if they have the same binaries a_j for $j \leq 0$. Physically, they both describe the same trajectory which collides with the origin of the (x,y) space at its intersection $j = 2$ with the x-axis.

The sequence ...., (U_{-1}, X_{-1}), (U_o, X_o) , (U_1, X_1), ... can be regarded as the result of an area conserving map P of the rectangle into itself. The corresponding map QPQ^{-1} of the square itself is the well known baker's transformation. It also conserves the measure. Therefore, we have two invariant measures, but they are not related by a differentiable transformation just as one would expect from an ergodic system.[8]

A periodic orbit is now given by a periodic binary sequence of even length (because the sign of the momentum v alternates at consecutive intersections with the x-axis). Only the two sequences with $a_j = +1$ for all j and with $a_j = -1$ for all j are not realized physically. A periodic binary sequence is characterized by a sequence of finite length which repeats itself. E.g., the finite sequence (.+−+−) determines a periodic orbit, but it is not primitive because it is the double pass of the po $(+-)$. The number $P(L)$ of po with length L is given by $P(L) = 2^L - 2$, where we now see an explicit example of the exponential growth in $P(L)$. It is clearly not easy to single out the primitive periods. Yet, this is what formulas (3) and (4) require.

The number of conjugate points turns out to equal the length L (which is an even number). It is tempting to transform (4) with the help of $T_o = dS_o/dE$ and the fact that β_o is fairly large, say $\sim L_o$, for the mass ratio 5. It seems legitimate to approximate $[2\sinh(m\beta_o/2)]^{-1} \sim \exp(-m\beta_o/2)$. Since β_o is independent of E we can write

$$\int_{-\infty}^{E} \tilde{g}(E)dE \cong \sum_{ppo} \log \frac{1-\exp\left[\frac{i}{h}S_o(E)-\frac{1}{2}(\beta_o+iL_o\pi)\right]}{1-\exp\left[-\frac{1}{2}(\beta_o+iL_o\pi)\right]} , \tag{9}$$

where the argument of the logarithm has been normalized so as to give zero in the limit $E \to -\infty$. The similarity with the so-called ζ-function which mathematicians have discussed in connection with abstract dynamical systems, is quite striking.[9] The theorems about the singularities of these ζ-functions may finally turn out to be statements about the approximate energy levels in Hamiltonian systems with stochastic behavior. For the time being, it would be important to evaluate either (3), (4), or (9) for the special case of the anisotropic Kepler problem.

References

1. A. Einstein, Verhandlugen der Deutschen Phsikalischen Gesellschaft **10**, 82-92 (1917).

2. M. C. Gutzwiller, J. of Math. Phys. **12**, 343-358 (1971); **14**, 139-152 (1973); **18**, 806-823 (1977).

3. R.P. Feynman and A.R. Hibbs, Quantum Mechanics and Path Integrals, McGraw Hill, New York, 1965.

4. W. H. Miller, J. Chem. Phys. **62**, 1899 (1975).

5. M. C. Gutzwiller, J. of Math. Phys. **10**, 1004-1020 (1969) and **11**, 1791-1806 (1970).

6. J. Hadamard, J. Math. Pure Appl. **4** , 27 (1898).

7. Cf. ref. 2.

8. Rufus Bowen, Equilibrium States and the Ergodic Theory of Anosor Diffemorphisms Springer-Verlag, Lecture Notes in Mathematics No. 470, 1975.

9. Cf. D. Ruelle, Bulletin Am. Math. Soc. **82**, 153 (1976).

SEMI-CLASSICAL ERGODICITY OF QUANTUM EIGENSTATES IN THE WIGNER REPRESENTATION

André Voros*
Service de Physique Théorique
Centre d'Etudes Nucléaires de Saclay
BP. n°2 - 91190 - Gif-sur-Yvette
FRANCE

ABSTRACT

We propose a new definition of ergodicity for a quantum system, pertaining to the semi-classical behavior of its energy eigenstates analyzed in the Wigner representation. This notion of ergodicity, stronger than the usual ones, leads to observable properties of the eigenfunctions. We discuss the possible existence (yet to proved) of such ergodic systems : we conjecture that "semi-classical ergodicity" is implied by the ergodicity of the limiting classical system found at $\hbar = 0$.

It has been known since Einstein that familiar semi-classical methods like the WKB method, Bohr-Sommerfeld quantization rules, ..., are inoperant to study the eigenstates and energy levels of a quantum system whose classical limit is ergodic [1,2]. Research on such systems has been reactivated rather recently and is far from being achieved, but the various studies ([1-7], and other contributions to this Conference) already suggest a few definite conclusions, which still remain to be mathematically confirmed.

Our analysis is based on the idea that the Wigner representation of quantum states [8] leads to the most powerful available semi-classical method (i.e. to which any other method can be reduced), so that it might remain useful in cases where other semi-classical methods have failed. Specifically, we shall use the expansion in $\hbar$ of the Wigner functions over a space of generalized functions, a method introduced in [5] and mathematically formalized in [6] (Berry [7] has also studied the Wigner function, but in a very different spirit). Here we shall emphasize the physical aspects of the problem and suppress the mathematical details.

Notations : let $Q = \mathbb{R}^\ell$ be the position space of a (non-relativistic, spinless) system with ℓ degrees of freedom, $P = \mathbb{R}^\ell$ be the momentum space ; $X = Q \oplus P$ is the classical phase space, and $\mathcal{H} = L^2(Q)$ is the representation space of the Heisenberg quantum commutation relations :
$[\hat{q}_j,\hat{p}_k] = i \hbar \delta_{jk} \mathbb{1}$.

---

* Member of CNRS

1. SEMI-CLASSICAL MECHANICS IN THE WIGNER FORMALISM

Semi-classical mechanics studies the behavior of quantum mechanics as a function of the parameter $\hbar > 0$ in the limiting transition to classical mechanics ($\hbar \to 0$). In particular it deals with the expansion of quantum mechanics in (formal) power series of $\hbar$.

To our knowledge, the most general framework for such semi-classical expansions is the Wigner formalism, in which an $\hbar$-dependent operator $A_\hbar$ acting on the Hilbert space $\mathcal{H}$, is represented by a function on phase space called its Wigner symbol [8] :

$$A_w(q,p)_\hbar = \int \langle q - \frac{r}{2} | A_\hbar | q + \frac{r}{2} \rangle \exp(ipr/\hbar)\, dr \; .$$

Usual manipulations on operators can be translated in terms of their symbols :

- conjugation : $(A^\dagger)_w = (A_w)^*$ (1.1)

- operator product (in power series of $\hbar$) :

$$(A_\hbar B_\hbar)_w\,(q,p)_\hbar = A_w(q,p)_\hbar \left[\exp \frac{i\hbar}{2}(\overleftarrow{\nabla}_q \overrightarrow{\nabla}_p - \overleftarrow{\nabla}_p \overrightarrow{\nabla}_q)\right] B_w(q,p)_\hbar \tag{1.2}$$

- Hilbert-Schmidt inner product :

$$\mathrm{Tr}\, B_\hbar^\dagger A_\hbar = (2\pi\hbar)^{-\ell} \int_X B_w^*(q,p)_\hbar\, A_w(q,p)_\hbar\, dq dp \; . \tag{1.3}$$

Wigner has shown that if the operator A is a quantum observable, the classical observable $A_{cl}(q,p) = \lim_{\hbar \to 0} A_w(q,p)_\hbar$ should be interpreted as the classical limit of A.

Considered as a linear form on observables, a quantum state is also represented by a (positive) operator ρ (such that $\rho(A) = \mathrm{Tr}\,\rho\, A$) : by Eq. (1.3), the limit $\rho_{cl}(q,p)$ as $\hbar \to 0$ of $\rho^w(q,p)_\hbar = (2\pi\hbar)^{-\ell}\, \rho_w(q,p)_\hbar$ considered as a measure on X, appears as the classical state corresponding to ρ (in all cases, the limits are assumed to exist).

Furthermore, we can define the semi-classical expansion of an operator $A_\hbar$ as the series expansion of its symbol around $\hbar = 0$:

$$A_w(q,p)_\hbar \sim \sum_{n=0}^{\infty} A_n(q,p)\, \hbar^n \qquad (A_o = A_{cl}) \; .$$

If we want more than a formal theory, we must impose that the coefficients A_n belong to an appropriate function space. "Nice" quantum observables (e.g. polynomials in the quantum operators $\hat{q}$ and $\hat{p} = -i\hbar \frac{\partial}{\partial q}$, or Hamiltonians $-\frac{\hbar^2}{2m}\Delta + V(\hat{q})$ with C^∞ potentials) can be expanded in terms of C^∞ functions $A_n(p,q)$: we refer to them as "admissible observables" (they are rigorously defined in [5-6]). To expand interesting quantum states however, apart from the different normalization suggested by (1.3) :

$$\rho^w(p,q)_\hbar = (2\pi\hbar)^{-\ell}\, \rho_w(q,p)_\hbar \sim \sum_{n=0}^{\infty} \rho_n(p,q)\, \hbar^n \tag{1.4}$$

(and $\rho_o = \rho_{cl}$), we need to enlarge considerably the space of allowed coefficients.

The example (ℓ = 1) of the coherent state at q = p = 0

$$\rho^{w}(q,p)_{\hbar} = (\pi\hbar)^{-1} \exp-(q^2+p^2)/\hbar \sim \sum_{n=0}^{\infty} \frac{\hbar^n}{4^n n!} \left(\frac{d^2}{dq^2} + \frac{d^2}{dp^2} \right)^n \delta(q)\ \delta(p)$$

shows that the semi-classical expansion is only possible in the sense of distributions. In [5] we have suggested to let the coefficients ρ_n be tempered distributions on the phase X ; we refer to such quantum states as "admissible states". The resulting semi-classical theory of quantum states is more general and more regular than the usual approach (where one tries to find a limiting form for the wave functions of a pure state, as in the WKB method).

2. THE ADMISSIBLE EIGENSTATE PROBLEM

We are now interested in the eigenstates of a given quantum Hamiltonian $H_{\hbar}$, i.e. the solutions in $\mathcal{H}$ of the equation :

$$H_{\hbar}\ \psi_{E,\hbar}(q) = E\ \psi_{E,\hbar}(q) \tag{2.1}$$

for real values E of the energy (we want the bound states, to fix the ideas).

A typical Schrödinger Hamiltonian, with non-singular potentials, is an admissible observable (indeed, its Wigner symbol coincides with the classical limit H_{cl} , and does not even depend on $\hbar$). The eigenstate problem then has a classical analog : a classical eigenstate of H_{cl} is a positive measure on phase space, carried by an energy surface $H_{cl}^{-1}(E)$, and invariant under the Hamiltonian motion (the Liouville flow of H_{cl} on X).

It is physically plausible (but by no means proved) that the quantum eigenstates of problem (2.1) tend to their classical analogs as $\hbar \to 0$. The simplest formulation of this conjecture is that the projector $\rho_{E,\hbar} = |\psi_E\rangle \langle\psi_E|$ (for any fixed E) is an admissible state as $\hbar \to 0$, i.e. admits a semi-classical expansion :

$$\rho_{E,\hbar} \sim \sum_{n=0}^{\infty} \rho_{E,n}(q,p)\ \hbar^n \qquad \left(\rho_{E,n} \in \mathcal{S}'(X)\right) . \tag{2.2}$$

Our problem is now to find the expansions (2.2) that satisfy the eigenprojector equations :

$$H_{\hbar}\rho_{E,\hbar} = \rho_{E,\hbar}\ H_{\hbar} = E\ \rho_{E,\hbar} \tag{2.3}$$

order by order in the symbols (operator products are to be evaluated using Eq.(1.2)). Thus, the lowest order term $\rho_{E,0}$ precisely satisfies the classical eigenstate equations at energy E.

Remarks : - if the quantum spectrum is discrete, the left hand side of (2.2) is defined only for the discrete values of $\hbar$ such that E is an eigenvalue of $H_{\hbar}$; nevertheless these values of $\hbar$ should accumulate around $\hbar$ = 0, thus making the right-hand side expansion possible.

- Eqs. (2.3) may have <u>several</u> admissible state solutions, describing distinct subfamilies of eigenstates as $\hbar \to 0$.

Our main result here is that one solution (2.2) to Eqs. (2.3) is explicitly known [6] ; it is the admissible state $\delta(H_{\hbar} - E)$, which is defined (and computed) by its semi-classical expansion :

$$\delta(H_{\hbar}-E)^{W} \sim (2\pi i)^{-1} \lim_{\varepsilon \to 0} \ [(H_{\hbar} -E-i\varepsilon)^{-1} - (H_{\hbar}-E+i\varepsilon)^{-1}]_{W} \tag{2.4}$$

the symbol in brackets being expanded in $\hbar$ <u>before</u> letting $\varepsilon \to 0$. The classical limit of (2.4) is $\delta(H_{cl}-E)$, the ergodic (Liouville) measure on $H_{cl}^{-1}(E)$; hence we call $\delta(H_{\hbar}-E)$ the <u>ergodic eigenstate</u> of H. The explicit evaluation of (2.4) yields :

$$\delta(H_{\hbar} - E)^{W} \sim \delta(H_{cl} - E) + \hbar^{2} \Big[\sum_{j,k} (\partial_{q_j}\partial_{p_k}H_{cl}.\partial_{p_j}\partial_{q_k}H_{cl} - \partial_{q_j}\partial_{q_k}H_{cl}.\partial_{p_j}\partial_{p_k}H_{cl}) \frac{\delta''(H_{cl}-E)}{8}$$

$$- \sum_{j,k} (\partial_{q_j}H_{cl}.\partial_{q_k}H_{cl}.\partial_{p_j}\partial_{q_k}H_{cl} - 2\partial_{q_j}H_{cl}.\partial_{p_k}H_{cl}.\partial_{q_k}\partial_{p_j}H_{cl} + \partial_{p_j}H_{cl}.\partial_{p_k}H_{cl}.\partial_{q_j}\partial_{q_k}H_{cl})$$

$$\times \frac{\delta'''(H_{cl}-E)}{24} \Big] + \mathcal{O}(\hbar^{4}) .$$

The interpretation of the ergodic eigenstate readily follows from Eq.(2.4) :

$$(2\pi\hbar)^{-\ell}\delta(H_{\hbar}-E) = \lim_{\varepsilon\to 0} \sum_{m} \frac{\varepsilon}{\pi[(E-E_m)^2+\varepsilon^2]} \cdot |\psi_{E_m}> <\psi_{E_m}| \quad . \tag{2.5}$$

This is the spectral measure of $H_{\hbar}$: $\sum_m |\psi_{E_m}> <\psi_{E_m}|\delta(E-E_m)$, averaged by a Lorentzian weighting function of width $\varepsilon \to 0$.

3. THE SEMI-CLASSICAL ERGODIC HYPOTHESIS

We propose here a new type of ergodic hypothesis, which concerns the semi-classical behavior of the quantum system under consideration, and provides a link betwen classical and quantum ergodic theories ; it precises certain related assumptions found in the recent litterature [1-3].

We recall that ergodicity of the classical Hamiltonian H_{cl} on the energy surface $H_{cl}^{-1}(E) \subset X$ (equality of time and phase averages) is equivalent to the statement that the only non-singular invariant measure on $H_{cl}^{-1}(E)$ is the Liouville measure $\delta(H_{cl}-E)$[9] (singular invariant measures may exist, but they are deemed unessential by the fact that the subset of the initial conditions, that generate them under time evolution, has measure zero). In the language of § 2, ergodicity implies that $\delta(H_{cl}-E)$ is the unique meaningful classical eigenstate. However the previous argument is deceptive, since the words "non-singular", "measure zero" ... hold with respect to the Liouville measure itself : so that the classical theory implicitly assumes for this measure a privileged status at the start.

On the other hand, quantum ergodicity for the Hamiltonian operator H, in von Neumann's sense, is equivalent to the statement that the energy spectrum is non-degenerate [9]. In contrast with the classical case, we restrict the dynamics much less, but we also get no prediction about the form of the eigenstates. Neither do we see a clear relation between quantum and classical ergodicity.

We propose the following new definition : if the quantum Hamiltonian $H_{\hbar}$ is an admissible observable (§2), we call it semi-classically ergodic [5a,6] if it has a non-degenerate spectrum (for all $\hbar > 0$), and if the equality :

$$|\psi_{E,\hbar}\rangle \langle\psi_{E,\hbar}|^{w} = \delta(H_{\hbar} - E)_{w} \tag{3.1}$$

holds in the sense of semi-classical expansions of the symbols, whenever E is an eigenvalue of $H_{\hbar}$ for the eigenfunction $\psi_{E,\hbar}$ (not normalized).

We can assess the relevance of our definition against the previous ones :

- to the usual quantum ergodicity condition for the admissible $H_{\hbar}$ we have only added the requirement that the equality (2.5) should hold sharply, i.e. with a smoothing width $\varepsilon = 0$. This is guaranteed iff the eigenprojectors $|\psi_E\rangle \langle\psi_E|$ behave as regular functions of the classical energy E (and of no other parameter) as $\hbar \to 0$ (this requirement is for instance violated when the system admits quantum constants of the motion besides the energy),

- in the classical limit, $\delta(H-E)_w$ tends to the Liouville measure $\delta(H_{cl} - E)$, which thus appears as the only physically realistic classical eigenstate, in contrast with the other (singular) eigenstates of H_{cl} . In that sense (not necessarily in Birkhoff's sense [9]), the classical system defined by H_{cl} also exhibits an ergodic behavior : it tends to occupy the whole energy surface in phase space.

Of course the whole difficulty now lies in proving that a given admissible Hamiltonian $H_{\hbar}$ is ergodic in our sense, and this is an unsolved problem for us, at present. But at least, this is a well-formulated question within the theory of pseudo-differential equations, in which we avoid the usual considerations about the physical measurement process to explain the observed ergodicity of the system.

4. DISCUSSION

Do there exist ergodic systems in our sense ? We shall conjecture here that in general, an admissible Hamiltonian $H_{\hbar}$ should be ergodic whenever its limit H_{cl} is classically ergodic. Our discussion will depend on the number ℓ of degrees of freedom :

- if $\ell = 1$: classical ergodicity means that the curve $H_{cl}^{-1}(E)$ is connected, i.e. consists of a single orbit. In that case the eigenfunctions of an arbitrary admissible $H_{\hbar}$ can be computed in the form of a WKB expansion $\psi_{E,\hbar}(q) = a(q,\hbar) \exp(iS(q)/\hbar)$. This allowed us to check, up to the order $\hbar^4$, that Eq.(3.1) is true [6] : it is then probably true to all orders (conversely, Eq.(3.1) provides a new way of computing the WKB eigenfunctions from the semi-classical expansion of $\delta(H_{\hbar}-E)_w$). We also note that the mathematical validation of the WKB method is well under way [10]. We can thus assert, not quite rigorously but with strong confidence, that one-dimensional systems are semi-classically ergodic, provided $H_{cl}^{-1}(E)$ is connected.

- if $\ell > 1$: the main obstacle here is that no semi-classical form is known for the eigenfunctions. The WKB method applies to integrable systems but certainly not to ergodic systems, since the classical limit of a WKB state is the Dirac distribution on a manifold of dimension ℓ (called Lagrangian manifold, or invariant torus), whereas we want to obtain the measure $\delta(H_{cl}-E)$ on the energy surface, which has dimension $2\ell - 1 > \ell$! More generally, we ignore of what analytic form in $\hbar$ could be a function $\psi_\hbar \in \mathcal{H}$ such that the measure : $\lim_{\hbar \to 0} (|\psi_\hbar\rangle \langle\psi_\hbar|^W)$ has a support of dimension $> \ell$! So we shall only list a few consequences of our conjecture, suggesting that they should be tested on examples in the future.

We first mention that if the conjecture was false, then Eq.(3.1) would hold only in the averaged form (2.5), meaning that eigenfunctions of neighboring energies must behave very differently from one another. Since, due to ergodicity, this could not be accounted for by additional contants of the motion, this situation would render meaningless any semi-classical analysis of individual eigenfunctions (in our language : the eigenstates of $H_\hbar$ would not be admissible).

Our conjecture expresses the hope that this pessimistic view is not realized. If semi-classical ergodicity holds, we obtain indeed a partial knowledge about the semi-classical behavior of the eigenfunctions $\psi_{E,\hbar}$. For instance :

- for any admissible observable $A_\hbar$, the expression $\langle\psi_{E,\hbar}|A_\hbar|\psi_{E,\hbar}\rangle$ has an expansion in $\hbar$ given by Eq.(1.3), with B replaced by $(2\pi\hbar)^\ell |\psi_{E,\hbar}\rangle\langle\psi_{E,\hbar}| = \delta(H_\hbar - E)$.

- integration of Eq.(3.1) in momentum space yields :

$$|\psi_{E,\hbar}(q)|^2 = \int_P \delta(H_\hbar - E)_w \, (q,p) \, dp \,, \tag{4.1}$$

hence the modulus of the eigenfunction is known, apart from its short wavelength oscillations, which are smoothed away in the semi-classical expansion of $\delta(H_\hbar-E)_w$. The irrelevance of the WKB method also suggests that the phase of the eigenfunction is the component that behaves very irregularly as $\hbar \to 0$.

- In the WKB case, we know that when the various complex branches of the wave function are superposed to form the real eigenfunction, it is the phase function that produces the short wavelength oscillations of $|\psi_{E,\hbar}|$ and that also determines the semi-classical spectrum of the operator $H_\hbar$ via single-valuedness conditions (the Bohr-Sommerfeld rules). A contrario, in the ergodic case, with no computable phase function, we should expect the short wavelength behaviour of $\psi_{E,\hbar}$ to be very irregular as $\hbar \to 0$, and the same for the spectrum. Berry has suggested that $\psi_{E,\hbar}$ is a gaussian random function [7] . The conjecture that the spectrum is very irregular and has no semi-classical limit, is otherwise due to Percival [1].

- the integration in (4.1) becomes singular at points of $H_{cl}^{-1}(\Xi)$ where $|\vec{\nabla}_p H_{cl}| = 0$, and in particular on the boundary surface between the classically allowed and forbidden regions. In the classical limit of (4.1), we found with Berry that $|\psi|$ behaves like $d^{(\ell-2)/4}$ as the distance d to this boundary tends to zero from the allowed side. For instance if $\ell = 2$, $|\psi|$ tends to a constant and if $\ell = 3$ it tends to zero, hence the name "anticaustics" suggested in [7] (in the separable case instead, the WKB eigenfunctions blow up at the caustics like $d^{-1/4}$). This curious effect should be observable, although the corrections in $\hbar$ should induce higher singularities.

- the autocorrelation function of $\psi_{E,\hbar}$, which is the average C(r;q) of $\psi_{E,\hbar}(q-\frac{r}{2})\,\psi^*_{E,\hbar}(q+\frac{r}{2})$ in a small region around the point q, also exhibits a semi-classical behaviour :

$$\langle\psi(q-\tfrac{r}{2})\,\psi^*(q+\tfrac{r}{2})\rangle \sim (2\pi\hbar)^{-\ell}\int_P \delta(H_{cl}(q,p)-E)\,\exp(-ipr/\hbar)\,dp$$

$$\Rightarrow C(r;q) = \frac{\langle\psi(q-\frac{r}{2})\psi^*(q+\frac{r}{2})\rangle}{\langle|\psi(q)|^2\rangle} \sim \int_P \delta(H_{cl}-E)\exp(-ipr/\hbar)\,dp \Big/ \int_P \delta(H_{cl}-E)\,dp. \quad (4.2)$$

For instance [7] , if $H_{cl} = \frac{p^2}{2m} + V(|q|)$:

$$C(r,q) = \Gamma\left(\frac{\ell}{2}\right)\frac{J_{\ell/2-1}(\xi)}{(\xi/2)^{\ell/2-1}} \quad \text{with } \xi = |r|\,\sqrt{2m(E-V(|q|))}/\hbar .$$

5. CONCLUSIONS

Although our work does not draw firm conclusions, we hope to have convinced the reader that certain well-defined statements about ergodic eigenstates should be, if not proved, at least tested on examples ; and this is not an easy task, since the semi-classical behavior only appears at large quantum numbers.

Several numerical investigations on model Hamiltonians already exist [1,3,4] and seem to confirm ergodic assumptions like ours (but less precise). We stress however that the corresponding classical Hamiltonians are of the quasi-integrable type, satisfying the Kolmogorov-Arnold-Moser theorem [11] : roughly speaking, the phase space splits into "regular" regions filled with invariant tori and "irregular regions" containing everywhere dense orbits [1]. Whereas several extensions of the WKB method seem to apply to quantum eigenstates associated with the regular regions [6b,12], the suggestion made in [1,3,4] and also in [2,7] is that the quantum eigenstates associated with the irregular regions should be ergodic in some sense. However, it is not even known if the classical motion in these regions is ergodic in Birkhoff's sense (it might be more irregular instead). Hence we feel that it would be worth testing our conjecture on a truly ergodic system such as the toric billiard [13]. A very encouraging result has been announced by Schnirelman [14] : "almost all" eigenstates of the Laplace-Beltrami operator on a compact riemannian manifold M are ergodic (in a sense close to ours) in the large eigenvalue limit, when the geodesic flow on the constant energy surface S^*M is ergodic.

Physical applications of these theories should be found in complex many-body (e.g. atomic or molecular) systems.

ACKNOWLEDGEMENTS

We are grateful to Profs. M. Berry, I. Percival and M. Tabor for very fruitful discussions at the Conference, which have notably influenced the present form of this text, and to Prof. Y. Pomeau for many enlightening discussions.

REFERENCES

1. I.C. Percival, J. Phys. B6 (Letters) (1973), L229 ; Adv. Chem. Phys. 36 (1977),1.
2. M. Berry, Phil. Trans. Roy. Soc. A287 (1977), 237.
3. K. Nordholm, S. Rice, J. Chem. Phys. 61 (1974), 203 and 768.
4. N. Pomphrey, J. Phys. B7 (1974), 1909 ; I.C. Percival and N. Pomphrey, Molec. Phys. 31 (1976), 97.
5. A. Voros, a) Comptes-Rendus RCP 25 vol. 22 (1975), 142 (IRMA, University of Strasbourg) ; b) Ann. Inst. H. Poincaré 24A (1976), 31.
6. A. Voros, a) Ann. Inst. H. Poincaré 26A (1977), 343 ; Thèse d'Etat, University of Paris-Sud, Orsay (May 1977).
7. M. Berry, J. Phys. A10 (1977),2083.
8. E. Wigner, Phys. Rev. 40 (1932), 749.
9. I. Farquhar, Ergodic Theory in Statistical Mechanics, Interscience (1964).
10. V.P. Maslov, Théorie des Perturbations et Méthodes Asymptotiques, Dunod (1972) ; J.P. Eckmann, R. Seneor, Arch. Rational Mechanics 61 (1976), 153.
11. A. Kolmogorov, Dokl. Ak. Nauk. 98 (1954), 527 ; V. Arnold, Usp. Mat. Nauk.18-5 (1963), 13 ; J. Moser, Nachr. Ak. Wiss. Göttingen 1 (1962), 1.
12. M. Gutzwiller, J. Math. Phys. 12 (1971), 343 ; W. Eastes and R. Marcus, J. Chem. Phys. 61 (1974), 4301 ; D. Noid and R. Marcus, J. Chem. Phys. 62 (1975), 2119 ; A. Voros, in : Géométrie Symplectique et Physique Mathématique (CNRS, Paris 1975) ; I. Percival and N. Pomphrey, J. Phys. B9 (1976), 3131.
13. V. Arnold, A. Avez, Problèmes Ergodiques de la Mécanique Classique, Gauthier-Villars, 1967 (English translation : Benjamin, 1968) ; J. Sinai, Usp. Mat. Nauk 30-1 (1966), 15 ; G. Gallavotti, in : Dynamical Systems (Theory and Apllications), Springer Lecture Notes in Physics, Vol. 38 (1975).
14. A. Schnirelman, Usp. Mat. Nauk 29-6 (1974), 181.

STOCHASTIC BEHAVIOR OF A QUANTUM PENDULUM UNDER A PERIODIC PERTURBATION

G. Casati

Istituto di Fisica, Via Celoria 16, Milano, Italy

and

B. V. Chirikov and F. M. Izraelev

Institute of Nuclear Physics, 630090 Novosibirsk 90, U.S.S.R.

and

Joseph Ford

School of Physics, Georgia Institute of Technology
Atlanta, Georgia 30332, U.S.A.

ABSTRACT

This paper discusses a numerical technique for computing the quantum solutions of a driven pendulum governed by the Hamiltonian

$$H = (p_\theta{}^2/2m\ell^2) - [m\ell^2\omega_o{}^2\cos\theta]\ \delta_p(t/T),$$

where p_θ is angular momentum, θ is angular displacement, m is pendulum mass, ℓ is pendulum length, $\omega_o{}^2 = g/\ell$ is the small displacement natural frequency, and where $\delta_p(t/T)$ is a periodic delta function of period T. The virtue of this rather singular Hamiltonian system is that both its classical and quantum equations of motion can be reduced to mappings which can be iterated numerically and that, under suitable circumstances, the motion for this system can be wildly chaotic. Indeed, the classical version of this model is known to exhibit certain types of stochastic behavior, and we here seek to verify that similar behavior occurs in the quantum description. In particular, we present evidence that the quantum motion can yield a linear (diffusive-like) growth of average pendulum energy with time and an angular momentum probability distribution which is a time-dependent Gaussian just as does the classical motion. However, there are several surprising distinctions between the classical and quantum motions which are discussed herein.

Since we have not yet developed a completely adequate explanation for all these distinctions, this paper should be regarded as a progress report describing work on a highly interesting and numerically solvable model.

I. INTRODUCTION

Noteworthy recent progress[1-3] has been made in illustrating the truly stochastic behavior which can occur for the strictly deterministic systems of classical mechanics. Indeed, quite simple classical Hamiltonian systems can exhibit precise phase space trajectories so chaotic in their phase space wanderings that slightly imperfect observation cannot distinguish this deterministic motion from completely stochastic motion. Even though some of these model systems have only a few degrees of freedom, they nonetheless illustrate a generic type of wild classical mechanical trajectory behavior which is of great significance in the study of dynamical stability and of statistical mechanics; moreover, the nature of such wild behavior is now being studied for more general systems of widespread physical interest. Thus, even at this early stage, it becomes highly desirable to establish the effect introduced by quantum mechanics on this classical mechanical stochastic behavior. For even more than in classical mechanics, most of the exact work in quantum mechanics has been devoted to integrable (solvable) systems, with the remaining more difficult problems being left either to generally divergent perturbation theory or to quantum statistical mechanics whose foundations are less well understood than those of classical statistical mechanics. As a consequence, there is not only a dearth of quantum models exhibiting stochastic behavior, there is also some ambiguity even concerning the criteria for and the definition of quantum stochastic behavior.

Thus a number of recent papers have appeared, using a variety of techniques, which seek to develop viable criteria for and a definition of quantum stochastic behavior. To establish their criterion, Pukhov, et. al.,[4] generalize the notion of local instability (exponential separation) of initially close orbits valid for classical stochastic systems. In particular, they investigate the time growth of the change in the wave function $\delta\psi$ due to a small external perturbation added to the original Hamiltonian, and they establish criteria sufficient to insure an exponential growth in $\delta\psi$. Along somewhat related lines, Percival and Pomphrey[5] treat quantum mechanically a Hamiltonian system known to exhibit a classical transition from near-integrable to stochastic behavior. They show that the quantum energy levels for this Hamiltonian become highly sensitive to very small added external perturbations only above the classical stochastic threshold, and they argue that their study thus reveals a transition to quantum stochastic behavior. Both of these studies investigate quantum stochasticity as revealed through small but explicit outside perturbations.

The works of Nordholm and Rice[6,7] and of Shuryak[8] are more directly concerned with the case of isolated Hamiltonian systems for which the Hamiltonian splits naturally into dominant terms describing independent degrees of freedom which are coupled by small nonlinear interaction terms. Nordholm and Rice[6] also present a brief but clear review of the older literature on quantum ergodic theory for isolated systems, pointing out that this problem is still very much an open one. Both Nordholm and Rice as well as Shuryak regard their isolated quantum systems as behaving stochastically when the unperturbed eigenstates of the dominant, independent modes are strongly coupled by the weak interaction in the sense that the expansion of the exact quantum state is a sum of many unperturbed, independent mode states. The approach of Nordholm and Rice is, in principle, exact but in practice is forced to rely heavily on numerical (computer) computations whereas the work of Shuryak is strictly analytic but only approximate since it is based on a generalization of the order of magnitude resonance-overlap estimates of Chirikov.[3] For further details on work in this area, the reader is referred to the papers listed as Reference 9. Additional discussions of quantum stochasticity appear, of course, in several of the companion papers in this volume.

But strangely enough, to our knowledge, no quantum investigation has previously been made for the simplest possible Hamiltonian system known to exhibit chaotic trajectories, namely a periodically driven (i.e., time-dependent) one degree of freedom Hamiltonian system. Perhaps this is because such driven systems are, in general, no easier to solve than the conservative two degree of freedom systems whose quantum behavior has been previously studied. However, there is one notable exception to this general rule, and it is this exception which we seek to exploit in this paper. In particular, both the classical and quantum equations of motion for pendulum Hamiltonians of the type

$$H = (p_\theta/2m\ell^2) + V(\theta)\ \delta_p(t/T)\ , \tag{1}$$

where θ is angular displacement, p_θ is angular momentum, m is pendulum mass, ℓ is pendulum length, $V(\theta)$ is angular potential energy, and

$\delta_p(t/T)$ is a periodic delta function of period T, can be reduced to mappings and solved numerically as we shall show. Indeed, the classical motion for Hamiltonian (1) using $V(\theta) = -mg\ell\cos\theta = -m\ell^2\omega_o^2\cos\theta$ has already been extensively investigated,[3] and it is for this reason that we chose to begin our quantum studies using this particular model.

In Section II, we derive the classical mapping equations for our model and discuss the nature of its solutions. In Section III, we derive and discuss the quantum mapping equations. Section IV presents our numerical results for the quantum problem and these results are then discussed in Section V. Brief concluding remarks appear in Section VI. This paper represents a progress report describing calculations on an exceptional type of driven quantum system which can be solved, at least numerically. It is our hope that future studies on this or related model systems may lead to a broader understanding of quantum chaotic behavior.

II. DISCUSSION OF THE CLASSICAL MODEL

The specific Hamiltonian we choose to study is given by

$$H = (p_\theta/2m\ell^2) - [m\ell^2\omega_o^2\cos\theta]\ \delta_p(t/T) \ , \tag{2}$$

which is merely Hamiltonian (1) specialized to $V(\theta) = -mg\ell\cos\theta = -m\ell^2\omega_o^2\cos\theta$, where obviously $\omega_o^2 = g/\ell$. Here the periodic delta function, which may be expressed as

$$\delta_p(t/T) = \sum_{j=-\infty}^{\infty} \delta[j-(t/T)] = 1+2\sum_{n=1}^{\infty}\cos(2n\pi t/T) \ , \tag{3}$$

"turns on" the gravitational potential for a brief instant during each period T. The classical equations of motion for Hamiltonian (2) are

$$\dot{p}_\theta = -(m\ell^2\omega_o^2\sin\theta)\ \delta_p(t/T) \tag{4a}$$

$$\dot{\theta} = p_\theta/m\ell^2 \tag{4b}$$

where the dot notation indicates time derivative. Letting θ_n and $(p_\theta)_n$ be the values of θ and p_θ just before the nth delta function "kick", we may integrate Eq. (4a,b) to obtain the mapping equations

$$(p_\theta)_{n+1} = (p_\theta)_n - m\ell^2\omega_o^2 T\sin\theta_n \ , \tag{5a}$$

$$\theta_{n+1} = \theta_n + (p_\theta)_{n+1}T/m\ell^2 \quad . \tag{5b}$$

As an aside, let us observe that, when T tends to zero allowing us to replace T by dt, Eq. (5a,b) becomes

$$dp_\theta = -(m\ell^2\omega_o{}^2\sin\theta)\ dt \tag{6a}$$

$$d\theta = (p_\theta/m\ell^2)\ dt \quad , \tag{6b}$$

which are precisely the equations of motion for the conservative, gravitational pendulum Hamiltonian

$$H = (p_\theta{}^2/2m\ell^2) - mg\ell\cos\theta \tag{7}$$

recalling that $\omega_o{}^2 = g/\ell$. Thus as the time T between delta function "kicks" tends to zero, the gravitational potential is "turned on" continuously and the motion generated by Hamiltonian (2) becomes identical with that generated by the integrable[1] Hamiltonian (7). As T increases away from zero, the orbits of Hamiltonian (2) increasingly deviate from those of the integrable pendulum, eventually exhibiting chaotic or stochastic behavior. Consequently, one may adopt either of two viewpoints regarding Hamiltonian (2). As written above in Eq. (2), it obviously describes a free rotator perturbed by delta function "kicks". The remarks of this paragraph show that Hamiltonian (2) may also be regarded as describing a gravitational pendulum perturbed by a periodic driving force. This latter viewpoint becomes clearer if we rewrite Hamiltonian (2), using Eq. (3), as

$$H = [(p_\theta/2m\ell^2) - mg\ell\cos\theta] - 2mg\ell\cos\theta \sum_{n=1}^{\infty} \cos(2n\pi t/T) \tag{8}$$

In particular, it is the viewpoint represented by Hamiltonian (8) that led to the title of this paper.

Returning to Eq. (5a,b), let us now define the dimensionless angular momentum P_n via the equation $P_n = (p_\theta)_n T/m\ell^2$, and then let us write Eq. (5a,b) in the so-called standard form [3]

$$P_{n+1} = P_n - K\sin\theta_n \tag{9a}$$

$$\theta_{n+1} = \theta_n + P_{n+1} \quad , \tag{9b}$$

where $K = (\omega_o T)^2$ is the only remaining mapping parameter. As mentioned earlier when K tends to zero, this mapping becomes precisely integrable with all orbits lying on (or forming) simple invariant curves of the mapping. When K is small, the KAM theorem[2] insures that the mapping is near-integrable, meaning here that most mapping orbits continue to lie on simple invariant curves. Numerical computations[3,10] show that at least some of these simple invariant curves persist as K approaches unity. Moreover, for the K region $0<K<1$, the momentum variation is bounded with $|\Delta P| \approx K^{1/2}$. As K reaches and exceeds unity, all the previously existing simple invariant curves completely disappear and most mapping orbits become chaotic point sets. For $K>>1$, numerical evidence shows that the P-motion is characterized by a simple, random walk diffusion equation having the form

$$\overline{P^2} \approx (K^2/2)t \ , \tag{10}$$

where $\overline{P^2}$ is the average of the squared angular momentum at integer time $t = n$ measured in units of the period T and where initially P is taken to be zero. The average here is over many orbits having distinct θ_o initial conditions or over many sections of the same orbit (normalizing P to zero at the beginning of each segment). In order to indicate the source of Eq. (10), let us note that from Eq. (9a) we may obtain

$$(P_n - P_o)^2 = K^2 \sum_{j,k=o}^{(n-1)} (\sin\theta_j \sin\theta_k) \ . \tag{11}$$

Averaging Eq. (11) over θ_j and θ_k, taking both to be uniformly distributed random variables, yields Eq. (10); this procedure is validated by numerical iteration of chaotic orbits. But in addition to Eq. (10), the empirically observed angular momentum distribution itself has the time-dependent Gaussian form

$$f(P) = [K(\pi t)^{1/2}]^{-1} \exp[-P^2/K^2 t] \ , \tag{12}$$

as would be expected from the central limit theorem provided we regard P_n in the equation $P_n = P_o - K \sum_{j=o}^{(n-1)} \sin\theta_j$ as being a sum of random variables. It is precisely this stochastic momentum (or energy) diffusion we shall seek to empirically observe in the quantum description of this model to which we now turn.

III. DISCUSSION OF THE QUANTUM MODEL

We now seek to solve Schrodinger's equation

$$H\psi(\theta,t) = i\hbar\, \partial\psi(\theta,t)/\partial t \tag{13}$$

for the system governed by Hamiltonian (2). In this section, we regard the system as a free rotator perturbed by delta function "kicks" and we expand the wave function $\psi(\theta,t)$ in terms of the free rotator eigenfunctions $(2\pi)^{-1}e^{in\theta}$. In particular, we write

$$\psi(\theta,t) = (2\pi)^{-1}\sum_{n=-\infty}^{\infty} A_n(t)e^{in\theta} . \tag{14}$$

Over any period T between delta function "kicks", the $A_n(t)$ evolve according to

$$A_n(t+T) = A_n(t)e^{-E_nT/\hbar} = A_n(t)e^{-in^2\tau/2} , \tag{15}$$

where

$$E_n = n^2\hbar^2/2m\ell^2 \tag{16}$$

are the free rotator energy eigenvalues and where

$$\tau = \hbar\, T/m\ell^2 . \tag{17}$$

During the infinitesimal time interval of a "kick", we may write Eq. (13) as

$$i\hbar\partial\psi/\partial t = -m\ell^2\omega_o^2\cos\theta\,\delta_p(t/T)\psi . \tag{18}$$

Integrating Eq. (18) over the infinitesimal interval (t+T) to $(t+T^+)$, we find

$$\psi(\theta,t+T^+) = \psi(\theta,t+T)\, e^{ik\cos\theta} , \tag{19}$$

where

$$k = (m\ell^2\omega_o^2T)/\hbar) . \tag{20}$$

Now expanding both sides of Eq. (19) in free rotator eigenfunctions yields

$$\sum_{n=-\infty}^{\infty} A_n(t+T^+)\ e^{in\theta} = \sum_{r,s=-\infty}^{\infty} A_r(t+T)b_s(k)\ e^{i(r+s)\theta}, \tag{21}$$

where we have used the expansion

$$e^{ik\cos\theta} = \sum_{s=-\infty}^{\infty} b_s(k)\ e^{is\theta} \tag{22}$$

with $b_s(k) = i^s J_s(k) = b_{-s}(k)$ and with $J_s(k)$ being the ordinary Bessel function of the first kind. Because of the orthogonality of the $e^{in\theta}$, we may use Eq. (21) to establish that

$$A_n(t+T^+) = \sum_{r=-\infty}^{\infty} A_r(t+T)b_{n-r}(k) \quad . \tag{23}$$

Finally using Eq. (15), we obtain the quantum mapping

$$A_n(t+T^+) = \sum_{r=-\infty}^{\infty} A_r(t)b_{n-r}(k)\ e^{-ir^2\tau/2} \tag{24}$$

giving the A_n at time $(t+T^+)$ in terms of the A_n at time t. Using Eq. (24) and Eq. (14), we could obtain $\psi(\theta,t+T^+)$ in terms of $\psi(\theta,t)$, but in the calculations of interest here, the A_n momentum-representation is more useful. Before turning to these calculations however, let us briefly discuss the quantum mapping of Eq. (24).

First, let us note that the mapping of Eq. (24) can actually be numerically iterated to obtain the time development of a quantum solution provided only one or a few of the initial $A_n(0)$ are non-zero. Even though the sum on r is infinite, the $b_s(k) = i^s J_s(k)$ coefficients become negligible[11] outside the range $s \lesssim 2k$. Thus for reasonable sized k-values starting from only one (or a few) initially non-zero $A_n(0)$, we may iterate through many periods T before the increasing number of non-zero $A_n(t)$ exceeds the practical limitations of a large computer. Moreover, accuracy can be monitored via the normalization condition $\Sigma\ |A_n|^2 = 1$. Next, let us observe that the classical mapping depends on the single parameter $K = (\omega_o T)^2$. Thus one might, at first glance, expect the quantum mapping to depend only on the product $(k\tau)$ since $K = (k\tau)$ as one may immediately verify from Eq. (17,20). Indeed Ehrenfest's theorem makes it reasonable to suppose that the product $K = k\tau$ is particularly significant since it would control the time evolution of the center of a wave packet; however, the additional quantum spreading of the packet might not depend on this product alone. Not only does the numerical evidence presented later indicate that the quantum mapping does indeed depend separately on k and τ, but examination of Eq. (24) itself shows that this must be the case. First, independent of the value of k, Eq. (24) is invariant to the replacement of τ by $(\tau+4\pi)$; thus contrary to the classical case, the quantum motion places an upper bound on τ (or T), since in Eq. (24) there is no loss of generality in restricting τ to the interval $0\leqslant\tau\leqslant4\pi$. This quantum anomaly arises because the wave function for the unperturbed free rotator is periodic in τ independent of initial conditions. Also in Eq. (24) when $k<1$, only one or a very few $b_s(k)$ coefficients will be appreciably different from zero. For both these reasons, one would expect the classical and quantum motion to differ greatly when K is large due to a small k and a large τ. This is only one of a variety of perhaps interrelated classical-quantum distinctions. Finally since from Eq. (19) it is clear that k is the parameter which controls the amount of energy absorbed by the rotator due to the driving delta function "kicks", it may be worthwhile to use Eq. (20) and write Hamiltonian (2) as

$$H = (p_\theta/2m\ell^2) - (\hbar k T^{-1}\cos\theta)\ \delta_p(t/T)\ , \tag{25}$$

thus revealing k as an explicit coupling parameter.

IV. NUMERICAL RESULTS FOR THE QUANTUM MODEL

For the quantum model discussed in Sec. III, we have numerically iterated Eq. (24) to obtain the time evolved $\{A_n(t)\}$ starting from various initial $\{A_n(0)\}$ sets. For each computer run, starting with a definite $\{A_n(0)\}$ set, we calculated at each iteration the probability distribution

$$\rho(n) = |A_n|^2\ , \tag{26}$$

the average energy (in units of $\hbar^2/m\ell^2$)

$$\langle E\rangle = \sum_n (n^2/2)\ \rho(n)\ , \tag{27}$$

and the average angular momentum (in units of $\hbar$)

$$\langle p_\theta \rangle = \sum_n n\rho(n) \quad . \tag{28}$$

In all runs after fixing k and τ, we chose only one $A_n(0)$ or a few ($\sim$10) adjacent $A_n(0)$ to be non-zero. Surprisingly, the computed final state $\rho(n)$ appeared to be independent of the precise initial state, a point to which we return later. Typical results for four runs are discussed below. Each of these four runs was started with only the ground state free rotator $A_o(0)$ being non-zero.

In order to investigate the extent to which the numerically computed quantum distribution $\rho(n)$ of Eq. (26) mimics the classical stochastic distribution of Eq. (12), let us write Eq. (12) in terms of the quantum variables. Using the definitions of τ, P, and K and the fact that $p_\theta = n\hbar$, we may write the quantum version of Eq. (12) as

$$f(n) = [k(\pi t)^{1/2}]^{-1} \exp(-n^2/k^2 t) \tag{29}$$

where t is integer time measured in multiples of the "kick" period T. Note in Eq. (29) the dependence on k rather than K ($=k\tau$). If the quantum system indeed mimics the classical motion for $K = k\tau >> 1$, then we would expect the $\rho(n)$ of Eq. (26) to equal the f(n) of Eq. (29), since loosely speaking the quantum solution is an "automatic" average over many classical orbits. For ease of graphical comparison, let us introduce the normalized variables $X = n^2/k^2t$, $f_N(n) = k(\pi t)^{1/2} f(n)$, and $\rho_N(n) = k(\pi t)^{1/2}\rho(n)$; in these variables we need only determine the validity of the simple equation

$$\rho_N(n) = f_N(n) = e^{-X} \quad . \tag{30}$$

For k = 40 and $\tau = 1/8$, we plot the numerically determined ($\ln\rho_N$) versus X at t = 25 in Fig. 1. The straight line $\ln(e^{-X}) = \ln f_N(n)$ is graphed for comparison. One notes here that the comparison is rather good indicating that the quantum motion indeed appears to be stochastic here. It must be noted in Fig. 1 that each plotted point for $\ln\rho_N(n)$ actually represents an average value of this quantity taken over ten adjacent energy levels; this averaging reduces but does not eliminate fluctuations. In Fig. 2 we present a graph of $\rho_N(n)$ and $f_N(n)$ for the parameter values k = 10 and $\tau = 1/2$. In Fig. 2, contrary to Fig. 1, one notes that the quantum system is not behaving stochastically despite the fact that the classical value of K is five for both cases and, classically, stochasticity would be expected for both. This is yet another of the several

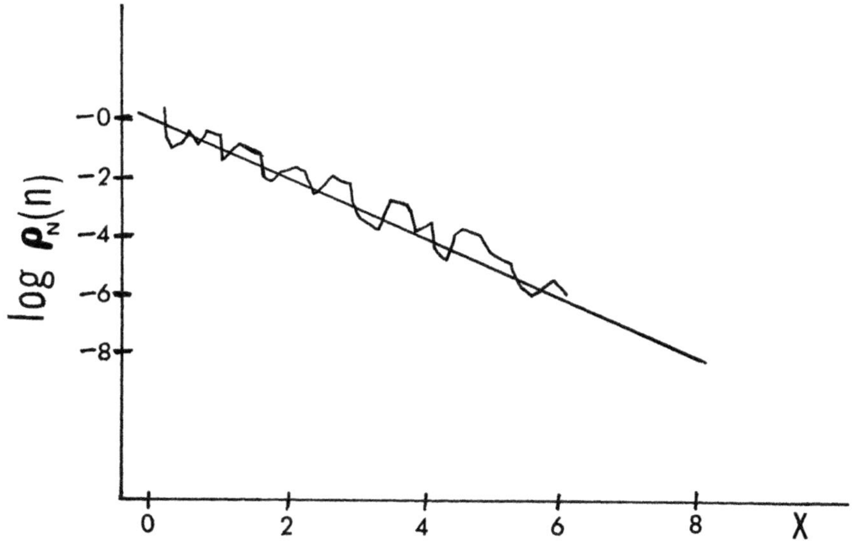

Fig. 1. A plot of the logarithm of the quantum probability distribution $\rho(n)$ versus the normalized variable $X = n^2/k^2t$ at time $t = 25$ for $k = 40$ and $\tau = 1/8$ ($K = 5$). The straight line is a plot of $\ell n(e^{-X})$. Were the quantum motion stochastic, these two curves should be identical; the fact that they are quite close indicates a great similarity between the classical and quantum motion for this case.

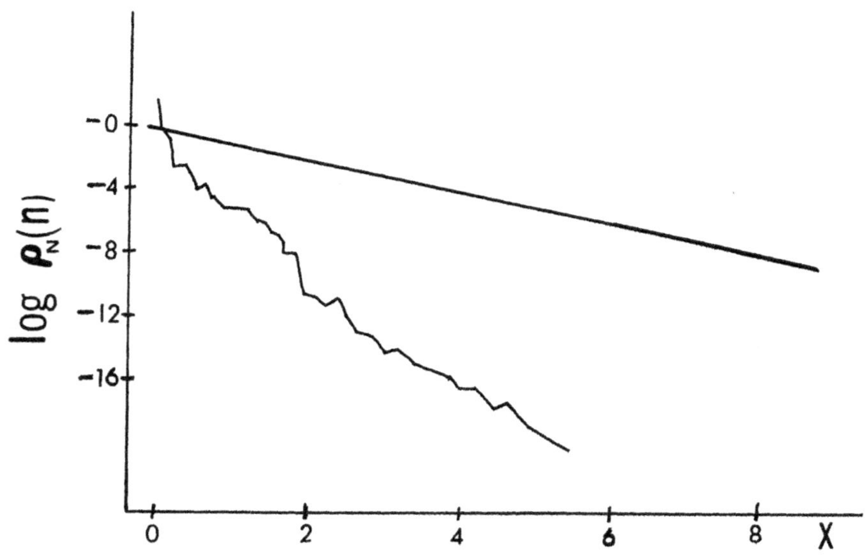

Fig. 2. A plot of the same variables as in Fig. 1 but for $k = 10$ and $\tau = 1/2$. Here the curves are quite distinct indicating a lack of stochasticity in the quantum motion even though $K = 5$ as in Fig. 1. We do not yet understand this quantum anomaly.

quantum anomalies which we shall discuss in the next section.

In order to summarize all the computed data for the above two runs plus two additional ones, we list the various computed parameters in Tables I-IV which appear at the end of this section. In each table, the first column lists the value of the integer time t. The second column lists the normalized average energy $\langle E\rangle_N$ at each time normalized in such a way that were the system motion stochastic exhibiting the Gaussian distribution (30) then the table values for $\langle E\rangle_N$ would sequentially read 1, 2, 3, 4, and 5; in particular $\langle E\rangle_N$ is given by

$$\langle E\rangle_N = \langle E\rangle/(k^2 t_1/4) \ , \tag{31}$$

where t_1 is the number of iterations per output. The third column lists a parameter B determined by a least squares fit of the experimental data to the formula

$$\rho_N(n) = A\ e^{-BX} \ . \tag{32}$$

If the motion were purely stochastic, B would equal unity. The fourth column lists a parameter W_d given by

$$W_d = [A/k(\pi t)^{1/2}] \int_{-\infty}^{\infty} \exp(-Bn^2/k^2t)\ dn = A/B^{1/2} \tag{33}$$

loosely speaking, W_d is related to the percentage of stochastic energy diffusion in the motion. For purely stochastic motion, we would find $W_d = 1$. The fifth column lists (W_d/B) which is the average energy computed using the fitted Eq. (32) and given by

$$\langle E\rangle = A \int_{-\infty}^{\infty} (n^2/2)\ e^{-BX}\ dn = W_d/B \ . \tag{34}$$

The sixth column lists the ratio ξ given by

$$\xi = \langle E\rangle\ /\ \langle E\rangle_d \ , \tag{35}$$

TABLE I. A listing of stochastic energy absorbtion parameters as a function of time for one computer solution of the driven quantum pendulum. The parameters are defined in the text. Here $k = 40$, $\tau = 1/8$, and $K = k\tau = 5$. The number of $b_s(k)$ values used in Eq. (24) was 101.

t	$\langle E\rangle_N$	B	W_d	W_d/B	ξ	R_o	$\lvert a_o\rvert^2$
5	1.55	3.67	331	90.3	1.55	0.38	0.0024
10	2.79	2.49	92.6	37.3	1.39	1.36	0.0061
15	3.77	0.76	0.90	1.18	1.26	1.01	0.0037
20	5.01	0.75	0.87	1.16	1.25	1.67	0.0053
25	5.66	0.83	0.87	1.05	1.13	1.43	0.0040

TABLE II. A listing of stochastic energy absorbtion parameters as a function of time for one computer solution of the driven quantum pendulum. Here $k = 40$, $\tau = 1/40$, and $K = k\tau = 1$. The number of N of $b_s(k)$ values used in Eq. (24) was 101.

t	$\langle E\rangle_N$	B	W_d	W_d/B	ξ	R_o	$\lvert a_o\rvert^2$
30	0.089	29.1	0.28	0.0096	0.0056	16.50	0.043
60	0.100	53.2	0.27	0.0051	0.0029	6.70	0.012
90	0.068	76.4	0.24	0.0031	0.0018	9.48	0.014
120	0.077	94.2	0.14	0.0015	-0.0005	4.89	0.0063
150	0.048	116.0	0.15	0.0013	-0.0002	25.4	0.028

where $\langle E\rangle$ is given by Eq. (27) and

$$\langle E\rangle_d = (k^2/4)t \tag{36}$$

is essentially Eq. (10) expressed in terms of the quantum parameters or, alternatively, it is $\langle n^2/2\rangle$ computed using Eq. (29). If the motion were completely stochastic, the fifth and sixth columns of each table would be identical. The seventh column lists the ratio $R_o=|A_o|^2k(\pi t)^{1/2}$ of the actual probability of the zeroth energy level to that expected from the Gaussian distribution of Eq. (29). The last column lists

$|A_o|^2$ itself. The number of Bessel functions (or $b_s(k)$) needed in Eq. (22) to accurately compute each run is listed in the table captions. Finally, in all four runs at least 1,000 A_n-values ($-500 \leq n \leq 500$) were computed at each iteration and, as mentioned earlier, accuracy was checked by verifying normalization of the A_n-sum.

We now turn to a discussion of these experimental results.

V. DISCUSSION OF NUMERICAL RESULTS

We have presented results for four runs selected to illustrate typical behavior in a modest variety of k and τ ranges. The parameter values $k = 40 \gg 1$ and $\tau = 1/8 \ll 1$ (corresponding to a classical $K = 5$) used in the computations yielding Fig. 1 and Table I are those for which one would expect quantum stochastic behavior, since the corresponding classical system is certainly highly stochastic for this case. A survey of Fig. 1 and Table I reveal that these expectations are verified reasonably well. However, it must be emphasized that our calculations yield a solution only over a finite time interval, that the computed $\rho(n)$ distribution contains large fluctuations, and that the various parameters in Table I are only crude indices which do deviate from their expected values. Pending further study, our results here must be regarded as providing only an indication of

TABLE III. A listing of stochastic energy absorbtion parameters as a function of time for one computer solution of the driven quantum pendulum. Here $k = 1$, $\tau = 5$, $K = k\tau = 5$. The number N of $b_s(k)$ values used in Eq. (24) was 23.

t	$\langle E\rangle_N$	B	W_d	W_d/B	ξ	R_o	$\|a_o\|^2$
150	0.0108	13.1	0.023	1.78×10^{-3}	3.18×10^{-4}	12.6	0.58
300	0.0092	29.5	0.029	9.83×10^{-4}	-7.50×10^{-6}	27.9	0.91
450	0.0079	40.0	0.015	3.70×10^{-4}	-4.60×10^{-5}	22.9	0.61
600	0.0057	55.4	0.015	2.80×10^{-4}	-2.50×10^{-5}	38.2	0.88
750	0.0079	66.2	0.020	2.96×10^{-4}	1.90×10^{-6}	35.9	0.74

TABLE IV. A listing of stochastic energy absorbtion parameters as a function of time for one computer solution of the driven quantum pendulum. Here $k = 10$, $\tau = 1/2$, and $K = k\tau = 5$. The number of N of $b_s(k)$ values used in Eq. (24) was 41.

t	$\langle E \rangle_N$	B	W_d	W_d/B	ξ	R_o	$\lvert a_o \rvert^2$
90	0.27	1.12	0.055	0.049	0.27	1.24	0.073
180	0.27	1.71	0.066	0.039	0.13	4.44	0.019
270	0.32	2.43	0.077	0.032	0.10	4.55	0.016
360	0.26	2.93	0.066	0.023	0.066	4.74	0.014
450	0.30	3.75	0.070	0.024	0.061	4.86	0.013

quantum stochastic behavior for the above parameter values.

The parameter values $k = 40$ and $\tau = 1/40$ (classical $K = 1$) used for Table II are those which classically lie on the border of stochasticity, and the data presented in Table II clearly indicates that the quantum behavior for this case is also non-stochastic. Table III which presents results for $k = 1$ and $\tau = 5$ (classical $K = 5$) indicates that the quantum motion is non-stochastic even though the corresponding classical motion is stochastic. However, as mentioned earlier, this difference between the quantum and classical case is understandable and to be expected. For small k values, the impulses or "kicks" do not give rise to many $e^{in\theta}$ "harmonics" in Eq. (22), and thus energy cannot be absorbed into many energy levels as is the case for $k \gg 1$ where 2k "harmonics" are involved. This dependence of quantum stochasticity on k and not just the product $(k\tau)$ as in the classical case receives support from the calculations resulting in Table III, but more work will be required to establish the approximate k-value determining the border of quantum stochasticity.

The results presented in Table IV and Fig. 2 which involve $k = 10$ and $\tau = 1/2$ constitute a true puzzle, since here $k \gg 1$ and the corresponding classical $K = 5$ just as for Table I and Fig. 1. Thus one might have expected stochastic behavior rather than the stable, non-stochastic data which actually appears. It is possible that the stochastic border occurs for higher k-values than previously anticipated, or it is possible that one here is observing a totally new and unexpected effect; only further study can reveal the appropriate and correct alternative. In this regard, let us mention that there are further unique quantum effects. For example regardless of k-value or initial $\psi(\theta,0)$, the free rotation mapping of Eq. (15) is the identity map when $\tau = 4\pi$; for this τ-value after an integer number t of "kicks", we have from Eq. (19) that

$$\psi(t) = \psi(0)\ \exp[itk\cos\theta] \tag{37}$$

which yields the average energy growth given by

$$\langle E\rangle = -(\hbar^2/2m\ell^2)\int d\theta\psi^*(\partial^2/\partial\theta^2)\psi \sim t^2 \tag{38}$$

that is proportional to t^2, corresponding to resonant rather than diffusive energy absorption. Moreover, when $\tau = 2\pi$, one may use Eq. (24) and a well-known Bessel function identity[11] to rigorously prove that $A_n(t+2T^+) = A_n(t)$ for all n; that is, the full driven quantum solution is strictly periodic independent of $\psi(\theta,0)$ or the value of k. Finally, resonant energy growth proportional to t^2 has also been observed for $\tau = 4\pi/m$, where m = 3, 4, 8, and 32, although the τ-widths of these resonances become increasingly narrow as m increases. Apparently, these peculiar resonant (or anti-resonant at $\tau = 2\pi$) effects are due to the strictly periodic nature of the unperturbed free rotator wave function; while all unperturbed classical rotator orbits are periodic, there is no period common to all solutions as occurs in the quantum case.

Subsequent to the Como Conference, we continued the above Fig. 1-run for k = 40 and $\tau = 1/8$ and discovered that the diffusive quantum energy absorption obeys Eq. (29) up to a time t_B (break-time) after which diffusive energy absorption appears to continue but at a much slower rate. Empirically, we find t_B proportional to k in sequences of runs for which $K = (k\tau)$ is held fixed. This result further "explains" the lack of any diffusive energy absorption in the data of Table III where k = 1 and $\tau = 5$ which is classically stochastic. A very crude, but possible explanation for the appearance of this break-time t_B may lie in the uncertainty relationship $\Delta E\Delta t \sim \hbar$. Note in Eq. (17) and Eq. (18) that the classical limit $\hbar\to 0$ is equivalent to $k\to\infty$ and $\tau\to 0$. Thus for large k and small τ ($\hbar$ fixed), we might expect the quantum behavior to mimic the classical at least for a time interval Δt during which the discrete nature of the free rotator energy level spectrum is insignificant. From Eq. (15), one notes that τ is a measure of the "effective" energy level spacing. Thus taking $\Delta E \sim \tau$ and $\Delta t \sim t_B$ in $\Delta E \Delta t \sim \hbar$, we have $t_B \sim \tau^{-1} \sim k$ for fixed $K(=k\tau)$, where t_B is the time required for the quantum system to "notice" that its energy levels are discrete. Further numerical study will be required to verify this possible explanation.

In closing this section, let us mention that we have sought to verify a sensitive dependence of final quantum state upon initial state similar to that implied by the exponential separation[3] of initially close classical orbits. Holding $K = k\tau$ fixed, we have tried k-values from $k = 1$ to $k = 100$ using various pairs of initially close $\{A_n(0)\}$ initial states only to find that the final state probability distribution $\rho(n) = A_n^* A_n$ was identical for each member of a pair to within numerical error. Even starting from non-close initial states such as $A_n = A_o \delta_{no}$ and $A_n = e^{-n^2}$ yielded the same final $A_n^* A_n$ distribution. Each of these initial quantum states (for large k and small τ) appeared to be approaching the unique final state probability distribution given by Eq. (29), at least for times $t < t_B$. Moreover, Eq. (29) is being approached from each of these definite initial states without the need for a time or an ensemble average. These rather startling results tempt one to speculate that the unique final state probability distribution for an isolated (chaotic) quantum system might be the microcanonical distribution, however premature such a speculation might be. Regardless of such speculations however, our present computations reveal no sensitive dependence of final state upon initial state, indeed they indicate a surprising lack of such dependence. Certainly in the classical limit of sufficiently large k and sufficiently small $\tau (K = k\tau >> 1)$, initially close wave packets must exponentially separate, but apparently even $k = 100$ and $\tau = 0.05$ does not lie in the classical parameter range.

VI. CONCLUDING REMARKS

This progress report has been presented in order to expose an example of a whole category of classically chaotic Hamiltonian systems whose exact quantum behavior can be investigated, at least numerically. It is our belief that future studies of the type Hamiltonian models revealed here can provide substantial information regarding the nature of chaotic behavior in deterministic quantum systems. Certainly the calculations for the specific pendulum Hamiltonian system considered herein provide at least an initial indication of the surprises and the possibly significant results which may await future investigators.

Regardless of final outcome, it now appears that a new doorway to quantum chaos may have opened and this progress report is an invitation for others to join us in crossing over its threshold.

In closing, we wish to express our profound appreciation to Ya. G. Sinai, E. V. Shuryak, G. M. Prosperi, G. M. Zaslavsky, D. Shepelyansky, and F. Vivaldi for many enlightening discussions regarding these problems. During the period of final editing for this paper, we received a specially prepared, handwritten preprint describing some splendid related work from Michael Berry, N. L. Balazs, M. Tabor, and A. Voros concerning "kicked" free particle systems. We have enormous admiration for the willingness of these authors to share their independent discoveries with us prior to publication.

REFERENCES

1. V. I. Arnold and A. Avez, Ergodic Problems of Classical Mechanics (W. A. Benjamin, Inc., New York, 1968); J. Moser, Stable and Random Motions in Dynamical Systems (Princeton Univ. Press, Princeton, 1973); Z. Nitecki, Differentiable Dynamics (MIT Press, Cambridge, 1971).
2. G. M. Zaslavsky, Statistical Irreversibility in Nonlinear Systems (Nauka, Moskva, 1970, in Russian); J. Ford in Fundamental Problems in Statistical Mechanics, III, Edited by E. G. D. Cohen (North-Holland, Amsterdam, 1975).
3. B. V. Chirikov, "A Universal Instability of Many-Dimensional Oscillator Systems," Physics Reports (to appear 1979).
4. N. M. Pukhov and D. S. Chernasvsky, Teor. i. Matem. Fiz. 7, 219 (1971).
5. I. C. Percival, J. Phys. B6, 1229 (1973); J. Phys. A7, 794 (1974); N. Pomphrey, J. Phys. B7, 1909 (1974); I. C. Percival and N. Pomphrey, Molecular Phys. 31, 97 (1976).
6. K. S. J. Nordholm and S. A. Rice, J. Chem. Phys. 61, 203 (1974).
7. K. S. J. Nordholm and S. A. Rice, J. Chem. Phys. 61, 768 (1974).
8. E. V. Shuryak, Zh. Eksp. Teor. Fiz. 71, 2039 (1976).
9. G. M. Zaslavsky and N. N. Filonenko, Soviet Phys. JETP 38, 317 (1974). K. Hepp and E. H. Lieb in Lecture Notes in Physics, V. 38, Edited by J. Moser (Springer-Verlag, New York, 1974). A. Connes and E. Stormer, Acta Mathematica 134, 289 (1975).
10. John M. Greene, "A Method for Determining a Stochastic Transition," Preprint, Plasma Physics Laboratory, Princeton, New Jersey.

11. M. Abramowitz and I. A. Stegun, Handbook of Mathematical Functions (Dover Publications, Inc., New York, 1972), p. 363 and 385.

PERIODIC SOLUTIONS OF ARBITRARY PERIOD, VARIATIONAL METHODS

Robert H. G. Helleman and Tassos Bountis

School of Physics, Georgia Tech., Atlanta, Georgia 30332

Applied Mathematics, Cal. Tech., Pasadena, California 91125

ABSTRACT

We have constructed by a rapidly convergent variational method the periodic solutions, analytic in the time t and with *'arbitrary'* (long) *period*, of the well known *Hénon-Heiles System*, described by the Hamiltonian: $H = \frac{1}{2}(\dot{x}^2 + \dot{y}^2 + x^2 + y^2) + x^2y - y^3/3$. We obtain, what appears to be, a *dense* set of 1-parameter families of periodic solutions, wherever there exist bounded solutions, even in the so called 'stochastic regions' of this nonintegrable system. The 'stochastic regions' manifest themselves in our results as "regions of confluence" of the 1-parameter families, i.e. these families all come together there. Thus the period [i.e. the Poincaré Recurrence Time] and the orbits, exhibit a very "sensitive dependence" on the initial conditions there, going from one family to the "next". These regions show up already in our *analytic*, "zeroth order" (a-priori-), approximations of the converged results. The variational method is of Newton form and the -asymptotically- quadratic rate of convergence does not deteriorate inside the 'stochastic regions' or at very high nonlinearity.

CONTENTS

1. INTRODUCTION

"D'ailleurs, ce qui nous rend ces solutions périodiques si précieuses, c'est qu'elles sont, pour ainsi dire, la seule brèche par où nous puissions essayer de pénétrer dans une place jusqu'ici reputée inabordable."

HENRI POINCARÉ,

Vol. 1, chapt. 3, artl. 36 of ref. 17, note the end of our section 1.2.

In this paper we construct a large number of those 'precious' periodic solutions, of 'arbitrary' (long) period, for a well known 'non-integrable' [≈nonseparable] [*36,3,35*] system of nonlinear oscillators, the Hénon-Heiles system [*7,3,11,13,8*], cf. (1.1). We obtain these solutions, by a rapidly convergent variational method, as Fourier series which are 'analytic' in the time: t. Within the so called 'stochastic regions' of a nonintegrable system the orbits display an extremely 'sensitive dependence' on the initial values x(0), y(0), etc. [*36,3,35, 4,7*]. More-familiar examples of this 'sensitive dependence' on the initial conditions are: Throwing dice, playing pin-ball machines, etc. While completely deterministic these processes are usually referred to as 'stochastic'. In these examples the cause of the sensitive dependence is obvious: the singular edges of the dice and the pins! Yet Siegel tells us that 'most' Hamiltonian systems have such singularities, even if the Hamiltonian is non-singular [*3,11,35,12*]. We use the periodic solutions, following the (above) advice of Poincaré, as the breach through which to penetrate those 'stochastic regions'.

We obtain, what appears to be, a *dense* set of 1-parameter families of periodic solutions even in the 'stochastic regions', cf. Fig. 2 in Section 1.1. We do find that these families all "bunch together" in the stochastic regions thus creating a 'sensitive dependence', of the period and the orbits, on the initial conditions. Some results for the Hénon-Heiles system are presented in Section 1.1.

The variational method is of 'Newton form' [*27,28*] and avoids the usual divergence problems of classical mechanics [*3,11,12,35,36,6*]. Its rate of convergence does not deteriorate in the 'stochastic regions' or at very high energy [*24,23*]. It is essential in our methods to abandon the 'initial value problem' [*6,23,26*], i.e. the usual specification of x(0), y(0) values. Instead we specify the period, or: 'recurrence time', of the orbit and the number of zero's for the velocities, within one period. The essential parts of this method are discussed in Section 1.2.

Global *analytic* "zeroth-order" formulae can be obtained from it which require no iterations and are quite accurate [8]. In Section 2.1 we derive the simplest "zeroth-order" formulae for the Hénon-Heiles system (1.1). The ('stochastic') regions in which the 1-parameter families "bunch together" already show up in these zeroth-order results. In Sections 2.2 and 2.3 we discuss practical short-cuts in the methods and give the actual recursion relations for the Hénon-Heiles system.

Variational methods are seldom used, in practice, to solve a classical N-body problem even though mechanics is formally derived from a celebrated variational principle: Hamilton's Principle [16]. The problem lies in the fact that the Action is not a local minimum under variations, but a saddlepoint [6,16] as we shall see. We hope to convert it here, and elsewhere [6], into a less formal principle:

> *"... it is easy to conjecture that the significance of the analogous variational principles of dynamics is largely formal."*
>
> G. D. BIRKHOFF,
>
> chapter 2, ref. 31.

1.1 Results for the Hénon-Heiles System.

Our example here is the Hénon-Heiles system

$$\ddot{x} = -x - 2xy \tag{1.1a}$$

$$\ddot{y} = -y + y^2 - x^2 \quad , \tag{1.1b}$$

a famous, if not notorious, system [7,3,11,13,8,6,35-38] which is non-integrable, by most indications. After this, the extension of these methods to other systems will be obvious [6,8,9,23-26].

Consider some exact periodic solution $\hat{x}(t)$, $\hat{y}(t)$ of (1.1), with period T_r: the Poincaré - "Recurrence Time" [and "Recurrence Frequency": $\nu_r \equiv 2\pi/T_r$]. Exceedingly important to us are the numbers of zero's in the velocities $\dot{\hat{x}}$, $\dot{\hat{y}}$, i.e. the integers m_1 and m_2 defined by:

$$\left.\begin{aligned} 2m_1 &\equiv \text{"the number of zero's of } \dot{\hat{x}}(t) \\ 2m_2 &\equiv \text{"the number of zero's of } \dot{\hat{y}}(t) \end{aligned}\right\} \text{ within } [0,T_r)\text{",} \tag{1.2}$$

cf. refs. 6, 23-26. This is schematically illustrated in Figure 1:

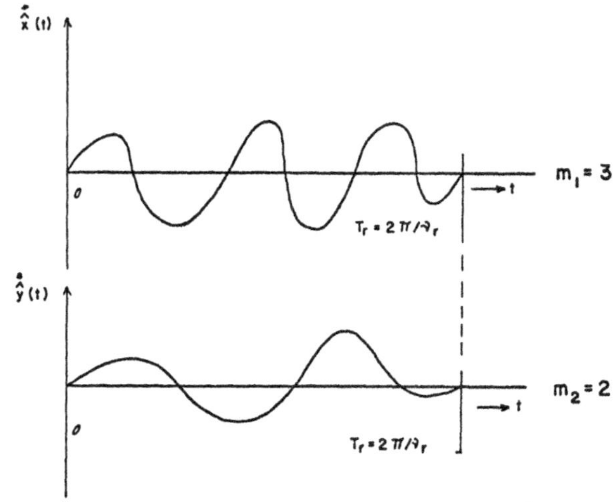

Figure 1 - The number of zero's of $\dot{\hat{x}}$ is: $2m_1$, and that of $\dot{\hat{y}}$ is: $2m_2$, within $[0,T_r)$, cf. (1.2). Here: $m_1=3$ and $m_2=2$.

for simplicity's sake we use solutions here which are initially at rest, i.e. $\dot{\hat{x}}(0) = 0 = \dot{\hat{y}}(0)$, cf. Fig.1. Ordinarily one would now specify the solution by its initial displacements $\hat{x}(0)$, $\hat{y}(0)$. *Here however we specify such periodic solutions by the* <u>*integers*</u>*: m_1 and m_2 and the period:* T_r. These uniquely specify the solution of (1.1) (within a "symmetry sector" of (1.1)), cf. Section 2.2 and ref. 6, and allow us to find the Fourier series representation of the solution $\hat{x}(t)$, $\hat{y}(t)$ by a variational (iterative) method which avoids the divergence problems of classical mechanics. In Sections 1.2 and 2.2 we explain how a knowledge of m_1, m_2, T_r enables one to avoid those problems and to ensure monotone convergence (quadratic convergence, asymptotically). From the $\hat{x}(t)$, $\hat{y}(t)$ thus obtained we finally recover the "initial" displacements $x_o \equiv \hat{x}(0)$, $y_o \equiv \hat{y}(0)$, a "backward approach" if ever there was one...!

It is convenient to present some of our results for the Hénon-Heiles system (1.1) by plotting the above initial displacements x_o, y_o at *constant* values of m_1, m_2, for a range of T_r values. This is done in Figure 2 on the other page. We obtain smooth non-intersecting curves (wherever there exist bounded solutions), parametrized by a rational number σ, defined as:

$$\sigma \equiv m_2/m_1 \quad , \tag{1.3}$$

the 'winding number'; the 'rotation number' is $\omega \equiv 2\pi\sigma$ [*7-14*][for Fig. 2 we only use the "primary" periodic solutions, i.e. the ones with m_1, m_2 relative primes [*10*] whence σ suffices to recover m_1, m_2]. These (constant-) "σ-curves" have been superimposed in Fig. 2 on the dashed equipotential lines of (1.1), separately displayed in Fig. 3 [*7,13*]. We find the known 'simple' [*5*] periodic solutions discussed by Rod, Pecelli and Churchill [*13*] and an infinity of new ones, bifurcating off the "(integrable-) $\sigma=1$ solutions", <u>along</u> the symmetry axes. Along each σ-curve the value of T_r changes *continuously* [*6*]. The

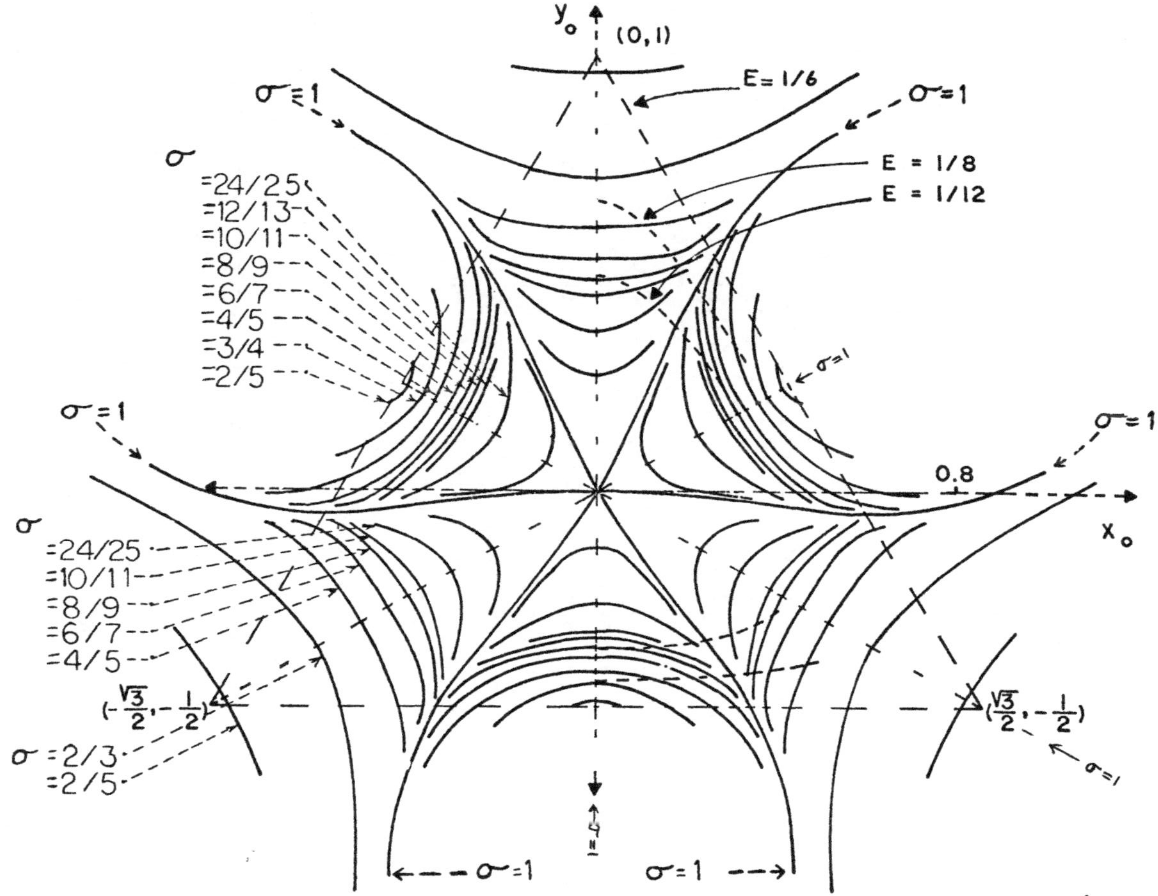

Figure 2 - Converged initial displacements x_o, y_o of the periodic solutions of the Hénon-Heiles system (1.1) (with $\dot{x}(0)=0=\dot{y}(0)$) at different values of $\sigma \equiv m_2/m_1$;only the "primary" solutions are used (i.e. m_2, m_1 relative prime). Note the confluence of σ-curves(near σ=1) resulting in a 'sensitive dependence on initial conditions' there, yielding 'stochastic regions' [*7,3,35-38*].

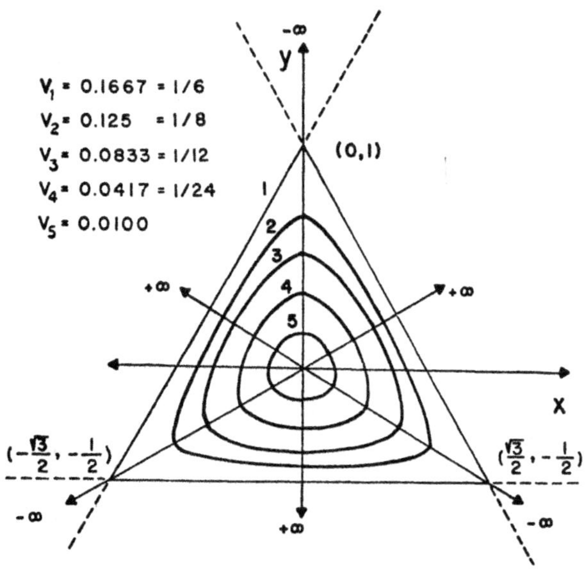

Figure 3 - Equipotential lines V=E of the Hénon-Heiles system, with $V(x,y) = \frac{1}{2}(x^2+y^2) + x^2y - y^3/3$.

existence of such continuous 1-parameter families is of course well known in celestial mechanics [*14-18,13*]. The value of T_r changes *dis*continuously as we go from one σ-curve to the "next" : Take two σ-curves in Fig. 2, one with $\sigma=2/3$ and another with $\sigma = 2{,}000{,}000/3{,}000{,}001$. Both curves intersect a symmetry axis, e.g. the y-axis, in *nearby* points. Orbits along such axes can be solved exactly, by an elliptic cosine [plus constant] [*13,19,16*] whose period changes *continuously* with y_0; so we expect our "y-period", T_r/m_2, cf. Fig. 1, to change continuously as well. However, the m_2 for the second curve is 10^6 larger than for the first one. Hence the T_r for the second curve is $\approx 10^6$ *smaller* than for the first one, *nearby*!

The values of T_r on the symmetry axes of Fig. 2, can be read-off in this manner and change continuously from that value, along each σ-curve. Some values are tabulated in Ref. 8.

Adding more σ-curves to Fig. 2 would merely clutter up the picture; yet it appears that the σ-curves do cover the x_0, y_0 plane *densely*: We can construct a σ-curve arbitrarily close to any point of the plane, by (rational) σ-interpolation since the ordering of the nested σ-curves varies "continuously" with σ, in this plane [often requiring larger m_1, m_2 values with a corresponding increase in the algebra- or computer - work]. The threefold symmetry of the Hénon-Heiles potential [*7,13,8*], cf. Fig. 3, is clearly reflected in our Figure 2.

It is interesting to note from Fig. 2 that *the σ-curves "bunch together"* along the $\sigma = 1$ curves here . In these regions of *σ-confluence* small changes in x_0, y_0 can apparently cause large relative changes in σ(and m_1, m_2). Since the subsequent time evolution $\hat{x}(t)$, $\hat{y}(t)$ depends very sensitively on the value of σ [e.g. compare two "Lissajous Figures"] we have here "regions of sensitive dependence on x_0, y_0" [*35,36,3,11,12*] as well. These coincide of course with the well known *'stochastic regions'* of the Hénon-Heiles system [*7,35,3,11*]. At higher values of the energy E the σ-confluence regions, near σ = 1, gather many more σ curves and this sensitive dependence on the initial x_o, y_o becomes very pronounced indeed. This can be seen clearly from Figure 4 where we graphed the σ, along some equipotential lines. In our calculations we can usually deduce the degree of 'stochasticity' already from the slopes of such graphs or from the presence or absence of σ-confluence in certain regions: Similar regions of σ-confluence are abundantly present in another nonintegrable system, the Duffing Equation [*23*]; however they are absent "nearly everywhere" [*24*] in the 'Fermi-Pasta-Ulam chain', a 'nearly-integrable' system [*2*]. Even the "zeroth-order" *analytic* approximations to our σ-curves (and to Figs. 2 and 4), in Section 2.1, usually suffice to show the approximate location of these σ-confluence regions.

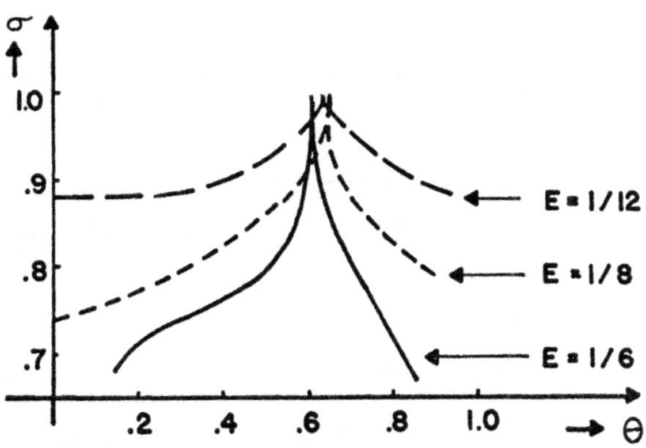

Figure 4 - Graph of the 'winding number' $\sigma \equiv m_2/m_1$ (1.3) along an equipotential line at energy E, as a function of the polar angle θ about the origin of the x_o, y_o plane in Fig. 2 [θ in radians; θ = 0 is chosen at 30°]. Note that the "sensitive dependence" on θ increases sharply with E.

1.2 A Variational Method of Newton Form.

> *"I cannot pass quietly over the relations between the theory of solutions of the second kind [i.e. of arbitrarily long period] and the Principle of Least Action; and it is even because of these relations that I have written chapter 29."*
>
> HENRI POINCARÉ
>
> Vol. 3, chapt. 31, artl. 371 of ref. 1.

The Action S(x,y) is stationary, but not necessarily an extremum, under variations about the exact (periodic) solution $\hat{x}(t)$, $\hat{y}(t)$, according to Hamilton's Principle or one of its variants [*16,33*]. So we must find the functional "roots" $\hat{x}$, $\hat{y}$ of $\delta S(x,y) = 0$ ... . Since root-finding is easier in algebraic equations we reduce the problem, as usual, to a *purely algebraic* one by Fourier decomposition:

$$x(t) = \sum_{n=-\infty}^{\infty} A_n e^{in\nu_r t} \; ; \; y(t) = \sum_{n=-\infty}^{\infty} B_n e^{in\nu_r t}, \text{ with } \nu_r \equiv 2\pi/T_r . \quad (1.4)$$

The action (the Lagrangian integrated over one period T_r) is a purely algebraic function $\overline{S}(A,B)$ of the infinite vectors A, B in (1.4)(and ν_r) and Hamilton's Principle becomes

$$\partial\overline{S}/\partial A_n^*\Big|_{\hat{A},\hat{B}} = 0 \text{ and } \partial\overline{S}/\partial B_n^*\Big|_{\hat{A},\hat{B}} = 0 \; , \quad (1.5)$$

for all integers n. Hence "all" that is left to do is find the roots $\hat{A}$, $\hat{B}$ of (1.5)... . If this stationary point were a local *minimum* of $\overline{S}(A,B)$ any variational method minimizing ('optimizing') $\overline{S}$ would suffice [*19-22*]. However, *the stationary point is a saddlepoint* [*6,16,20*], whose type changes *discontinuously* with (x_0, y_0 and) $\hat{A}$, $\hat{B}$, i.e. it can change *during* the variations (iterations)! The saddlepoint for the present Hénon-Heiles system is demonstrated and discussed in Ref. 6; here we merely reproduce Jacobi's elegant example in Section 2.2. Difficulties in locating such a pathological saddlepoint are at the heart of the classic divergence problems of mechanics [*26,23,6,3,12,11*]. Our unusual specification (1.2) of the periodic orbits enables us to pre-determine this saddlepoint and avoid those problems.

If (1.5) were a scalar "root-finding" problem Newton's method would be preferred [*3,27,28*] in which one uses the slope evaluated at a trial root A to calculate a correction, ΔA, to A, etc. The vector-version uses the gradient of (1.5), i.e. one would calculate corrections ΔA, ΔB from

$$\left.\begin{bmatrix} \partial^2\overline{S}/\partial A^*\partial A & \partial^2\overline{S}/\partial A^*\partial B \\ \partial^2\overline{S}/\partial B^*\partial A & \partial^2\overline{S}/\partial B^*\partial B \end{bmatrix}\right|_{A,B} \begin{pmatrix} \Delta A \\ \Delta B \end{pmatrix} = - \left.\begin{pmatrix} \partial\overline{S}/\partial A^* \\ \partial\overline{S}/\partial B^* \end{pmatrix}\right|_{A,B} \; , \quad (1.6)$$

in what, we hope, is an obvious notation for the matrix of second derivatives, evaluated at the trial vectors, A, B. Imagine, for argument's sake, this (Hermitian-) matrix diagonalized, by a transformation to a new vector c. The correction Δc might then be solved from *the diagonal version of* (1.6):

$$\lambda_n(c)\Delta c_n = -\partial\overline{S}/\partial c_n^*\Big|_c \; , \text{ with } \lambda_n(c) \equiv \partial^2\overline{S}/\partial c_n \partial c_n^*\Big|_c \; , \quad (1.7)$$

where we order the new variables c_n according to the value of the action - curvature, λ_n; whence $n \geq 0$ (here) and

$$\lambda_0 \leqslant \lambda_1 \leqslant \lambda_2 \leqslant \dots \quad . \tag{1.8}$$

Since $\bar{S}$ has a saddlepoint at the exact solution $\hat{c}$ [6,16] the curvatures λ_n change their sign, e.g. at $n = m$,

$$\lambda_n(\hat{c}) \begin{cases} < 0 & \text{for } n < m \\ = 0 & \text{for } n = m \longleftarrow \text{(cf. ref. 6)} \\ > 0 & \text{for } n > m \end{cases} , \tag{1.9}$$

i.e. $\bar{S}(\dots,c_n,\dots)$ has a local maximum for $0 \leqslant n < m$ and a minimum for $n > m$ [6]. Going from c to $c + \Delta c$ the corresponding change in the action would be

$$\Delta\bar{S}_{(n)} \approx \partial\bar{S}/\partial c_n \, \Delta c_n \overset{(1.7)}{\overset{\downarrow}{=}} -|\partial\bar{S}/\partial c_n|^2/\lambda_n(c) \quad , \tag{1.10}$$

per component ($n \neq m$), per iteration. If the λ values were available at the exact solution $\hat{c}$, rather than at the present c, eq.(1.9) would guarantee $\Delta\bar{S}_{(n)} > 0$ for $n < m$, and <0 for $n > m$ [$n=m$ requires a single special calculation], i.e. there would be monotone convergence to the saddlepoint of $\bar{S}$. However, the actual "levels" $\lambda_n(c)$ do change with c, at each iteration of (1.7), and several λ_n (near $n=m$) can easily have the wrong sign compared to the exact levels (1.9). Thus some of the Δc_n [and $\Delta\bar{S}_{(n)}$] acquire the wrong sign from (1.7) [and (1.10)] and the new $c + \Delta c$ diverges away from $\hat{c}$. In the usual 'initial value problem' [$x(0) = x_o$, $y(0) = y_o$] this *"Level Crossing Problem"* [or: 'Small Denominator Problem', cf. (1.10)] is hard to avoid since the crucial index $n=m$ (1.9) is unknown and changes discontinuously with the trial function (and x_o, y_o) [26,23].

Our obvious remedy is to *specify* the index $n=m$, at which the curvature (1.9) changes sign, (plus T_r) instead of the usual initial values x_o, y_o. This immediately allows us to modify ["relocate"] Newton's method (1.7) into:

$$[\lambda_n(c) - \lambda_m(c)] \, \Delta c_n = -\partial\bar{S}/\partial c_n^* \Big|_c \quad , \text{ whence} \tag{1.11}$$

$$\Delta\bar{S}_{(n)} \approx -|\partial\bar{S}/\partial c_n|^2/[\lambda_n(c) - \lambda_m(c)] \quad , \tag{1.12}$$

$n \neq m$, cf. (1.10). These $\Delta\bar{S}_{(n)}$ have the correct sign by (our) definition of the orbit and $\bar{S}$ *now approaches its saddlepoint <u>monotonically</u>.* Note that this method, of Newton form [27], reverts back to Newton's method, as $\lambda_m(c)$ approaches its final value: 0, cf. (1.9), i.e. we have "quadratic convergence" asymptotically [27,28].

The usual constraint on A,B imposed by the initial values $\Sigma A_n = x_o$, $\Sigma B_n = y_o$ is replaced here by *a new constraint*

$$\partial\bar{S}/\partial c_m^*\Big|_c = 0 \quad , \tag{1.13}$$

i.e. (1.7) - (1.11) at n = m, to be satisfied by *each* trial vector c, including the first one: the "starting vector". Eq. (1.13) can usually by satisfied by solving for $|\Delta c_m|^2$ [or, linearly..., for a scaling factor [*24,23*]]. Since $\partial^2\bar{S}$ is diagonal, cf. (1.7), and the constraint (1.13) must be satisfied at c as well as at c + Δc we have

$$0 = \partial\bar{S}/\partial c_m^*\Big|_{c+\Delta c} - \partial\bar{S}/\partial c_m^*\Big|_c \approx \lambda_m(c)\ \Delta c_m \ , \qquad \therefore\ \lambda_m(c) \approx 0, \tag{1.14}$$

ensuring the change of sign (1.9) and asymptotic "quadratic convergence" of (1.11). These deceptively simple remedies (1.11), (1.13) can only be used of course if one knows *beforehand* the n = m at which the exact levels λ_n in (1.9) change sign, as we do by definition. It is easy to check that the velocities $\dot{\hat{x}}$, $\dot{\hat{y}}$, defining the previous m_1, m_2 in (1.2), satisy the (second) 'variational equations' [arising from $\delta^2\bar{S} = 0$] exactly [*15,16*] [the equations of motion (1.1) arise of course from $\delta\bar{S} = 0$]. Hence the Fourier coefficients of $\dot{\hat{x}}$, $\dot{\hat{y}}$ constitute the eigenvector of (1.6) with eigenvalue zero, i.e. belonging to the m-th eigenvalue (1.9) of (1.6): $\lambda_m(\hat{c}) = 0$ [*6*]. We could recover the value of m from m_1, m_2 but in practice we usually employ algorithms directly in terms of the A,B rather than c, as discussed in Section 2.2, so the m_1, m_2 are preferred.

It seems that the arguments of this section can easily be repeated for the 'quasi-periodic' solutions as well, i.e. solutions of the form:

$$x(t) = \Sigma_{n_1} \Sigma_{n_2} A_{n_1,n_2} \exp[i(n_1\nu_1 + n_2\nu_2)t], \tag{1.15}$$

with ν_2/ν_1 *irrational,* and a similar expansion for y(t). However, in that case the λ's can become dense and could cluster about λ=0, creating practical problems in (1.12), (1.14) and (1.7). On the other hand when we restrict ourselves to periodic solutions this does not happen and there are no clusterpoints in the level-spectrum. Asymptotic estimates for the levels usually show a level separation of the order of $1/T_r^2$(2.27) [*20-22,6*]. Confining ourselves to periodic solutions, of 'arbitrarily' large T_r, does not create a serious practical restriction since *for 'most' Hamiltonian systems such periodic solutions are 'dense' among all bounded solutions*, cf. theorems 28.5 - 31.5 and § 33 of ref. 29. This, in turn, allows all bounded solutions of these systems to be approximated *arbitrarily well* by such periodic solutions, including the solutions in the 'stochastic regions' of a nonintegrable system, cf. our Section 1.1. Until the very last year of his life, Poincaré repeatedly emphasized these facts and returned to work towards a proof. Later results were established by Birkhoff [*31*], Moser, Pugh, Markus and others [*29*]. While these facts are not established for every particular orbit of every particular system we have, in our particular systems [*23-25,8,9*], not seen any evidence to the contrary and subscribe fully to Poincaré's views on the subject:

> *"We shall see further on that ... the (3-body) problem still contains an infinity of periodic solutions...*
>
> *It seems at first that this fact can be of no interest whatsoever in practice. Indeed, there is zero probability for the initial conditions of the motion to be precisely those corresponding to a periodic solution. But it can happen that they differ very little from them, and that takes place just in the case where the old methods are not applicable anymore. One can then advantageously take the periodic solution as first approximation, ...*
>
> *There is even more: here is a fact which I have not been able to demonstrate rigorously, but which seems to me very much like the truth nevertheless.*
>
> *Given equations of the form defined in artl. 13 ['General Problems of Dynamics'] and some particular solution of these equations, one can always find a periodic solution (whose period, it is true, may be very long) such that the difference between the two solutions is as small as one wants, over as long a time as one wants. Moreover, what renders these periodic solutions so precious to us is that they are, so to speak, the only breach through which we could try to penetrate into an area, hitherto deemed inaccessible."*

HENRI POINCARÉ,

in: Periodic Solutions, Non-Existence of Uniform Integrals, Asymptotic Solutions, Vol. 1, chapt. 3, artl. 36 of ref. 17.

II. DETAILS OF THE RESULTS AND METHODS

Having discussed the essentials we can now present practical algorithms and iterative procedures to implement the variational method of Section 1.2. In Section 2.3 we give the actual recursion relations for the Fourier vectors A,B, used to obtain the "Hénon-Heiles results" of Section 1.1. Before we start iterating these we need to calculate "starting vectors" within the region of convergence. These analytic "zeroth-order" approximations require no iterations and only a minimal knowledge of Section 1. Moreover, they can be very accurate, i.e. close to the converged results, depending on the amount of algebra one is willing to do. In any case, since they are explicit (analytic) functions of all the parameters and variables they provide an 'a-priori' *global* overview of the converged results. In Section 2.1 we derive the simplest ones for the Hénon-Heiles system (1.1). In Section 2.2 we reproduce Jacobi's example of an action saddlepoint and find several practical short-cuts to the variational method.

2.1 Zeroth-Order Results for the Hénon-Heiles System

Since the velocities $\dot{\hat{x}}$, $\dot{\hat{y}}$ have $2m_1$, $2m_2$ zero's (1.2) by (our) definition, it is clear that their main Fourier coefficients are $A_{\pm m_1}$ and $B_{\pm m_2}$. It is obvious also that the x^2, y^2 terms in (1.1) will generate A_0, B_0 terms from them. So to "zeroth-order" (1.4) becomes

$$\begin{aligned} x(t) &= A_0 + 2A_{m_1} \cos(m_1 \nu_r t) \\ y(t) &= B_0 + 2B_{m_2} \cos(m_2 \nu_r t) \end{aligned} \quad , \qquad (0.\text{it.}), \quad (2.1)$$

with $A_{-n} = A_n$, to obtain $\dot{\hat{x}}(0)=0=\dot{\hat{y}}(0)$ as before. According to (1.13) we must satisfy the m-th Fourier components of the equations of motion (1.1), i.e.

$$0 = (m_1^2\nu_r^2 - 1)\, A_{m_1} - 2B_0\, A_{m_1} \qquad [n = m_1 \text{ in } (1.1a)], \quad (2.2)$$

$$0 = (m_2^2\nu_r^2 - 1)\, B_{m_2} + 2B_0\, B_{m_2} \qquad [n = m_2 \text{ in } (1.1b)], \quad (2.3)$$

$m_2 \neq m_1$, obtained by substituting (2.1) in (1.1) and equating like-exponentials. Similarly we find for the constant terms

$$0 = -A_0 - 2A_0\, B_0 \qquad [n = 0 \text{ in } (1.1a)], \quad (2.4)$$

$$0 = -B_0 + B_0^2 + 2B_{m_2}{}^2 - 2A_{m_1}{}^2 - A_0^2 \qquad [n = 0 \text{ in } (1.1b)], \quad (2.5)$$

since $A_{-n} = A_n$.

The initial values are related to A,B by (2.1)

$$x_o = A_0 + 2A_{m_1} \qquad (2.6)$$

$$y_o = B_0 + 2B_{m_2} \qquad (2.7)$$

Now, given the values of m_1, m_2, ν_r , as discussed in Section I, eqs. (2.2) - (2.7) are six equations which can be solved for the four Fourier coefficients plus x_o, y_o.

Eliminating B_0 from (2.2), (2.3) yields a (zeroth-order) "restriction" on our 'perturbed' frequencies $m_1\nu_r$, $m_2\nu_r$ in terms of the 'unperturbed' frequencies: 1,1 , cf. (1.1),

$$m_1^2\nu_r^2 + m_2^2\nu_r^2 \approx 2 \qquad \text{(0.it.), (2.8)}$$

This relation already allows us to estimate the range of m_1, m_2, ν_r values encountered in the periodic solutions of the Hénon-Heiles system. While (2.8) cannot survive as an identity in the converged results [m_1, m_2, ν_r can be *specified*, independently], those results do show (2.8) to be satisfied to within $\approx$ 7% [better near the symmetry axes, worse in the regions of "σ-confluence", in Fig. 2] [*8*].

The calculation is simplest if we set:

$$A_0 = 0 \ , \qquad \text{(0.it.), (2.9)}$$

thus limiting ourselves to results with $x_o \approx 0$ (or near the other two symmetry axes) since the potential is symmetric about x = 0, cf. Fig. 3. Eliminating all but B_0 from (2.2) - (2.7), eq. (2.5) becomes a quadratic equation, with the solution

$$B_0 = \{1+y_o \pm\sqrt{1+2y_o-2y_o^2+3x_o^2}\} /3 \ , \qquad \text{(0.it.), (2.10)}$$

which, even at $x_o = 0$, yields a real B_0 for $-0.36 \lesssim y_o \lesssim 1.35$, cf. Figs. 2 and 3. The 'perturbed' frequencies solved from (2.2), (2.3) are

$$m_1^2\nu_r^2 = 1 + 2B_0 \text{ and } m_2^2\nu_r^2 = 1 - 2B_0 \ , \qquad \text{(0.it.)} \quad (2.11)$$

cf. (2.40), (2.41) of ref. 6. Combined with (2.10) we have the frequencies as a function of x_o, y_o. Since $B_0 > 0$ for any choice of sign in (2.10) we find for all x_o, y_o

$$m_1\nu_r \geqslant 1 \text{ and } m_2\nu_r \leqslant 1 \ , \qquad (2.12)$$

while the 'unperturbed' frequencies are both equal to 1, and from (1.3):

$$\sigma^2 \equiv m_2^2/m_1^2 = (1-2B_0)/(1+2B_0) \leq 1 \quad , \tag{2.13}$$

cf. Fig. 2. The relations (2.12), (2.13) are satisfied by all our converged results as well. Solving for B_0 from (2.13) we obtain

$$B_0 = \tfrac{1}{2}\mu \ , \text{ with } \mu \equiv (1-\sigma^2)/(1+\sigma^2) \tag{2.14}$$

Substituting this in (2.10) and squaring we finally arrive at

$$y_o^2 - x_o^2 = \mu\, y_o - 3\mu^2/4 + \mu \quad , \tag{2.15}$$

a zeroth order approximation for the σ-curves, of Fig. 2, near the symmetry axes. These are *hyperbolae*, with $y_o = \pm x_o$ as the

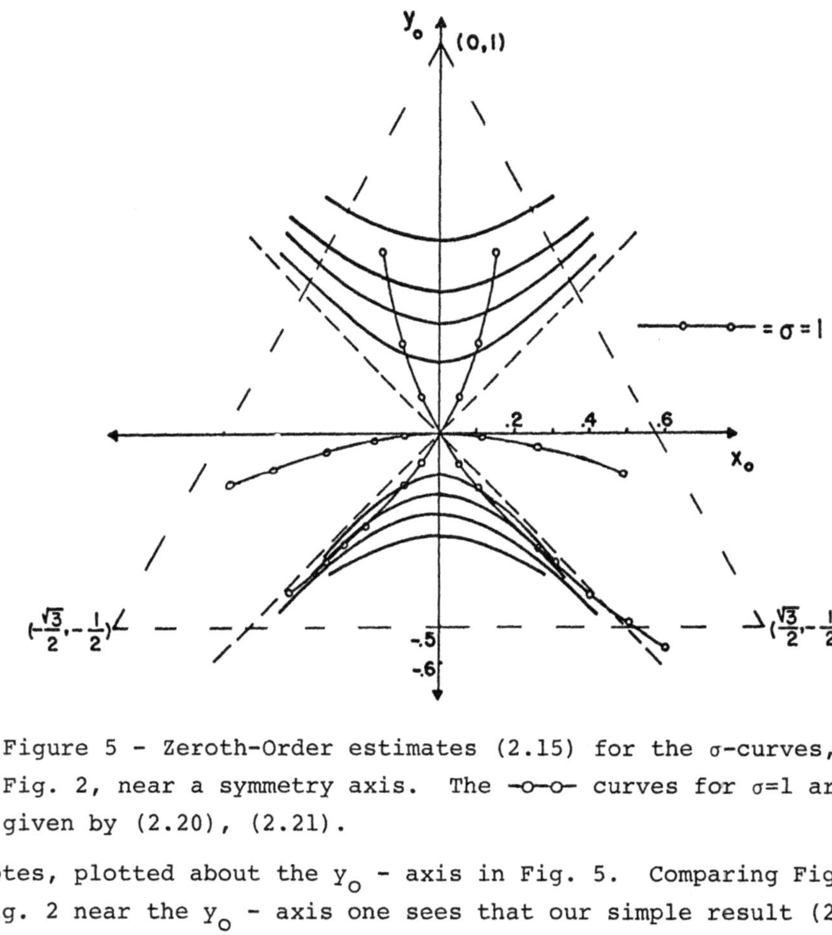

Figure 5 - Zeroth-Order estimates (2.15) for the σ-curves, of Fig. 2, near a symmetry axis. The -o-o- curves for σ=1 are given by (2.20), (2.21).

asymptotes, plotted about the y_o - axis in Fig. 5. Comparing Fig. 5 with Fig. 2 near the y_o - axis one sees that our simple result (2.15) is a very good approximation of the converged results there. When we plot these simple hyperbolae about all three symmetry axes, in Fig. 6, we notice that they "overlap" in the "σ-confluence" - or 'stochastic' regions of Section 1.1, cf. Fig. 2. Thus the result (2.15) provides bounds on its own region of validity, with the aid of Fig. 6. Worst-off in our simple estimate (2.15) are the $\sigma=1$ curves emanating from the origin [where the unperturbed frequencies are equal], hardly a surprise since we assumed $m_2 \neq m_1$ in (2.2), (2.3). Yet the $\sigma=1$ curves are interesting since they are at the center of the regions of overlap in Fig. 6 and of the "σ-confluence" -, or 'stochastic', regions in Fig. 2. A better estimate is obtained by allowing $A_0 \neq 0$ in (2.2) - (2.7), cf. (2.9). In the $\sigma=1$ case we have

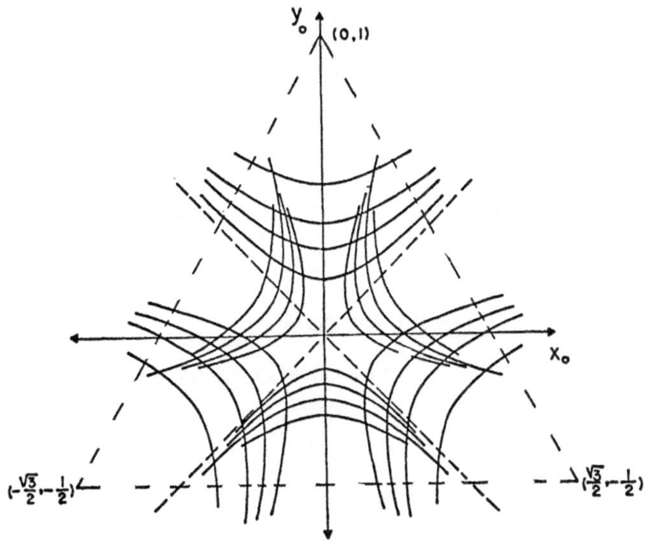

Figure 6 - The "zeroth-order" estimates (2.15) about all three symmetry axes. They overlap in the "σ-confluence" regions, or 'stochastic regions' of Fig. 2.

$$m_1 = m_2 = 1 \ , \quad \sigma = 1 \text{ and } \nu_r = \nu \ , \tag{2.16}$$

for the primary solutions [when m_2, m_1 are not relative primes we can find the "islands about islands" [7, 35]] and equations (2.2) - (2.4) change into

$$0 = (\nu^2-1)\ A_1 - 2B_0A_1 - 2A_0B_1 \qquad [n = 1 \text{ in } (1.1a)], \tag{2.17}$$

$$0 = (\nu^2-1)\ B_1 + 2B_0B_1 - 2A_0A_1 \qquad [n = 1 \text{ in } (1.1b)], \tag{2.18}$$

$$0 = -A_0 - 2A_0B_0 - 4A_1B_1 \qquad [n = 0 \text{ in } (1.1a)], \tag{2.19}$$

Eqs. (2.5) - (2.7) still apply and similar algebra yields three expressions for the $\sigma=1$ curves

$$5x_o^2 - y_o^2 + 4y_o \pm 4\sqrt{3}\ x_o \pm 6\sqrt{3}\ x_oy_o = 0 \ , \qquad (0.\text{it.}), \tag{2.20}$$

$$x_o^2 - 2y_o^2 + 2y_o = 0 \ , \qquad (0.\text{it.}), \tag{2.21}$$

cf. (2.15), each of which is a hyperbola with one blade passing through the origin, as plotted in Fig. 5. Now however, these 3 hyperbolae do transform into each other under 120° and 240° rotations. They do reveal the features of the converged $\sigma=1$ curves at the center of the "σ-confluence" regions of Fig. 2, analytically and before undertaking any variational calculations. At low energy they are quite accurate as well, cf. Figs. 5 and 2. While calculating (2.20) one finds the frequency as a function of x_o, y_o

$$\nu^2 = 1 - y_o \mp \sqrt{3}\, x_o \quad , \qquad (0.\text{it.}),\ (2.22)$$

along the $\sigma=1$ curves of (2.20)

2.2 Scaling the Action Saddlepoint without Sliding

Here we find some short-cuts to the general method of Section 1.2. But let us first establish, using Jacobi's example [*30*], that there *exists* an action-saddlepoint somewhere. The saddlepoint for the Hénon-Heiles system (1.1), and others, is discussed at length in ref. 6.

Consider a particle constrained to move along the surface of a sphere, at *constant* velocity v. Its action,

$$S(\tau) \equiv \int_0^{\tau} \mathcal{L}(\vec{x}(t), \dot{\vec{x}}(t))\, dt = \tfrac{1}{2}v^2\tau = \tfrac{1}{2}vs \quad , \qquad (2.23)$$

is proportional to the distance, s, travelled along the sphere. The particle completes a great circle in some period T and returns to its original position, which we call the "north pole". When $\tau < \frac{1}{2}T$ the actual orbit, along a great circle, is the shortest path between the initial - and final points. So $S(\tau)$ has a local minimum under all variations of the actual orbit if $\tau < \frac{1}{2}T$. On the other hand when we pass the "south pole", i.e. when $\tau > \frac{1}{2}T$, there are nearby paths on the sphere which are *shorter* than the actual orbit which goes from north pole via south pole to some further point Q [a shorter nearby path is obtained by taking another nearby great circle but making a short-cut to Q just before arriving at the south pole]. Since the previous, short-term, variations are still possible there are *longer* paths also. Hence $S(\tau)$ has a local maximum under the long term variations and a minimum under the short term ones, i.e. $S(\tau)$ *has a saddlepoint for* τ *large enough*. In terms of the Fourier [or Stürm-Liouville [*6*]] coefficients (cf. Section 1.2) this means that $\bar{S}(\ldots, c_n, \ldots)$ has a local maximum for the lowest n values, $n < m$, and a local minimum for all larger ones $n > m$, cf. (1.9).

Similar arguments could be made for an ellipsoid with periods T_1 in the x-direction and T_2 in the y-direction. The $S(\tau)$ then has a saddlepoint under variations in $x(t)$ if τ exceeds some fraction of T_1 and under variations in $y(t)$ if τ exceeds some fraction of T_2. It is clear that our "y-period" T_r/m_2 discussed in Section 1.1, cf. Fig. 1, functions as the T_2, and our "x-period" T_r/m_1 as the T_1, in the Hénon-Heiles system. This emphasizes again the role of the m_1, m_2 (1.2) in our periodic solutions [6].

Specifying m_1, m_2 and T_r as we do, we cannot extract more information than is contained in the *velocities* $\dot{\hat{x}}$, $\dot{\hat{y}}$ since nothing else is used in the definition of m_1, m_2 and T_r, cf. (1.2). Suppose we do indeed obtain the velocities exactly: Then we would recover the solution $\hat{x}(t)$, $\hat{y}(t)$, up to integration constants x_o, y_o ... , by integrating these velocities. [even in practice this would not be difficult, having obtained the velocities as Fourier series]. The 'primitive' functions thus acquired would be substituted in the equations of motion (1.1), and evaluated at $t = 0$, to solve for the "integration constants" x_o, y_o. If these nonlinear equations for x_o, y_o allow several solutions for x_o, y_o we obtain several orbits at once. Our methods are, in some sense, equivalent to the above [6] and as a result we obtain several solutions at once, depending on the symmetries inherent in the equations of motion: Solving for x_o from the Hénon-Heiles equations (1.1) we may choose either $-x_o$ or $+x_o$ since the equations are invariant under $x \to -x$. Hence we do not have to repeat our calculations to check whether our solutions to the left of the y-axis in Fig. 2 are, in fact, the 'mirror-image' of the ones to the right. The same holds for the threefold rotational symmetry of the Hénon-Heiles equations, cf. Fig. 3. [7,13,8] and (2.20), (2.21) for example.

In (1.11), (1.13) we found an algorithm yielding monotone convergence at a high rate of convergence. Here we introduce short-cuts which reduce the required labor and still avoid "sliding out of the saddlepoint". Transformed back into the original A,B Fourier coefficients our "relocated" recursion relation (1.11) is:

$$\left\{ \left[\begin{matrix} \partial^2\overline{S}/\partial A^*\partial A & \partial^2\overline{S}/\partial A^*\partial B \\ \partial^2\overline{S}/\partial B^*\partial A & \partial^2\overline{S}/\partial B^*\partial B \end{matrix} \right] \Bigg|_{A,B} - \lambda_m(A,B)\ I \right\} \begin{pmatrix} \Delta A \\ \Delta B \end{pmatrix} = - \begin{pmatrix} \partial\overline{S}/\partial A^* \\ \partial\overline{S}/\partial B^* \end{pmatrix} \Bigg|_{A,B} , \qquad (2.24)$$

cf. (1.6), a linear system which we solve for $\Delta A, \Delta B$. It does not require extra work to evaluate this matrix of second derivatives after each iteration, in our approach, since: "$\partial^2\overline{S}$ is (one power) less nonlinear than $\partial\overline{S}$", i.e. evaluating the <u>rhs</u> of (2.24) requires more effort anyway, cf. Section 2.3. However, calculating the m-th eigenvalue λ_m

(≈ 0) after each iteration, *does* require extra labor. While we do it on occasion, we prefer short-cuts to decrease that labor. These short-cuts usually have a slower rate of convergence and the number of (computer-...) iterations increases.

In the arguments following (1.14) we noted that the m-th eigenvector contains the Fourier coefficients of the $\dot{\hat{x}}$, $\dot{\hat{y}}$ [6], which in turn define m_1, m_2 (1.2). Hence $\overline{S}(\ldots, A_n, \ldots; \hat{B})$ has a maximum for $n < m_1$ and a minimum for $n > m_1$; the same holds for $\overline{S}(\hat{A}; \ldots, B_n, \ldots)$ about m_2. We may therefore replace the constraint (1.13) by the new constraints:

$$\partial\overline{S}/\partial A^*_{m_1}\Big|_{A,B} = 0 \text{ and } \partial\overline{S}/\partial B^*_{m_2}\Big|_{A,B} = 0 \quad , \tag{2.25}$$

cf. (1.5), to be satisfied by the A,B at *each* iteration and by the starting vectors. The above argument *separates* the B-variation from the A-variations. We similarly separate the lhs of (2.24).

$$\left\{\begin{bmatrix} \partial^2\overline{S}/\partial A^*\partial A & 0 \\ 0 & \partial^2\overline{S}/\partial B^*\partial B \end{bmatrix}\Bigg|_{A,B} - \begin{bmatrix} \lambda'_{m_1} I \\ \lambda''_{m_2} I \end{bmatrix}\right\} \begin{pmatrix} \Delta A \\ \Delta B \end{pmatrix} = -w\begin{pmatrix} \partial\overline{S}/\partial A^* \\ \partial\overline{S}/\partial B^* \end{pmatrix} \quad , \tag{2.26}$$

where the λ' are the eigenvalues of $\partial^2\overline{S}/\partial A^*\partial A$ and the λ'' those of $\partial^2\overline{S}/\partial B^*\partial B$. This separation of the *iterations* may be justified locally [27,28], in a small enough neighborhood of A,B. To avoid large corrections ΔA, ΔB and leave that neighborhood we have introduced the scalar factor w on the rhs, which can be chosen as small as needed $[0 < w \leq 1]$. In linear systems the w is known as a 'relaxation parameter' [27,28].

Finally, we note that the matrices $\partial^2\overline{S}/\partial A^*\partial A$ and $\partial^2\overline{S}/\partial B^*\partial B$ are not too far from being diagonal: First, it follows from Stürm-Liouville Theory [21,22,6] that the n-th element of the n-th eigenvector of $\partial^2\overline{S}/\partial A^*\partial A$ is the larger one in that eigenvector; second, consider that,

$$\partial^2\overline{S}/\partial A_n\partial A^*_k \underset{n\to\infty}{\sim} \delta_{n,k}\, n^2\nu_r^2 \quad , \tag{2.27}$$

due to the $\ddot{x}$ term in the first equation of motion, cf. (1.1a), etc. So the matrices in (2.26) are asymptotically diagonal and (2.27) also provides a well known [20-22,6] asymptotic estimate of the eigenvalues λ'_n. The corresponding diagonal, asymptotic, version of the recursion relations (2.26) is:

$$\begin{aligned} (n_1^2 - m_1^2)\,\nu_r^2\,\Delta A_{n_1} &= -\,w\,\partial\overline{S}/\partial A^*_{n_1} \\ (n_2^2 - m_2^2)\,\nu_r^2\,\Delta B_{n_2} &= -\,w\,\partial\overline{S}/\partial B^*_{n_2} \end{aligned} \quad , \tag{2.28}$$

for all integers n_1, n_2. The forefactors on the lhs. are actually good approximations of $(\lambda'_{n_1} - \lambda'_{m_1})$ and $(\lambda''_{n_2} - \lambda''_{m_2})$ since m_1, m_2 and ν_r are *specified* in our approach. Of all the above versions eq.(2.28), with (2.25), clearly is the simplest one to implement but requires the smallest values of w to stay within the region of validity of all approximations. In Section 2.3 and ref 8 we describe the use of (2.28), (2.25) for the Hénon-Heiles system (1.1). Results for the Duffing equation can be found in refs. 23 and 24, for Hénon's mapping in ref. 25 and for the Fermi-Pasta-Ulam chain in ref. 24. Typically the iterations of (2.28), (2.25) converge rapidly at any nonlinearity up to infinity. The same holds in the 'stochastic regions', cf. Section 1.1, and even for the *unbounded* orbits. In the latter case we take a real-exponential basis, i.e. ν_r imaginary in (2.14) [*24,23*]. The rate of convergence does decrease considerably, very close to a point where the orbits change from bounded to unbounded [e.g. where our $\sigma \to 0$ or $\sigma \to \infty$ cf. Section 1.1] i.e. the "escape" points on the potential surface or the "breaking" points of a chain.

2.3 Practical Recursion Relations

In this Section we describe the specific recursion relations used to calculate the Hénon-Heiles results of Section 1.1 and ref. 8. It is helpful to note that since $\delta\overline{S}/\delta x = 0$ yields the first equation of motion (1.1a), the term $\partial\overline{S}/\partial A_n^*$ in (2.28) yields the n-th Fourier component of that equation of motion (1.1a), found by substitution of the Fourier series (2.14). We rewrite the corrections ΔA, ΔB as

$$\Delta A \equiv A' - A \text{ and } \Delta B \equiv B' - B \tag{2.29}$$

where the primes indicate the "corrected" version of A,B, after the iteration. Thus the explicit recursion relations (2.28) for the Hénon-Heiles system (1.1) are:

$$A'_n = w\left[(m_1^2\nu_r^2-1)\,A_n - 2\sum_{k=-\infty}^{\infty} A_k\,B_{n-k}\right]/(m_1^2-n^2)\,\nu_r^2 + (1-w)\,A_n, \tag{2.30}$$

$$B'_n = w\left[(m_2^2\nu_r^2-1)\,B_n + \sum_{k=-\infty}^{\infty} B_k\,B_{n-k} - \sum_{k=-\infty}^{\infty} A_k\,A_{n-k}\right]/(m_2^2-n^2)\,\nu_r^2+(1-w)\,B_n\ , \tag{2.31}$$

for $n \geq 0$ $[A_{-n} = A_n,\ B_{-n} = B_n]$, $n \neq m_1$ in (2.30) and $n \neq m_2$ in (2.31); i.e. the old A,B are substituted in the rhs and the new A',B' are solved from the lhs., etc. Similarly, the constraints (2.25) are:

$$0 = (m_1^2\nu_r^2 - 1)\, A_{m_1} - 2 \sum_k A_k\, B_{m_1-k} \tag{2.32}$$

$$0 = (m_2^2\nu_r^2 - 1)\, B_{m_2} + \sum_k B_k\, B_{m_2-k} - \sum_k A_k\, A_{m_2-k} \tag{2.33}$$

where we dropped the primes. Ordinarily we would solve (2.32) for A'_{m_1} and (2.33) for B'_{m_2} after each iteration [*23,24*]. However guided by the "zeroth order" results in Section 2.1 we elected to change this slightly: The first equation of motion (1.1a) is formally "linear" in x, causing the approximate "restriction" (2.8) on the frequencies $m_1\nu_r$ and $m_2\nu_r$. Hence a narrow range of ν_r results in a large range of x_o, y_o values. This "linearity" causes (2.32) to be nearly proportional to A_{m_1}, cf. (2.2). So we preferred to solve (2.32) linearly for ν_r^2 and recover A_{m_1} linearly from $x_o = \Sigma A_n$, cf. (1.4), i.e. we specified m_1, m_2 plus x_o, rather than m_1, m_2 plus T_r, to obtain Figs. 2 and 4.

Noting the zeroth order eqs. (2.3), (2.5), we preferred to solve (2.33) linearly for B_0 while finding $B_{m_2}^2$ linearly from (2.30) at n=0, i.e. (1.5b) at n=0, cf. [*25*].

More results and (even) more details may be found in ref. 8. In this paper we did not yet touch upon the stability - type of these periodic solutions [*10,13*]. However in references 8 and 9 we provide a method, based on the use of Hill's determinants, to find the 'Floquet Multipliers' of the solution, and determine which of our solutions, are 'elliptic' and which 'hyperbolic' [*8-13*]. This method is particularly simple in the case of the periodic solutions of an 'area-preserving mapping' [*10,25,35,8,9*].

The reader is referred to references 13-15,19,32-35 for research on other aspects of the same problems as well as to references 37 and 38 for newer developments.

III. ACKNOWLEDGEMENTS

Rarely have we attended a conference as interesting and enjoyable as this one. It was a delight to see people from such diverse areas of research actively interested in the same subjects. A great deal of credit goes to Giulio Casati, Joe Ford and the city of Como for making this a worthy successor to the other famous Como Conference .

In Como we benefitted greatly from Martin Kruskal's insights and enthousiasm as well as from the useful advice of Dr. Gutzwiller, Dr. Pasta, David Rod, Joe Ford, Dr. Contopoulos and Dr. Danby. We thank them for their correspondence and for sending us (p)reprints of refs.13-15,32,33.

We would like to compliment Audrey Ralston on the excellent typing of this paper.

IV. REFERENCES

1. Topics in Nonlinear Dynamics, ed. S. Jorna, Am. Inst. Phys. Conf. Proc., Vol. 46, A.I.P. New York (1978); this volume contains ref. 2-6, and other articles.
2. J. Moser, "Nearly Integrable and Integrable Systems", 1, ref. 1.
3. M. V. Berry, "Regular and Irregular Motion", 16, ref. 1.
4. J. Ford, "A Picture Book of Stochasticity", 121, ref. 1.
5. A. Weinstein, "Simple Periodic Orbits", 260, ref. 1.
6. R. H. G. Helleman, "Variational Solutions of Non-Integrable Systems", 264, ref. 1; addendum, below (2.37): "when m_1, m_2 are not relative primes, see the remark following (2.16) in the present paper and ref. 9."
7. M. Hénon and C. Heiles, Astron. J. 69, 73-79 (1964).
8. T. Bountis, Thesis, University of Rochester, N.Y., Physics (Sept. 1978); and University Microfilms, Inc.
9. T. Bountis and R. H. G. Helleman, to be submitted to J. Math. Phys. (1979).
10. J. Greene, "A Method for Determining a Stochastic Transition", to be submitted to J. Math. Phys. (1978); as well as: J. Math. Phys. 9, 760 (1968).
11. J. Moser, "Lectures on Hamiltonian Systems", Memoirs Am. Math. Soc. 81, 1-60 (1968).
12. J. Moser, Stable and Random Motions in Dynamical Systems, Princeton University Press (1973).
13. R. C. Churchill, G. Pecelli and D. L. Rod, in this volume.
14. G. Contopoulos, Astron. J. 75, 96(1970); 76, 147(1971) and this volume
15. J. M. A. Danby, Qualitative Methods in Celestial Mechanics, Publications of the Institute of Mathematics of the University of São Paulo, Brasil (1971); espec. sections 4 and 5.
16. L. A. Pars, Analytical Dynamics, John Wiley, Inc. (1965); espec. chapt. 26 (26.3), chapt. 23 (23.5, 23.6).
17. H. Poincaré, "Les Méthodes Nouvelles de la Mécanique Céleste", Gauthier-Villars, Paris (1892); Dover Press (1957); N.A.S.A. - Translation TT F-450/452, U.S. Federal Clearinghouse Springfield, Virginia (1967); espec. chapters 3, 29, 28.
18. H. Poincaré, Leçons de Mécanique Céleste, 3 volumes, Gauthier-Villars, Paris (1905-1910).
19. A. H. Nayfeh and D. T. Mook, Nonlinear Oscillations, J. Wiley (Interscience Series, Pure and Applied Math.) New York (1978).

20. P. Funk, Variationsrechnung und ihre Anwendung in Physik und Technik, 2nd Ed., Springer Verlag (1970); espec. chapters 2, 3, and 7.
21. R. Courant and D. Hilbert, Methods of Mathematical Physics, Interscience Publ. Inc., N.Y. (1965); espec. chapters 5 and 6 of Vol. 1.
22. S. H. Gould, Variational Methods for Eigenvalue Problems, Univ. of Toronto Press (1957); espec. chapters 2 and 3.
23. C. R. Eminhizer, R. H. G. Helleman and E. W. Montroll, "On a Convergent Nonlinear Perturbation Theory without Small Denominators or Secular Terms", J. Math. Phys. 17, 121-140 (1976); espec. sections 4 and 5.
24. C. R. Eminhizer, Thesis, Univ. of Rochester, N.Y., Physics (1975); and University Microfilms, Inc.
25. R. H. G. Helleman, "On the Iterative Solution of a Stochastic Mapping", p. 343-370, in Statistical Mechanics and Statistical Methods , ed. U. Landman, Plenum Publ. Co., N.Y. (1977); espec. section III.D.
26. J. Moser, p. 1-9 in: The Stability of the Solar System and of Small Stellar Systems, ed. Y. Kozai, Reidel Co., Dordrecht, Boston.
27. W. C. Rheinboldt, Methods for Solving Systems of Nonlinear Equations, S.I.A.M. Monographs (1974); espec. section 3.2
28. J. M. Ortega and W. C. Rheinboldt, Iterative Solution of Nonlinear Equations in Several Variables, Academic Press, N.Y. (1970).
29. R. Abraham and J. E. Marsden, Foundations of Mechanics, Benjamin, Inc., New York (1967).
30. C. G. J. Jacobi, Vorlesungen über Dynamik, ed. A. Clebsch, Reimer Verlag, Berlin (1866); espec. p. 48.
31. G. D. Birkhoff, Dynamical Systems, Am. Math. Soc., Providence, R.I. (1927); revised edition (1966); espec. chapt. 2.
32. M. Kruskal, J. Math. Phys. 3, 806-828 (1962).
33. M. C. Gutzwiller, J. Math. Phys. 18, 806-823 (1977); & this volume.
34. G. B. Whitham, Linear and Nonlinear Waves, J. Wiley, New York (1974); as well as chapters 5 and 6 in: Nonlinear Waves, eds. Leibovich and Seebass, Cornell Univ. Press, Ithaca (1974).
35. J. Ford, "The Statistical Mechanics of Classical Analytical Dynamics", 215-255 in: Fundamental Problems in Statistical Mechanics, Vol. 3, ed. E. Cohen, North Holland, Amsterdam (1975); and Adv. Chem. Phys. 24, 155-185 (1973).
36. R. H. G. Helleman, "Nonlinear Dynamics, a Glossary",400,ref. 1.

37. Nonlinear Dynamics and the Beam-Beam Interaction, eds. M. Month and H. Wiedemann, Springer Verlag (?) , to appear (1979); proceedings of a Brookhaven-symposium on accelerator dynamics, with articles by Moser, Ford, Laslett, Keil, Courant, et al.,
38. Proceedings of the (December) 1979 -'International Conference on Nonlinear Dynamics' [New York], ed. R. H. G. Helleman et al., Ann. N. Y. Acad. Sci. (1980).

Lecture Notes in Physics

Vol. 68: Y. V. Venkatesh, Energy Methods in Time-Varying System Stability and Instability Analyses. XII, 256 pages. 1977.

Vol. 69: K. Rohlfs, Lectures on Density Wave Theory. VI, 184 pages. 1977.

Vol. 70: Wave Propagation and Underwater Acoustics. Edited by J. Keller and J. Papadakis. VIII. 287 pages. 1977.

Vol. 71: Problems of Stellar Convection. Proceedings 1976. Edited by E. A. Spiegel and J. P. Zahn. VIII, 363 pages. 1977.

Vol. 72: Les instabilités hydrodynamiques en convection libre forcée et mixte. Edité par J. C. Legros et J. K. Platten. X, 202 pages. 1978.

Vol. 73: Invariant Wave Equations. Proceedings 1977. Edited by G. Velo and A. S. Wightman. VI, 416 pages. 1978.

Vol. 74: P. Collet and J.-P. Eckmann, A Renormalization Group Analysis of the Hierarchical Model in Statistical Mechanics. IV, 199 pages. 1978.

Vol. 75: Structure and Mechanisms of Turbulence I. Proceedings 1977. Edited by H. Fiedler. XX, 295 pages. 1978.

Vol. 76: Structure and Mechanisms of Turbulence II. Proceedings 1977. Edited by H. Fiedler. XX, 406 pages. 1978.

Vol. 77: Topics in Quantum Field Theory and Gauge Theories. Proceedings, Salamanca 1977. Edited by J. A. de Azcárraga. X, 378 pages 1978.

Vol. 78: Böhm, The Rigged Hilbert Space and Quantum Mechanics. IX, 70 pages. 1978.

Vol. 79: Group Theoretical Methods in Physics. Proceedings, 1977. Edited by P. Kramer and A. Rieckers. XVIII, 546 pages. 1978.

Vol. 80: Mathematical Problems in Theoretical Physics. Proceedings, 1977. Edited by G. Dell'Antonio, S. Doplicher and G. Jona-Lasinio. VI, 438 pages. 1978.

Vol. 81: MacGregor, The Nature of the Elementary Particle. XXII, 482 pages. 1978.

Vol. 82: Few Body Systems and Nuclear Forces I. Proceedings, 1978. Edited by H. Zingl, M. Haftel and H. Zankel. XIX, 442 pages. 1978.

Vol. 83: Experimental Methods in Heavy Ion Physics. Edited by K. Bethge. V, 251 pages. 1978.

Vol. 84: Stochastic Processes in Nonequilibrium Systems, Proceedings, 1978. Edited by L. Garrido, P. Seglar and P. J. Shepherd. XI, 355 pages. 1978

Vol. 85: Applied Inverse Problems. Edited by P. C. Sabatier. V, 425 pages. 1978.

Vol. 86: Few Body Systems and Electromagnetic Interaction. Proceedings 1978. Edited by C. Ciofi degli Atti and E. De Sanctis. VI, 352 pages. 1978.

Vol. 87: Few Body Systems and Nuclear Forces II, Proceedings, 1978. Edited by H. Zingl, M. Haftel, and H. Zankel. X, 545 pages. 1978.

Vol. 88: K. Hutter and A. A. F. van de Ven, Field Matter Interactions in Thermoelastic Solids. VIII, 231 pages. 1978.

Vol. 89: Microscopic Optical Potentials, Proceedings, 1978. Edited by H. V. von Geramb. XI, 481 pages. 1979.

Vol. 90: Sixth International Conference on Numerical Methods in Fluid Dynamics. Proceedings, 1978. Edited by H. Cabannes, M. Holt and V. Rusanov. VIII, 620 pages. 1979.

Vol. 91: Computing Methods in Applied Sciences and Engineering, 1977, II. Proceedings, 1977. Edited by R. Glowinski and J. L. Lions. VI, 359 pages. 1979.

Vol. 92: Nuclear Interactions. Proceedings, 1978. Edited by B. A. Robson. XXIV, 507 pages. 1979.

Vol. 93: Stochastic Behavior in Classical and Quantum Hamiltonian Systems. Proceedings, 1977. Edited by G. Casati and J. Ford. VI, 375 pages. 1979.

GPSR Compliance

The European Union's (EU) General Product Safety Regulation (GPSR) is a set of rules that requires consumer products to be safe and our obligations to ensure this.

If you have any concerns about our products, you can contact us on ProductSafety@springernature.com

In case Publisher is established outside the EU, the EU authorized representative is:

Springer Nature Customer Service Center GmbH
Europaplatz 3
69115 Heidelberg, Germany

Batch number: 10704763

Printed by Printforce, the Netherlands

Dictionary of SOIL SCIENCE

Dictionary of
SOIL SCIENCE

By
Dr. Subhash Chand
Assistant Professor
Division of Soil Science
S.K. University of Agricultural Sciences and Technology of Kashmir,
Shalimar, Srinagar – 191 121, J&K

2010
DAYA PUBLISHING HOUSE
Delhi - 110 035

ISBN 9789351240006

Published by : **Daya Publishing House**
A Division of
Astral International Pvt. Ltd.
– ISO 9001:2008 Certified Company –
4760-61/23, Ansari Road, Darya Ganj
New Delhi-110 002
Ph. 011-43549197, 23278134
E-mail: info@astralint.com
Website: www.astralint.com

Laser Typesetting : **Classic Computer Services**
Delhi - 110 035

Printed at : **Chawla Offset Printers**
Delhi - 110 052

PRINTED IN INDIA

Foreword

Soil science is a very important and complex subject. Often people find it difficult to understand various terminologies. Dr. Subhash Chand has made an attempt to prepare a dictionary of terms and terminology used in soil science by the soil scientists, agronomists and engineers in a very lucid and simple way. It will definitely help the students and agricultural scientists especially those deeply involved in aspects of soils and agronomy. It is a very good attempt and I compliment Dr. Subhash Chand for his work and hope that this publication will be well received by the students and practitioners of agriculture alike.

Prof. Anwar Alam

Vice-Chancellor

S.K. University of Agricultural Sciences and Technology of Kashmir, Shalimar, Srinagar – 191 121, J&K

Preface

Soil is a mixture of mineral and organic materials that capable of supporting plant life, hence a vital natural resource. Soil Science deals the study of natural resource on the surface of the earth including its formation, composition and uses. Present time is a time of knowledge and information. A sound knowledge about the soil science words provides a basis for better understanding about soils for their conservation and best uses. There are several branches of soil science like soil physics, soil chemistry, soil mineralogy, soil biology, pedology, soil survey and soil technology. Information about soil science words is wider and scattered, hence an effort has been made to compiled a book in the form of "Dictionary of Soil Science" in Alphabatic fasion "A to Z"

This dictionary is the result of the assimilation of knowledge and work of various scientists and researcher which have been appropriately compiled with suitable modification and explanations, wherever needed. The author express his gratitude to the writers of numerous books, dictionaries which may have been consulted for this compilation.

The present dictionary consists about 750 words involving nearly all sphere of soil science. It includes the words like Integrated Plant Nutrition System (IPNS), Organic farming, Precision agriculture, Sustainable agriculture and Geographical Information System (GIS) which have been directly or indirectly most popularity and widely accepted in Soil Science. Though the manuscript has been prepared in all the due cares, yet may lack in / errors because of the nature of the subject. For providing some important comprehensive and details informations, appendices (I to XI) has been given in the last pages of the manuscript.

I hope and wish that the Dictionary of Soil Science will be fruitful for agriculture graduate, post graduate students, researchers, environmentalist, soil scientist, faculty members of agriculture colleges and universities, aspirants of agriculture related competitive examinations, research institutes etc.

I am thankful to Dr. L.L. Somani, Ex-Director and Professor, MPUAT, Udaipur for encouragement and right advise at right time. Sincere thanks to Dr. A.R.Trag, Director of Research and Dr. G.M. Wani, Director of Extension, SKUAST-K, Srinagar. Thanks are always due to my colleagues Dr.D.Ram, Dr.Tahir Ali, Dr.M.A. Malik, SKUAST-K, Srinagar. Sincere thanks to Dr. Sunil Pabbi, Senior Scientist CCUBGA, IARI, New Delhi. The dedicated computerized composing and setting by M/S Aadil Computers, Shalimar, Srinagar and the help rendered in designing, printing this book by Daya Publisher, New Delhi is also greatly acknowledged.

Lastly the author wishes to expresses his sincere gratitude to his wife Ms. Anju Singh for her untired efforts. Thiscompilation is dedicated to my loving parents and each and every member of MADERNA family who inspired me to do something best among all.

Dr. Subhash Chand

A

A horizon: The surface horizon of a mineral soil having maximum organic matter accumulation, biological activity and/or eluviation of materials such as iron and aluminum oxides and silicate clays.

ABC soil: A soil with a distinctly developed profile, including A,B, and C horizons.

Absorption: The process by which a substance is taken into and included within another substance, *i.e.* intake of water by soil or intake of nutrients by plants.

Absorption, active: Movement of ions and water into the plant root as a result of metabolic processes against an activity gradient.

Absorption, passive: Movement of ions and water into the plant root as a result of diffusion along a gradient.

AC soil: A soil having a profile containing only A and C horizons with no clearly developed B horizon.

Accessory Structural Elements: The nutrients or elements forming parts of more active and vital living plant tissues particularly proteins. For example:

Nitrogen–a constituent of proteins, and constitutes 1-3 per cent of the plant tissue.

Phosphorus–a constituent of nucleoproteins, phytin and lecithin constitutes 0.05 to 1.0 per cent of plant tissues.

Sulphur–a constituent of many proteins; its content varies from 0.05 to 1.50 per cent in plant tissues.

Acid rain: Atmospheric precipitation with pH values less than about 5.6, the acidity being due to inorganic acids such as nitric and sulfuric that are formed when oxides of nitrogen and sulphur are emitted into the atmosphere.

Acid sulphate soils: Soils having sufficient sulphides (FeS_2 and others) and pH < 3 and when drained and aerated enough for cultivation also termed as cat clays.

Acidity active: The activity of hydrogen ion in the aqueous phase of a soil. It is measured and expressed as a pH value.

Acidity, residual: Soil acidity that can be neutralized by lime or other alkaline materials but cannot be replaced by an unbuffered salt solution.

Acidity, salt replaceable: Exchangeable hydrogen and aluminum that can be replaced from an acid soil by an unbuffered salt solution such as KCl or NaCl.

Acid soil: A soil that is acid in reaction throughout the root zone. Practically, this means a soil having pH less than 6.6; precisely, a soil with a pH value less than 7.0. Such a soil has more hydrogen (H) than hydroxyl (OH) ions in the soil solution. Acid soils are grouped into five categories.

Extremely acid: pH below 4.5

Very strongly acid: 4.5-5.0

Strongly acid: 5.1-5.5

Medium acid: 5.6-6.0

Slightly acid: 6.1-6.5.

Acid soils are reclaimed by addition of liming materials.

Acidity, total: The total acidity in a soil. It is approximated by the sum of the salt replaceable acidity plus the residual acidity.

Acid-forming: A term applied to commercial fertilizers which leave an acid residue in the soil. The amount of calcium carbonate required to neutralize the acid residue is referred to its equivalent acidity. Common fertilizers along with their equivalent acidity. Common fertilizers along with their equivalent acidity.

Actinomycetes: A group of microorganisms intermediate between the bacteria and the true fungi that usually produce a characteristic branched mycelium. Includes many, but not all, organisms belonging to the order of Actinomycetales.

Activity index: Index for measuring availability of water insoluble nitrogen in fertilizers. Activity index as per AOAC method is determined by the following formula:

$$\text{Activity index (\%)} = \frac{(\%\text{CWIN} - \%\text{HWIN})}{(\%\text{CWIN})} \times 100$$

CWIN: Cold water insoluble nitrogen

HWIN: Hot water (98–100°C) insoluble N

Used as a measure of availability of nitrogen from slow release fertilizers. For urea formaldehyde, which is a slow release fertilizer, activity index should not test less than 40 per cent.

Additive: A material added to fertilizer to improve its physical condition.

Adhesion: It is process of molecular attraction that holds the surfaces of two substances (*e.g.* water and sand particles) in contact.

Adsorption complex: The group of organic and inorganic substances in soil capable of adsorbing ions and molecules.

Adsorption: It is a process of the attraction of ions or compounds to the surface of a solid. Soil colloids adsorb large amounts of ions and water.

Aerate: To impregnate with gas, usually air.

Aeration, soil: The process by which air in the soil is replaced by air from the atmosphere. In a well-aerated soil, the soil air is similar in composition to the atmosphere above the soil. Poorly aerated soils usually contain more carbon dioxide and correspondingly less oxygen than the atmosphere above the soil.

Aerobic: 1) Having molecular oxygen as a part of the environment 2) Growing only in the presence of molecular oxygen, as aerobic organisms 3) Occurring only in the presence of molecular oxygen (said of certain chemical or biochemical processes, such as aerobic decomposition).

Aggregate (soil): Many soil particles held in a single mass or cluster such as a clod, crumb, block, or prism.

Agricultural Chemistry: Includes almost all branches of chemistry, namely, analytical chemistry, inorganic chemistry, organic chemistry, bio-chemistry, radiation chemistry, physical chemistry, as applicable to agriculture system. It is an application of chemistry to soil, water, plant and animal systems. It involves the use of chemistry in fertilizers, amendments, pesticides and insecticides, herbicides and growth hormones.

Agronomy: A specialization of agriculture concerned with the theory and practice of field crop production and soil management. The scientisific management of land.

Air porosity: The proportion of the bulk volume of soil that is filled with air at any given time or under a given condition such as a specified moisture potential, usually the large pores.

Air-dry: 1) The state of dryness (of a soil) at equilibrium with the moisture content in the surrounding atmosphere. The actual moisture content will depend upon the relative humidity and the temperature of the surrounding atmosphere. 2) To allow to reach equilibrium in moisture content with the surrounding atmosphere.

Alkali Soil: A soil that contains sufficient sodium salt to interfere with the growth of most crop plants.

Alkaline forming: A term applied to commercial fertilizers that leaves an alkaline or basic residue in the soil. The basic residual effect is expressed in terms of 'equivalent basicity'. More the basic residue, more will be the equivalent basicity.

Alkaline soil: A soil that is alkaline in reaction throughout the root zone or for a major part of the root zone. Precisely, any soil having a pH value greater than 7.0. Practically, a soil having a pH above 7.3 is called alkaline. An alkaline soil is reclaimed by addition of gypsum or sulphur. Soils having pH from 7 to 7.5 are sometimes referred to as alkali soils. They correspond to black alkali soils and occur in irregular patches. They can be reclaimed by adding calcium or magnesium.

Allophane: An aluminosilicate mineral that has an amorphous or poorly crystalline structure and is commonly found in soils developed from volcanic ash.

Alluvial soil: A soil developing from recently deposited alluvium and exhibiting essentially no horizon development or modification of the recently deposited materials.

Alluvium: A general term for the detrital material deposited or in transit by streams, including gravel, sand, silt, clay and all variations and unconsolidated mixtures of these.

Alpha particle: A positively charged particle (consisting of two protons and two neutrons) that is emitted by certain radioactive compounds.

Aluminosilicates: Compounds containing aluminum, silicon and oxygen as main constituents. An example is microcline, $KAlSi_3O_8$.

AM: Abbreviated terms for 2-ammino-4-chloro-6 methylpyridine used as a nitrification inhibitor. Toxic to bacteria *nitrosomonas* which oxidize ammonium ions to nitrate and hence control nitrification. Rate of application is 4-10 ppm (on soil basis) on mixing with fertilizer. Efficiently reduces nitrification of urea and can also be used as coating material for solid fertilizers using oil as a binder. The effect of AM can last upto 40 days.

Amendment, soil: Any substance other than fertilizers, such as lime, sulfur, gypsum, and sawdust, used to alter the chemical or physical properties of a soil, generally to make it more productive.

Amino acids: Nitrogen-containing organic acids that couple together to form proteins. Each acid molecule contains one or more amino groups ($-NH_2$) and at least one carboxyl group (-COOH). In addition, some amino acids contain sulphur.

Ammonification: It is biochemical process whereby ammoniacal nitrogen is released from nitrogen-containing organic compounds.

Ammonium fixation: The entrapment of ammonium ions by the mineral or organic fractions of the soil in forms that are insoluble in water and at least temporarily nonexchangeable.

Ammo-Phos: It is the trade name generally used in the USA and Canada to designate a series of products of commercial monoammonium phosphate, either alone or in combination with ammonium sulphate or other material. Two grades namely Ammo-Phos A and Ammo-Phos B are marketed. Ammo-Phos A is a product of phosphoric acid neutralization with ammonia. It contains 11 per cent N and 48 per cent available P_2O_5. Ammo-Phos B is manufactured by neutralizing phosphoric acid-sulphuric acid mixture with ammonia. It contains 16 per cent N and 20 per cent available P_2O_5.

Ammonium sulphate: Chemically produced sulphur containing nitrogenous fertilizer. A crystalline colourless salt. Due to impurities its colour sometime appears to be brown or grey. Highly soluble in water; not hygroscopic in nature and can be stored and handled without any difficulty. In addition to nitrogen, it also supplies 24 percent sulphur. It is manufactured by three processes. Its specifications as laid down in the Indian Fertiliser Legislation (FCO, 1985) are;

1.	Moisture percent by weight maximum	1.0
2.	Ammonical nitrogen percent by weight minimum	20.6
3.	Free acidity (as H_2SO_4) percent by wt., minimum (0.04 for material obtained from by-product ammonia and by product gypsum)	0.025
4.	Arsenic (as As_2O_3) percent by weight, maximum	0.01

All the nitrogen supplied by ammonium sulphate is in ammoniacal form. It is also a carrier of sulphur, hence, favoured for legume crops, tea and coffee. Its equivalent acidity is 110.

Ammonium nitrate: A synthetically produced nitrogenous fertilizer containing one half of its nitrogen in ammonia cal form and the other half in nitrate form. Commercial grade carries 33 per cent nitrogen. It is acid-forming in reaction with 60 equivalent acidity. This fertilizer needs to be stored carefully to avoid fire hazards.

Ammonium phosphate: A series of fertilizer materials containing a chemically combined form of ammonium nitrate and dicalcium phosphate. Two commercial grades of ammonium phosphate being manufactured in India during the Fourth Five Year Plan are 20-20-0 and 28-28-0. These fertilizers are acidic in residual effect. The 20-20-0 grade ammonium phosphate has 25 equivalent acidity.

Ammonium sulphate-nitrate: A synthetically produced nitrogenous fertilizer containing nitrogen in ammoniacal and nitrate form. Commercial grade of this fertilizer contains 26 per cent nitrogen. It is acid-forming in nature, with 93 equivalent acidity.

Anhydrous ammonia: Contains 82 per cent nitrogen in liquid form obtained by compressing ammonia gas. It is highly acidic, with 148 equivalent acidity. It requires special equipment to apply to the soil.

Anion: An ion carrying a negative charge of electricity.

Amphiboles: A group of aluminosilicate minerals having a structure containing double chains of link silica tetrahedral.

Anaerobic: Without molecular oxygen. The opposite of aerobic.

Anion: Negatively charged ion; during electrolysis it is attracted to the positively charged anode.

Anion sorption: The process by which an anion, such as phosphate, replaces OH^- groups on mineral surfaces and edges.

Antagonistic Effect: The application of one nutrient element may accentuate or induce deficiency of another nutrient due to related antagonism. For example, excess copper adversely affects the iron nutrition. Iron and manganese have an antagonistic effect on each other. Zinc and iron behave in a similar way. Heavy doses of phosphorus application induce zinc deficiency. High levels of potassium and calcium depress boron absorption.

AOAC: Abbreviation for Association of Official Analytical Chemists. The new name for the earstwhile American Association which was called the Association of Official Agricultural Chemists. The association was founded in Philadelphia on September 9, 1884. The membership of the association is institutional and includes State Departments of Agriculture, State Agricultural Colleges and Experimental Stations, the Federal Department of Agriculture and the Federal State and City officers charged with enforcement of food, feed, drug, fertilizer, insecticide and fungicide control law.

Ap: The surface layer of a soil disturbed by cultivation or pasturing.

Apatite: A phosphate bearing mineral contained in rock phosphate. There are five types of apatites.

1. Hydroxy apatite $Ca_{10}(PO_4)_6(OH)_2$

2 Chloroapatit $Ca_{10}(PO_4)_6Cl_2$

3. Fluoroapatite $Ca_{10}(PO_4)_6F_2$

4. Carbonatoapatite $Ca_{10}(PO_4)_6CO_3$

5. Sulphatoapatite $Ca_{10}(PO_4)_6SO_4$

Apatite is a source of phosphorus and phosphoric acid used in fertilizer and other industries. Also used for manufacturing bone chine. Fluorine rich apatite yields hydrofluosilicic acid as a by-product.

AR Grade: An abbreviation for Analytical Reagent Grade. AR grade chemicals are high purity chemical reagents which are used in any standard analytical work requiring high precision. Generally, on packings of such reagents, detailed assay including maximum impurities that can be found in the reagent are given.

Arid climate: Climate in regions that lack sufficient moisture for crop production without irrigation. In cool regions annual precipitation is usually less than 25 cm. It may be as high as 50 cm in tropical regions. Natural vegetation is desert shrubs.

Arnon's criteria of essential elements: Plants absorb a number of elements from soil during their growth. All these elements are not essential for plants. The criteria proposed by Arnon for deciding whether an element is essential for plants or not are;

1. The deficiency of the element makes it impossible for the plant to complete the vegetative or reproductive stages of its life cycle.
2. The deficiency is specific to the element in question: and as such, can be prevented or corrected only by supplying that particular nutrient element to the plant.
3. The element must have a direct influence on the plant and must be directly involved in the nutrition and metabolism of the plant quite apart from its possible effect on correcting some unfavourable microbiological or chemical conditions of the soil or other growing medium.

Atom: The smallest existing unit of an element. The term is derived from the Greak word meaning "individual". It consists of a nucleus which contains protons (positively changed particles) and neutrons (neutral particles). The nucleus is surrounded by electrons (negatively charged particles) which move in their respective orbits. The atom is electrically neutral as the number of electrons in an atom is equal to number of protons.

Atomic Absorption Spectrophotometer (AAS): An instrument used in analysis of a variety of products (plant, soil and water) including fertilizers based on the absorption of light by atoms. More than 60 elements can be determined by AAS.

The basic components of an atomic absorption spectrophotometer are: Atomizer, Light sources (cathode lamps) Optics, Monochromators and Detector.

Available nutrients in soils: A part of the plant nutrient in the soil that can be taken up by growing plants immediately. Available nitrogen is defined as the water soluble nitrogen plus the part that can be made soluble or converted into free ammonia. Available phosphoric acid is that part which is soluble in water or in a week dilute acid such as 2 per cent citric acid. Available potash is defined as that portion which is soluble in water or in a solution of ammonium oxalate.

Autotrophs: Group of micro organisms which obtain their energy from sunlight or by oxidation of inorganic compounds. If the energy is derived from sunlight, the group is called photo-autotroph. The other group which derives energy from oxidation of inorganic material is called chemo-autotroph. Algae are photo-autotrophs. Some autotroph bacteria which play an important part in availability of plant nutrients are:

1. *Nitrosomonas*–oxidize ammonium to nitrate

2. *Thiobacillus*–oxidize inorganic S to sulphate

3. *Nitrobacters*–oxidize nitrite to nitrate.

4. *T.ferroxidans-oxidise* ferrous iron to ferric form.

Azofication: Non-symbiotic nitrogen fixation by the *azotobacter* group of soil bacteria which use organic matter as a source of energy and are able to obtain nitrogen from the atmosphere to build up their body protein. After the death of the bacterial cell, this nitrogen is returned to the soil for use by higher plants.

Azolla: A small floating aquatic fern that is found in the tropics and sub-tropics and some temperate zones of the world. It is a member of order Salviniales and comprises of seven recognized living species, namely:

1. *Careliniana*
2. *Filiculoides*
3. *Mexicana*
4. *Microphylla*
5. *Nilotica*
6. *Pinnata*
7. *Rubra*

For rice cultivation azolla is applied as green manure both by basal application and as top dressing. Under optimum conditions an azola crop may produce 3 kg or more N/ha/day. Thus a 20 day old azolla crop can produce 60 kg N/ha. Two basal crops of 20 days each thus can produce almost 120 kg N/ ha for the following rice crop.

Azola provides nutrients and protective cavity to anabaena algae, which in exchange fixes atmospheric nitrogen. In addition to nitrogen fixation, the azolla is also used as fodder or feed for animals and as a weed suppressor in flooded rice. In some places like lakes, ponds, waterways and aquatic crops, azolla may be considered as a weed.

Azotobacter: Free living bacteria capable of utilizing atmospheric nitrogen and fixing it for their biosynthetic reaction. These can also utilize ammonium, nitrate, nitrite, urea and sometimes organic nitrogen.

Azotobacter are essentially aerobic. There are five species of azotobacter recognized on the basis of cell shape, pigmentation and mobility:

1. *Azotobacter chroococcum*
2. *Azotobacter beijerinckii*
3. *Azotobacter vinelandii*
4. *Azotobacter macrocytogenes*
5. *Azotobacter agilis.*

B

B horizon: Soil horizon usually beneath the A that is characterized by one or more of the following: q) a concentration of silicate clays, iron and aluminum oxides, and humus, alone or in combination b) a blocky or prismatic structure and c) coatings of iron and aluminum oxides that give darker, stronger, or redder color.

Bacterial culture: Any media enriched with any particular bacteria. The culture describes the bacteria contained in it *e.g.*

1. Rhizobium culture: is used for inoculating seeds of leguminous crops.
2. Azotobacter culture: are of a general nature and same culture can be used for inoculating seeds of a number of crops.

Band placement: The placement of fertilizers in the soil in continuous band or narrow ribbons in between the plant rows or around each plant. The fertilizer applied in band is covered by the soil but is not mixed with it. Phosphatic fertilizers are normally placed in bands to reduce phosphatic fixation.

Bar: A unit of pressure equal to one million dynes per square centimeter (10^6 dynes/ cm^2).

Base saturation percentage: The extent to which the adsorption complex of a soil is saturated with exchangeable cations other than hydrogen and aluminum. It is expressed as a percentage of the total cation exchange capacity.

$$\%BS = \frac{BC}{CEC} \times 100$$

where,

%BS: Base saturation

BC: Me of basic cations per 100 gram soil.

CEC: Total cation exchange capacity (me/100 g soil)

Basic slag: A by-product of steel industry obtained from phosphate iron ores. It contains about 6 to 18 per cent total phosphoric acid and 5 to 15 per cent in available form (soluble in two per cent citric acid). Basic slag produced in India being of low grade is not used as a commercial fertilizer.

BC soil: A soil profile with B and C horizons but with little or no A horizon. Most BC soils have lost their A horizons by erosion.

Bedrock: The solid rock derlying soils and the regolith in depths ranging from zero (where exposd by erosion) to several hundred feet.

Beer-Lambert law: Beer's law and Lambert's law combined together to form the Beer-Lambert Law. According to this law the magnitude of light absorption is proportional to concentration, thickness of media and intensity of light. Spectrophotometric instruments are based on the Beer-Lambert Law.

Mathematically it is expressed as:

$$\text{Log} \frac{lo}{It} = KCH$$

where,

lo: Intensity of incident light

It: Intensity of transmitted light

C: Concentration

H: Thickness of media

K: Equilibrium Constant depending on the nature of the substance.

Biogas: The gas produced from organic waste by microbiological reaction under anaerobic conditions. The organic wastes used in the production of biogas are generally cattle yard waste human waste, vegetative crop residues, unwanted aquatic plants and weeds. Biogas is sometimes termed after the material or method of gas production, *e.g.* marsh gas, sewage gas, sludge gas, digestor gas and gobar (cowdung) gas, etc. As a fuel, biogas is a potential source of energy. It is usually have the following composing.

Methane: 50-60 per cent

Carbondioxide: 30-40 per cent

Hydrogen: 5-10 per cent

Hydrogensulphide: traces

Water vapour: traces

Biological test: A technique for assessing nutrient status of soil using biological material. The following types of tests are commonly used:

1. Field test: Involves use of field crops in experimental or farmers' fields.
2. Laboratory or green house tests:
   (a) Mitscherlich pot culture
   (b) Lettuce pot culture
   (c) Neubauer seedling method
   (d) Sunflower pot culture
3. Microbiological tests:
   (a) Azotobacter culture
   (b) Sackett and Stewart technique
   (c) Aspergillus Niger test
   (d) Mehlich's Cunninghamella Plaque method.

Biomass: The amount of living matter in a given area.

Biuret ($C_2O_2N_3H_5$): A chemical compound formed by the combination of two molecules of urea with the release of a molecule of ammonia, when the temperature during the urea manufacturing process exceeds a certain level. Fertiliser grade urea contains variable amounts of biuret. Biuret is toxic to plants, particularly when it is applied through sprays. As per the Indian fertilizer Legislation (FCO, 1985) the biuret content in urea should not exceed 1.5 per cent.

Blood Meal: Dried blood obtained from a slaughter house. The blood lot is treated with copper sulphate at the rate of 2 oz per 100 lb of blood lot. It is then heated to evaporate from moisture and then dried under the sun. After drying it is powered, bagged and sold as blood meal. Various other treatments like tresting blood serum or clots with lime or very dilute hydrochloric acid or absorbing the blood in organic wastes and then drying in the sun, are also used in the preparation of blood meal. It is quite good organic fertilizer containing 10 to 12 per cent nitrogen and 1-2 per cent P_2O_3.

Blocky soil structure: Soil aggregates with block like shapes, common in B horizons of soils in humid regions.

Blown-out land: Areas from which all or almost all of the soil and soil material has been removed by wind erosion. Usually unfit or crop production. A miscellaneous land type.

Blue-Green Algae (BGA): Blue-green algae belong to the class Cyanophyceae or Myxophyceae. These are blue-green in colour because of the presence of pigments like chlorophyll, carotenes, xanthophylls, physcocyanin and phycoerythrin. The majority of blue-green algae species are found in fresh water and are often abundantly found in the standing water of rice fields. They reproduce vegetatively and multiply at a very fast rate. Blue green algae are divided into five orders:

1. *Chroococcales*
2. *Chmaesiphonates*
3. *Plerocapsales*
4. *Nostocales*
5. *Stigonematales.*

Important genera concerned with nitrogen-fixing activity are: Anabaena, Nostoc, Cylindrospermum. According to conservative estimates, blue-green algae can fix about 25-30 kg N/ha per cropping season.

The algae also give some additional benefits.

1. Because of photosynthetic activity they release oxygen which is a key factor in rice culture.
2. They increase the organic content of soil.
3. They improve the physical structure by increasing water stable aggregates.
4. They supply growth promoting substances like Vitamin B_{12}, IAA-IPA, Anthramilic acid and other allied compounds.

Bonemeal, raw: A fertilizer made of dried animal bones finely ground. It contains 20 to 24 per cent P_2O_5. The availability of its plant food depends largely upon how fine it is ground.

Bonemeal, steamed: A product made from grinding bones previously treated with steam under pressure. It contains one to two per cent nitrogen and 22 per cent phosphorus.

Borax: A chemical compound that contains approximately 11 per cent boron. Borax is applied to boron-deficient soil or sprayed on the plant's foliage which indicate boron deficiency.

Boron: One of the essential elements absorbed by plants as BO_3^{-3} and $HB_4O_7^-$ ions. It is required in a very small quantity, hence is classified as a micronutrient. It is primarily concerned with calcium uptake and its assimilation. It effects cell division, fruit formation, viability of pollen grains, carbohydrate metabolism and protein synthesis in plants. Its deficiency leads to symptoms like heart rot of sugarbeet, browning in cauliflower, corking in apple etc.

Bray's nutrient mobility concept: It states that as the mobility of a nutrient in the soil decreases, the amount of that nutrient needed in the soil to produce a maximum yield (the soil nutrient requirement) increases from a variable net value, determined principally by the magnitude of the yield and the optimum percentage composition of the crop, to an optimum percentage composition of the crop, to an amount whose value tends to be constant.

Breccia: A rock composed of coarse angular fragments cemented together.

Buffering capacity: The ability of a soil to resist changes in pH. Commonly determined by presence of clay, humus, and other colloidal materials.

Bulky organic manures: These manures are bulky in nature and supply plant nutrients in small quantities and organic matter in large quantities.

C

C horizon: A mineral horizon generally beneath the solum that is relatively unaffected by biological activity and pedogenesis and is lacking properties diagnostic of an A or B horizon. It may or may not be like the material from which the A and B have formed.

Calcareous soil: Soil containing sufficient calcium carbonate (often with magnesium carbonate) to effervesce visibly when treated with cold 0.1 N hydrochloric acid.

Calcium ammonium nitrate (CAN): This commercial nitrogenous fertilizer, prepared by mixing powdered limestone or dolomite with ammonium nitrate. It contain 25 percent nitrogen. One half of the nitrogen in calcium ammonium nitrate is in the nitrate form, the remaining half in the ammoniacal form. CAN is suitable for basal dressing as well as for top dressing.

Calcium sulphate or gypsum: Gypsum is hydrated calcium sulphate. The material can be manufactured by treating lime or calcium phosphate with sulphuric acid. It is applied to the soil to reclaim saline and alkaline soils or to bring down soil pH.

Caliche: A layer near the surface, more or less cemented by secondary carbonates of calcium or magnesium precipitated from the soil solution. It may occur as a soft, thin soil horizon, as a hard, thick bed just beneath the solum, or as a surface layer exposed by erosion.

Carbon: An essential plant element. Plants take their carbon requirement from atmospheric carbon dioxide. Since it is never deficient in the atmosphere to become a limiting factor in growth, it is not supplied in any form as fertilizer. Carbon present in organic compounds in soil cannot be used as a nutrient source of carbon.

Carbon cycle: The sequence of transformations whereby carbon dioxide is fixed in living organisms by photosynthesis or by chemosynthesis, liberated by respiration and by the death and decomposition of the fixing organism, used by heterotrophic species, and ultimately returned to its original state.

Carbon-nitrogen ratio: The ratio of the weight of total organic carbon to the weight of total nitrogen in a soil or in an organic material. C:N ratio of wheat straw is nearly 80:1 while that of soil is 10:1. When undecomposed straw with a high C:N ratio is applied to the soil, its C:N ratio is reduced through bacterial decomposition. To speed up bacterial decomposition, nitrogenous fertilizer is added to the soil at the time of turning under straw.

Catalyst: An element or compound that accelerates the rate of a chemical reaction, but itself does not take part in the reaction. Common characteristics of a catalyst are:

1. The quantity of catalyst required is very small.
2. It does not disturb or has no influence on state of equilibrium of a system.
3. It remains chemically unchanged at the end of a reaction.
4. It cannot start a reaction but can only accelerate the speed of reaction.
5. Catalysts are very specific in nature.

Catalysts are extensively used in the fertilizer industry. The most important example is the use of iron oxide catalyst promoted by metals like aluminium, potassium, calcium and magnesium in the synthesis reaction of ammonia. Nickel is used as a catalyst in steam reforming of natural gas or naptha.

Capillary water: The water held in the "capillary" or *small* pores of a soil, usually with a tension >60 cm of water.

Catena: A sequence of soils of about the same age, derived from similar parent material, and occurring under similar climatic conditions but having different characteristics because of variation in relief and in drainage.

Cation: An ion carrying positive charge of electricity. The common soil cations are calcium, magnesium, sodium, potassium and hydrogen.

Cation exchange: The exchange of cations held by the soil absorbing complex is rich in sodium (as is the case in alkali and alkaline soil), application of gypsum (calcium sulphate) causes calcium cations to exchange with sodium cations.

Cation exchange capacity: The sum total of exchangeable cations that a soil can adsorb. Sometimes called "total-exchange capacity" "base-exchange capacity", or "cation-adsorption capacity." Expressed in centimoles per kilogram (cmol/kg) of soil (or of other adsorbing material such as clay).

Channery: Thin, flat fragments of limestone, sandstone, or schist up to 15 cm (6 in.) in major diameter.

Chelates: An organic compound capable of holding the plant nutrient in a form which prevents it from getting tied with other elements in the soil, thus keeping it more or less in available form for the plant. The term refers to the claws of a crab illustrative of the way in which the atom is held. For examples EDTA, DTPA, HEEDTA and CDTA.

Chert: A structureless form of silica, closely related to flint, that breaks into angular fragments.

Chisel, subsoil: A tillage implement with one or more cultivator-type feet to which are attached strong knife like units used to shatter or loosen hard, compact layers, usually in the subsoil, to depths below normal plow depth.

Chlorite: A 2:1:1 type layer structured silicate mineral having 2:1 layers alternating with a magnesium dominated octahedral sheet.

Chlorosis: A condition in plants relating to the failure of chlorophyll (the green coloring matter) to develop. Chlorotic leaves range from light green through yellow to almost white.

Class, soil: A group of soils having a definite range in a particular property such as acidity, degree of slope, texture, structure, land use capability, degree of erosion, or drainage.

Clastic: Composed of brokwn fragments of rocks and minerals.

Clay: (1) A soil separate consisting of particles <0.002 mm in equivalent diameter (2) A soil textural class containing >40 per cent clay, <45 per cent sand, and <40 per cent silt.

Claypan: A dense, compact, slowly permeable layer in the subsoil having a much higher clay content than the overlying material, from which it is separated by a sharply defined boundary. Claypans are usually hard when dry and plastic and sticky when wet.

Clay humus complex: It is a complex formed due to interaction of the clay and humus colloids through different mechanisms and force like, cation bridge, hydrogen bonding, vander waals forces.

Clod: A compact, coherent mass of soil produced artificially, usually by such human activities as plowing and digging, especially when these operations are performed on soils that are either too wet or too dry for normal tillage operations.

Closed formula mixed fertilizers: The fertilizers grade is disclosed on each bag of such fertilizer mixture, but the ingredients or straight fertilizers used in formulating the mixture are not disclosed. In India, fertilizer mixtures sold to the cultivators are usually of the closed formula type.

Cohension: Holding together: force holding a solid or liquid together, owing to attraction between like molecules. Decreases with rise in temperature.

Colloid, soil (Greek, glue-like) Organis and inorganic matter with very small particle size and a correspondingly large surface area per unit of mass.

Colluvium: A deposit of rock gragments and soil material accumulated at the base of steep slopes as a result of gravitational action.

Complex fertilizers: The commercial fertilizers containing a least two or more of the primary essential nutrients. When such fertilizers contain only two of the primary nutrients, they are designed as incomplete complex fertilizers. While those containing all three primary nutrients are called as complex fertilizers. See multiple-nutrient materials for more information.

Compost: A mass of rotted organic matter made from waste.

Compost: Organic residues, or a mixture of organic residues and soil, that have been piled, moistened, and allowed to undergo biological decomposition. Mineral fertilizers are sometimes added. Often called "artificial manure" or "synthetic manure" if produced primarily from plant residues.

Composting: It is largely a biological process in which micro-organisms of both types namely aerobic and anaerobic, decompose the organic matter and lower the carbon-nitrogen ratio of the refuse. The final product of composting is a well-rooted manure known as compost.

Concretion: A local concentration of a chemical compound, such as calcium carbonate or iron oxide, in the form of grains or nodules of varying size, shape, hardness, and color.

Conditioner (of fertilizer)**:** A material added to a fertilizer to prevent caking and to keep is free-flowing.

Conduction: The transfer of heat by physical contact between two or more objects.

Consumptive use: The net surface charge of mineral particles, the magnitude of which depends only on the chemical and structural composition of the mineral. The charge arises from isomorphous substitution and is not affected by soil pH.

Contour: An imaginary line connecting points of equal elevation on the surface of the soil. A contour terrace is laid out on a sloping soil at right angles to the direction of the slope and nearly level throughout its course.

Convention: The transfer of heat through a gas or solution because of molecular movement.

Copper sulphate: It is the most common copper salt used in fertilizers, insecticides and fungicides. Used as a fertilizer, commercial grade of copper sulphate supplies 25 to 35 per cent copper (Cu).

Cotton seed cake: A by-product of the cotton seed crushing industry, when it is undecorticated, it is non-edible containing 3.9 per cent nitrogen, 18 1.8 per cent phosphoric acid and 1.6 per cent potash. When decorticated, cotton seed cake becomes edible oilcake containing higher percentages of plant nutrients. Decorticated cotton seed cake contains 6.4 per cent nitrogen, 2.9 per cent phosphoric acid and 2.2 per cent potash.

Crop rotation: A planned sequence of crops growing in a regularly recurring succession on the same area of land, as contrasted to continuous culture of one crop or growing different crops in haphazard order.

Crust: A surface layer on soils, ranging in thidkness from a few millimeters to perhaps as much as 3 cm, that is much more compact, hard, and brittle, when dry, than the material immediately beneath it.

D

Darcy's law: Darcy stated that the rate of flow increased with an increased depth of water above the bottom of the soil and decreased with an increased depth of soil through which water followed and

$$Qw = \frac{-K\,(\Delta dw)\,At}{\Delta ds}$$

where,

Qw: Quantity of water in c.c.

K: Hydraulic conductivity cm/ sec.

Δdw: Water of hydraulic head in cm.

A: Soil area in sq. cm.

t: Time in second

Δds: Soil depth in cm.

Deficiency: A condition of insufficient supply of essential plant nutrients required for various metabolic functions.

Plants cannot reach optimum growth, if any one of the essential elements is limited in supply. Deficiency may be caused due to any one of the following reasons:

1. The nutrient is in short supply in soil.
2. The nutrient is there in soil, but is not in the available form in which plants can absorb it.

3. There is the antagonistic effect of other elements. See also Antagonistic effect. This is called induced deficiency as the nutrient in question is not being metabolized because of the excess of other antagonistic elements.

Denitrification: The process by which nitrates or nitrites in the soil or in farmyard manure are reduced to ammonia or free nitrogen by bacterial action. The process results in the escape of nitrogen into the air and is therefore wasteful.

Desalinization: Removal of salts from saline soil, usually be leaching.

Diagnostic subsurface horizons: The following diagnostic subsurface horizons are used in *Soil taxonomy.*

Agric horizon: A mineral soil horizon in which clay, silt, and humus derived from an overlying cultivated and fertilized layer have accumulated. The wormholes and illuvial clay, silt, and humus occupy at least 5 per cent of the horizon by volume.

Argillic horizon: A mineral soil horizon characterized by the illuvial accumulation of layer lattice silicate clays.

Calcic horizon: A mineral soil horizon of secondary carbonate enrichment that is more than 15 cm thick, has a calcium carbonate equivalent of more than 15 per cent and has a least 5 per cent more calcium carbonate equivalent than the underlying C horizon.

Cambic horizon: A mineral soil horizon that has a texture of loamy very fine sand or diner, contains some weatherable minerals, and is characterized by the alteration or removal of mineral material. The cambic horizon lacks cementation or induration and has too few evidences of illuviation to meet the requirements of the argillic or spodic horizon.

Duripan: A mineral soil horizon that is cemented by silica, to the point that air-dry fragments will not slake in water or HCl.

Gypsic horizon: A mineral soil horizon of seconday calcium sulfate enrichment that is more than 15 cm thick.

Kandic horizon: A horizon having a sharp clay increase relative to overlying horizons and having low-activity clays.

Natric horizon: A mineral soil horizon that satisfies the requirements of an argillic horizon, but that also has prismatic, columnar, or blocky structure and a subhorizon having more than 15 per cent saturation with exchangeable sodium.

Oxic horizon: A mineral soil horizon that is at least 30 cm thick and characterized by the virtual absence of weatherable primary minerals or 2:1 latice clays and the presence of 1:1 lattice clays and highly insoluble minerals such as quartz sand, hydrated oxides of iron and aluminum, low cation exchange capacity, and small amounts of exchangeable bases.

Spodic horizon: A mineral soil horizon characterized by the illuvial accumulation of amorphous materials composed of aluminum and organic carbon with or without iron.

Diatomaceous earch: A geologic deposit of fine, grayish, siliceous material composed chiefly or wholly of the remains of diatoms. It may occur as a powder or as a porous, rigid material.

Diatoms: Algae having siliceous cell walls that persist as a skeleton after death, any of the microscopic unicellular or colonial algae constituting the class Bacillariaceae. They occur abundantly in fresh and salt waters and their remains are widely distributed in soils.

Diffusion: The transport of matter as a result of the movement of the constituent particles. The intermingling of two gases or liquids in contact with each other takes place by diffusion.

Dilution: The process of lowering the concentration of any component of mixture (say a solution, slurry, suspension or solid) by adding any other suitable substance called a diluant. It is very common practice undertaken in any chemical laboratory and is done to bring the elemental concentration within the working range of the equipment or method. For instance, zinc sulphate heptahydrate is guaranteed to contain 21 per cent or 210000 ppm zinc. To bring it to the working range (about 1 ppm Zn) of the atomic absorption spectrophotometery it is diluted 2,00,000 times by adding distilled water.

Dioctahedral: An octahedral sheet, or a mineral containing such a sheet, that has two thirds of the octahedral sites filled with trivalent ions such as aluminum or ferric iron.

Disintegration: The breakdown of rock and mineral particles into smaller particles by physical forces such as frost action.

Disperse (1) To break up compound particles, such as aggregates, into the individual component particles. (2) To distribute or suspend fine particles, such as clay, in or throughout a dispersion medium, such as water.

Diversion dam: A structure or barrier built to divert part or all of the water of a stream to a different course.

Dolomite: A natural mineral composed of calcium and magnesium carbonate widely used as liming material and as an ingredient in fertilizer mixtures. Dolomite is used in fertilizers to render them non-acid forming. It also supplies available magnesium.

Drain (1) To provide channels, such as open ditches or drain tile, so that excess water can be removed by surface or by internal flow (2) To lose water (from the soil) by percolation.

Drainage, soil: The frequency and duration of periods when the soil is free from saturation with water.

Drift: Material of any sort deposited by geological processes in one place after having been removed from another. Glacial drift includes material moved by the glaciers and by the streams and lakes associated with them.

Drumlin: Long, smooth cigar-shaped low hills of glacial till, with their long axes parallel to the direction of ice movement.

Dry farming: The system in which field crops are raised with an annual rainfall of less than 25 inches, without irrigation facilities. From plant nutrition point of view, lesser doses of fertilizers are recommended to dry farming areas for all field crops, compared to irrigated farming.

Dryland farming: The practice of crop production in lo rainfall areas without irrigation.

DTPA: It is abbreviated term of diethylene trinitrilopenta acetic acid. It is a chelating agent and used for making metal chelates keeping the different metellic cations available in the soil and also increases the translocation of metallic cations within the plant body.

Duff: The matted, partly decomposed organic surface layer of forest soils.

Duripan: A mineral soil horizon that is cemented by silica, to the point that air-dry fragments will not slake in water or HCl.

E

Edaphology: The science which deals with the influence of soil on living things, particularly plants. It also include the use of soil by men for plant growth.

Efficacy of fertilizer: The whole of the amount of nutrient from the applied fertilizer is not recovered by the crop as a part of it is lost or remains in the soil. The extent of recovery of applied nutrient by a crop or crop rotation indicates the efficacy of a fertilizer. Some factors for controlling efficiency of fertilizer application are listed below:

1. The nature of crop and its variety.
2. Method and time of application of fertilizer.
3. Crop management
4. Cropping system.
5. Chemical composition of soil and its pH.
6. Organic matter content of the soil.
7. Physical condition including drainage, aeration, etc.
8. Weather conditions.
9. Soil moisture.
10. Balance of nutrients.

E horizon: Horizon characterized by maximum illuviation (washing out) of silicate clays and iron and aluminum oxides, commonly occurs above the B horizon and below the A horizon.

Earth flow: The process of saturated soil moving down a slope under the force of gravity. The term is also used to describe the results of the process.

Effective precipitation: The amount of proportion of precipitation that infiltrates into soil.

Effluent: Liquid waste from either a septic tank or a sewage treatment plant.

Eutrofication: Pollution with unwanted nutrients.

Electromagnetic spectrum: Range of energy and wavelengths of electromagnetic radiation. The range from shortwave high energy radiation to longwave low energy radiation includes X-ray, ultraviolet, visible light, infrared and radio waves.

Electron: A small, negatively charged atomic particle.

Electrostatic (Attraction or repulsion): Interaction between electrically charged objects.

Element: Any substance that cannot be further separated except by nuclear disintegration.

Eluvial horizon: A soil layer (horizon) formed by the removal of constituents such as clay or iron.

Eluviation: Process of removing from a soil layer any soil constituents in suspension.

Emissivity: Relative measure of an objects ability to emit radiant energy at a given temperature.

Emulsion: The light-sensitive coating on photographic film, usually containing a silver salt such as silver chloride.

Enzymes: Protein catalysts produced in cells of living oranisms that direct and control the cells chemical reactions.

Eolian soil material: Soil material accumulated through wind erosion.

Epidermis: In plants, the outside layer of cells.

Epipedons: Surface layers of soil with specific characteristics used in classifying soils by soil taxonomy (*e.g.*, the mollic epipedon).

Equivalent: The unit formerly used to describe cation exchange capacity or quantities of ions. One gram atomic weight of hydrogen or the amount of any other ion that will combine with or displace this amount of hydrogen. The amount that provides 1 mol of charge. Equals 1 mol.

Equivalent acidity: A term used to express the calcium carbonate equivalent of acidic residue left by a fertilizer material.

Equivalent basicity: A term used to express the calcium carbonate equivalent of basic residue left by a fertilizer material See 'alkaline-forming' for further details.

Equivalent weight: The weight of an element that will combine with 7.9997 of oxygen or 1.0079 of hydrogen. Equivalent weight of molecules is calculated by dividing molecular weight with the number of hydrogen cations or protons/electrons that will take part in any reaction with the molecule, *e.g.* Equivalent weight of sulphuric acid = 98/ 2 = 49 where 98 is the molecular weight of sulphuric acid and 2 is the number of hydrogen cations that will take part in any reaction with sulphuric acid.

Erosion: The wearing away of the land surface by water, wind, ice, or gravity.

Essential plant nutrient: A nutrient essential for plants for proper growth and development. At present, there are sixteen essential plant nutrients recognized by plant physiologists. These are carbon, hydrogen, oxygen, nitrogen, phosphorus, calcium, potassium, magnesium, sulphur, manganese, boron, copper, zinc, iron, molybdenum and chlorine.

Eucaryotic: In biology, referring to the type of cells with a distinct nucleus with a nuclear membrane, characteristic of fungi, protozoa, algae, plants and animals.

Evapotranspiration: Evaporation plus transpiration.

Exchange complex: All the materials (clay, humus) that contribute to a soils exchange capacity.

Exchangeable base saturation: Exchangeable cations other than Al^{3+} and H^+, expressed as a percentage of cation exchange capacity measured at neutrality.

Exchangeable ions: Ions (charged atoms or molecules) held by electrical attraction at charged surfaces, can be displaced by exchange with other ions.

Exchangeable potassium: That form of potassium is soil exchange complex which is not recovered by water solution, but is recovered by the exchange process using a suitable reagent. The most commonly used reagent is ammonium acetate. Exchangeable potassium is the primary source of potassium for plants. The level of exchangeable potassium in soil is an important factor in determining the responsiveness of any crop to application of potassic fertilizers.

Exchangeable sodium percentage (ESP): Amount of exchangeable Na expressed as a percentage of total exchangeable cations. It can be expressed by given formula.

$$\text{ESP} = \frac{\text{Exchangeable sodium (Cmol/kg soil)}}{\text{Cation exchange capacity (Cmol/kg soil)}} \times 100$$

Extracellular: Outside the cell. Extracellular enzymes are excreted by some bacteria and fungi.

Extracellular enzyme: A protein substance that acts as an organic catalyst excreted outside the bacterial or plant cell. Also called an *exoenzyme.*

Extrusive rock: Any rock that forms from hot molten material by solidifying at the earths surface

F

Facultative organism: An organism capable of both aerobic and anaerobic metabolism.

Fallow: Cropland left idle in order to restore productivity, mainly through accumulation of water, nutrients, or both. Summer fallow is a common stage before cereal grain in regions of limited rainfall. The soil is kept free of weeds and other vegetation, thereby conserving nutrients and water for the next years crop.

Fallowing: Keeping the land free of a crop or weeds for a period of time. This is done to restore soil productivity mainly through accumulation of water, nutrients or both. Fallow during the kharif season is a common practice in wheat-growing regions which receive limited rainfall. During fallow, the field is cultivated to control weeds and to help the storage of moisture for the succeeding crop.

Family, soil: In soil classification, one of the categories intermediate between the great group and the soil series. Families are defined largely on the basis of physical and mineralogical properties of importance to plant growth.

Farmyard manure (FYM): It is a product of decomposition of the liquid and solid excreta of animals stored in the farm. In western countries, straw or other litter used as bedding is also included along with the animal excreta. In India, since straw is used mainly for fodder purposes, farmyard manure is made mainly from animal excreta. The composition of FYM varies with the nutrient content in excreta and with the method of preparation. On an average it contains 0.5 per cent N, 0.2 per cent P_2O_5 and 0.5 per cent K_2O.

Fauna: The animal life of a region.

Fermentation: A set of metabolic processes by which anaerobic organisms obtain energy by converting sugars to alsohols or acids and CO_2.

Ferrihydrite: A dark raddish brown poorly crystalline iron oxide that forms in wet soils.

Fertigation: The addition of soluble fertilizer through an irrigation system.

Fertility, soil: The quality of a soil that enables it to provide essential chemical elements in quantities and proportions for the growth of specified plants.

Fertility index: It is defined as the relative sufficiency expressed as a percentage of the amount of nutrient adequate for optimum yields. It is related with soil test values and crop response as follows:

Soil Test Rating	*Crop Response*	*Fertility Index*
Very low	Highly probable	0–10
Low	Probable	10–25
Medium	Possible	25–50
High	Unlikely	50–100
Very high	Highly unlikely	100+

Fertilizer (Control) order, (1985): An order issued by the Government of India under powers conferred by section-3 of the Essential Commodities Act, 1955. Under the powers, the Government controls the production, marketing, price and quality of fertilizers. There are 39 clauses of this order covering definitions, prices, registration of dealers and fertilizer mixtures, regulation on manufacture, sale, packing requirement, disposal of non-standard fertilizers, enforcement authorities, analysis of samples, etc. The Schedule I of the order includes detailed specifications of fertilizer covered by it. In Schedule II detailed procedure regarding sampling technique and methods of analysis is given.

Fertilizer grade: An expression that indicates the percentage of plant nutrients in a fertilizer, *e.g.* a 10-55 grade of fertilizer indicates 10 per cent nitrogen, 5 per cent phosphoric acid (P_2O_5) and 5 per cent potash (K_2O).

Fertilizer legislation: This refers to laws and regulations enforced by the Government to regulate the quality of fertilizers or fertilizer mixtures sold to consumers. According to the Fertilizer Control Order of the Govt.of India of 1957, each brand and grade of commercial fertilizer and fertilizer mixture is to be registered and each bag of fertilizer to be labeled with fertilizer grade clearly written on it.

Fertilizer ratio: The fertilizer ratio designates the relative proportion of three major plant nutrients, keeping the percentage of nitrogen as one of the ratio. Thus a 5-10-5 fertilizer mixture which contains 5 per cent total nitrogen. 10 per cent available P_2O_5 and 5 per cent soluble potash, has a nutrient ratio 1:2:1.

Fertilizer requirement: The quantity of certain plant nutrient elements needed, in addition to the amount supplied by the soil, to increase plant growth to a designated optimum.

Fertilizer: Any natural or manufactured material, dry or liquid, added to the soil in order to supply one or more plant nutrients. The term is generally applied to commercially manufactured materials other than lime or gypsum. When fertilizers are sold on a large scale, they are called commercial fertilizers. Fertilizers consumed in India on a large scale are of five types; nitrogenous fertilizers, phosphatic fertilizers, potassic fertilizers, complex fertilizers or fertilizers supplying more than one major plant nutrient and fertilizer mixtures.

Field capacity (field moisture capacity): The percentage of water remaining in a soil two or three days after its harving been saturated and after free drainage has practically ceased.

Field experiments: Experiments conducted in the field to determine the type and amount of fertilizers to suit particular soil type and crop.

Filler: Material added to a fertilizer mixture or fertilizer material to make up the difference between the weight of an added ingredient required and the total fertilizer that should be made according to the fertilizer formula. Materials commonly used as fillers are sand, dolomitic limestone, gypsum, furfural, vermiculite, rock salt, coal ashes, chalk grit, spent foundry sand, peanut hulls, saw dust and rice husk, etc.

Fixation (in soil): Conversion of a soluble material such as a plant nutrient like phosphorus, from a soluble or exchangeable form to a relatively insoluble form. To reduce fixation of phosphate, phosphatic fertilizers are brought in lesser contact with the soil particles and applied closer to the plant root through band placement.

Flocculation: Joining of colloidal particles to form clusters (flocs).

Foliar diagnosis: Diagnostic technique based on analyzing plant tissues for the total content of nitrogen, phosphorus, potassium, boron, etc. Low values of nutrient percentages in plants under test compared to nutrient percentage in high yielding plants indicate deficiency of nutrients. The nutrients commonly tested by this method are nitrogen, phosphorus and potash.

Foliar fertilization: Fertilization of plants, or feeding nutrients to plants, by applying chemical fertilizers to be foliage. This is also known as foliar feeding or spray fertilization.

Fortification: The process of putting an additional quantity of fertilizer to increase the nutrient content of manure/ compost, *e.g.* in the ADCO process of compost-making superphosphate is added to fortify the phosphate content of the manure.

Frictional resistance: Resistance to the movement of fluids or particles caused by the interaction (rubbing) of surfaces.

G

Gamma ray: A high energy ray (photon) emitted during radioactive decay of certain elements.

Geostatistics: Statistics that describe the variability of properties from place to place.

Geographical Information System (GIS): It is a computerised data base management system for capturing, storing, validation, analysing, displaying and managing spatially referenced data sources in addition to the primary data such as agro-climatic and soil characteristics. It can be used to generate map on erosion hazard, land suitability for a specified alternative land use type.

Genesis, Soil: The mode of origin of the soil, with special reference to the processes responsible for the development of the solum, or true soil, from the unconsolidated parent material.

Geological erosion: As opposed to accelerated erosion, erosion at natural rates, unaffected by human activity.

Geothite (FeOOH): A yellow-brown iron oxide mineral that account for the brown colour in many soils.

Gibbsite $Al(OH)_3$: An alluminium trihydroxide mineral most common in highly weathered soils such as oxisols.

Glacial outwash: Geological material moved by glaciers and subsequently sorted and depositied by streams flowing from melting ice. Also called *glaciofluvial deposits.*

Glacial till: Unsorted and unstratified geological material deposited directly by glacial ice.

Glaciation: The process of geological erosion by means of glacial ice.

Gleying: A process that produces reduction of iron and other elements under conditions of prolonged saturation.

Granular fertilizer: A fertilizer composed of particles of roughly the same composition, and about one-tenth of an inch in diameter. This kind of fertilizer is superior in efficiency compared to the fine or powdery fertilizer due to ease in handling and less fixation in soil.

Green leaf manuring: This refers to turning under of green leaves and tender green twigs collected from shrubs and trees grown on bunds, waste lands and nearby forest areas. The common shrubs and trees useful for this purpose are glyricidia *(Glyricidia maculate)*, *Sesbania speciosa*, karanj *(Pongamia pinnata)* etc.

Green manure crop: Any crop grown and buried into the soil for improving the soil condition by addition of organic matter. Such crops are legumes and non-legumes, but mostly legumes *e.g.* sannhamp *(Crotolaria juncia)*, Dhaincha–(*Sesbania aculeata)*

Green manuring: A practice of ploughing or turning into the soil undecomposed green plant material for improving the physical condition of the soil or for adding nitrogen when the green manure crop is legume. Two types of green manuring are being practiced by the cultivators of India. These are i) green manuring in situ and ii) Greenleaf manuring.

Green manuring in situ: A practice in which green manure crops are grown and buried in the same field which is to be green-manured. The crops are grown alone or intercropped with the main crop. Important green manure crops used in this system are sannhemp *(Crotalaria juncea)*, dhaincha *(Seshania aculeate)*, pillipesara *(Phaseolus trilobus)*, and guar *(Cyamopsis psoraloides* or *tetragonoloba)*.

Greenhouse effect: The warming of the earth's surface and atmosphere owing to absorption of out-going radiation by CO_2, CH_4 and H_2O (like absorption by glass).

Groundnut cake: A by-product of the oil industry. This is an edible oil cake. It contains 7.3 per cent nitrogen, 1.5 per cent phosphoric acid and 1.3 per cent potash. In the Deccan tract of Maharashtra. Karnataka. Andhra Pradesh and Tamil Nadu, groundnut cake is applied to sugarcane and bananas on a very large scale. Besides the nutrients, it supplies organic matter to the soil.

Guano: In some parts of America used synonymously with fertilizer. It is derived from the Spanish word meaning dung. Guano is made up of excrement of seafowl, together with their body remains. Deposits of guano found in many islands may vary from a few inches to over 200 feet in thickness. Peru is one of the pioneer producers and exporters of guano. It is also obtained from excrement and dead remains of sea creatures other than seafowl. It is also called bat guano, seal guano, fish guano, whale guano, etc. Some guanos produced from sheep and goat herds are called sheep guano or goat guano. The colour of guano may vary from grey to dark brown. The chemical composition of guano also varies. Its nitrogen content may vary from 4 to 16 per cent and total P_2O_5 may range between 12 and 26 per cent. When nitrogen content is very high, say, 8-16 per cent, it is termed as nitrogenous guano. When phosphorus content is very high, say, 20-25 per cent P_2O_5, it is termed as phosphatic guano. The potash (K_2O) content of guano is 2 to 3 per cent.

The product obtained after treatment of guano with sulphuric acid is called dissolved guano. The available phosphorus and nitrogen content increase as a result of this treatment. Some times any one or all of the three nutrients, namely, N, P_2O_5 and K_2O are added to guano to make it a balanced fertilizer. This is called rectified or fortified guano.

Gypsum: Hydrated calcium sulphate ($CaSO_4$, $2H_2O$) also known as land plaster. Agricultural grade gypsum contains 55 to 95 per cent gypsum. Gypsum is mined from sedimentary deposits. It is also obtained as a by product of phosphoric acid industry and is termed as phospo-gypsum. Gypsum is calcimined to produce plaster of paris which has wide industrial usage. It is also used in cement, pain and many other industries. In agriculture its main use is in reclamation of alkali soils. It is also a source of calcium, which is an essential plant element.

H

Halophyte: A plant that requires or tolerates a saline (high salt) environment.

Hardpan: A hardened soil layer, in the lower A or in the B horizon, caused by cementation of soil particles with organic matter or with materials such as silica, sesquioxides, or calcium carbonate. The hardness does not change appreciably with changes in moisture content and pieces of the hard layer do not slake in water.

Harrowing: A secondary broadcast tillage operation that pulverizes, smooths, and firms the soil in seedbed preparation, controls weeds, or incorporates material spread on the surface.

Heaving: The partial lifting of plants, buildings, road-ways, fenceposts, etc., out of the ground, as a result of freezing and thawing of the surface soil during the winter.

Heavy metals: Metals with particle densities > 5.0 Mg/m^3.

Heavy soil: A soil with a high content of the fine separates, particularly clay, or one with a high drawbar pull, hence difficult to cultivate.

Hematite, Fe_2O_3: A red iron oxide mineral that contributes red color to many soils.

Hemicellulose: A group of complex carbohydrates; polysaccharides that, unlike starch and cellulose, contain other sugars besides glucose. Important to plant cell walls.

Herbicide: A chemical that kills plants or inhibits their growth; intended for weed control.

Herbivore: A plant-eating animal.

Heterotroph: An organism capable of deriving energy for life processes only from the decomposition of organic compounds and incapable of using inorganic compounds as sole sources of energy or for organic synthesis.

High-analysis fertilizer material: The fertilizers containing a higher percentage of the plant nutrients. Such as urea containing 46 per cent nitrogen, triple superphosphate containing 45 to 47 per cent P_2O_5 and muriate of potash containing 60 per cent K_2O. High-analysis fertilizers have an advantage in the cost of bagging, handling and transportation per unit of plant.

Hill placement: This refers to applying fertilizer either in bands or near plants. This method is usually employed when relatively small quantities are to be applied. In Egypt, nitrogenous fertilizer is applied in pinches at the base of each plant of cotton. Maize, cauliflower, cabbage, brinjal, tomato, and other vegetable crops are fertilized by this method.

Horizon, soil: A layer of soil, approximately parallel to the soil surface, differing in properties and characteristics from adjacent layers below or above it.

Horn-and-hoof-meal: A manure prepared from horns and hoofs of animals. Such a meal contains about 13 per cent nitrogen.

Horticulture: The art and science of growing fruits, vegetables, and ornamental plants.

Humic acid: A mixture of variable or indefinite composition of dark organic substances, precipitated upon acidification of a dilute alkali extract from soil.

Humid climate: Climate in regions where moisture, when distributed normally throughout the year, should not limit crop production. In cool climate annual precipitation may be as little as 25 cm; in hot climates, 150 cm or even more. Natural vegetation in uncultivated areas is forests.

Humification: The process involved in the decomposition of organic matter and leading to the formation of humus.

Humin: The fraction of the soil organic matter that is not dissolved upon extraction of the soil with dilute alkali.

Humus: That more or less stable fraction of the soil organic matter remaining after the major portions of added plant and animal residues have decomposed. Usually it is dark in colour. The term is often used synonymously with soil organic matter.

Hydrated: Having water attached or incorporated as part of a chemical substance.

Hydration: The chemical combination of water with another substance.

Hydraulic conductivity: An expression of the readiness with which a liquid such as water flows through a solid such as soil in response to a given potential gradient.

Hydrologic cycle: The circuit of water movement from the atmosphere to the Earth and back to the atmosphere through various stages or processes, as precipitation, interception, runoff, infiltration, percolation, storage, evaporation, and transpiration.

Hydrolysis: A reaction that involves products (H^+ or OH^-) of the dissociation of water. A mineral weathering reaction that adds H^+ to a mineral structure.

Hydromulching: Technique of spraying a slurry of fiber, seed, fertilizer, and chemicals onto roadsides for erosion control.

Hydroscopic coefficient: The amount of moisture in a dry soil when it is in equilibrium with some standard relative humidity near a saturated atmosphere (about 98 per cent), expressed in terms of percentage on the basis of oven-dry soil.

Hydrous oxides, hydroxyoxides: Sesquioxides. Oxides of Fe, Al, or similar metals, with different proportions of water and hydroxyl in the structure.

Hydroxyl: A OH ion or group.

Hygroscopic: A term used to designate materials that absorb moisture from the air.

Hygroscopicity of fertilizer materials: Some fertilizer materials absorb moisture from the air, which causes them to become sticky. Such hygroscopic fertilizers are calcium ammonium nitrate, sodium nitrate, ammonium chloride and urea.

I

Igneous Rock: The earliest rocks formed by cooling of the earth's molten magma. Based on silicic acid content igneous rocks are classified in four groups:

1. Acid rocks: Silicic acid content is more than 65 per cent, *e.g.* granite.
2. Sub-Acid Rocks: Silicic acid content is 60-65 per cent, *e.g.* syenite.
3. Sub-Basic Rocks: Silicic acid content is 55-60 per cent, *e.g.* diorite.
4. Basic Rocks: Silicic acid content is 40-45 per cent, *e.g.* golomite, basalt.

Illuvial horizon: A soil layer or horizon in which material carried from an overlying layer has been precipitated from solution or deposited from suspension. The layer of accumulation.

Immature soil: A soil with indistinct or only slightly developed horizons because of the relatively short time it has been subjected to the various soil-forming processes. A soil that has not reached equilibrium with its environment.

Immobilization: The conversion of an element from the inorganic to the organic form in microbial tissues or in plant tissues, thus rendering the element not readily available to other organisms or to plants.

Imogolite: A poorly crystalline aluminosilicate mineral with an approximate formula $SiO_2Al_2O_3.5H_2O$; occurs mostly in soils formed from volcanic ash.

Impermeable: Unable to transmit water. A term of ten used with dense soil horizons through which water moves extremely slowly.

Impervious: Resistant to penetration by fluids or by roots.

Inactivated organisms: In waste disposal, harmful organisms made harmless by reaction with the soil and soil organisms.

Inclusion: One or more polypedons within a map unit, not identified by the map unit name.

Indicator plant: A plant that indicates, in general, and often in a specific manner, a deficiency of plant nutrient in the soil.

Indurated (soil): Soil material cemented into a hard mass that will not soften on wetting.

Inferred properties: Soil properties that are not seen or measured but are assumed from other properties. Suitability for a use is an inferred property.

Infiltration rate: A soil characteristic determining or describing the maximum rate at which water can enter the soil under specified conditions, including the presence of an excess of water.

Infrared: Refers to the electromagnetic radiation of wavelength longer than light but shorter than radio. Sometimes called heat radiation.

Inhibitor: The opposite of catalyst as it retards or reduces the rate of a chemical reaction. It is also known as a retarder. Antioxidants are well known inhibitors of oxidation reactions. Benzoic acid and salicylic acid are added to food products to prevent rancidity.

Inoculation: The process of introducing pure or mixed cultures of microorganisms into natural or artificial culture media.

Inorganic compounds: All chemical compounds in nature except compounds of carbon other than carbon monoxide, carbon dioxide, and carbonates.

In-place parent material: Parent material that has not been transported from its original location, for example, a bedrock is an in-place parent material.

Insecticide: A chemical that kills insects.

Insulator: A material (or space) that transmits heat or electricity poorly.

Interaction: A combined effect of two or more factors or variables such that the variables modify each other's effects.

Interaction, electrostatic: Adsorption caused by the electrical attraction of ions to a charged surface.

Intercrop: Two or more crops grown together on the same piece of land at the same time.

Intergrade: A soil that possesses moderately well developed distinguishing characteristics of two or more genetically related great soil groups.

Integrated plant nutrition system (IPNS): It is a concept proposed by FAO where the basic goal is the maintenance or adjustment and possibly improvement of soil fertility and of plant nutrient supply to an optimum level for sustaining the desired crop productivity through optimization of the benefits from all possible sources of plant nutrients in an integrated manner.

Interlayer (mineralogy): Materials between layers within a given crystal, including cations, hydrated cations, organic molecules, and hydroxide groups or sheets.

Interveinal: Between veins.

Interveinal chlorosis: Chlorosis only between leaf veins.

Intrusive rock: Any rock that forms by cooling from a hot molten mass below the earth's surface. Granite is an intrusive rock.

Ion exchange: The interchange between an ion in solution and another ion on the surface of any surface-active material such as clay or humus.

Ions: Electrically charged particles. As used in soils, an ion refers to an electrically charged element, resulting from the breaking up of many substances like acids, salts, and alkalines. Positively charged ions are called 'cations' and negatively charged as 'anions'. For further information see cation and anion.

Iron-pan: An indurated soil horizon in which iron oxide is the principal camenting agent.

Irrigation efficiency: The ratio of the water actually consumed by crops on an irrigated area to the amount of water diverted from the source onto the area.

Irrigation methods: Methods by which water is artificially applied to an area. The methods and the manner of applying the water are as follows.

border-strip: The water is applied at the upper end of a strip with earth borders to confine the water to the strip.

center pivot: Automated sprinkler irrigation achieved by automatically rotating the sprinkler pipe or boom, supplying water to the sprinkler heads or nozzles, as a radius from the center of the field to be irrigated.

check-basin: The water is applied rapidly to relatively level plots surrounded by levees. The basin is a small check.

corrugation: The water is applied to small, closely spaced furrows, frequently in grain and forage crops, to confine the flow of irrigation water to one direction.

drip: A planned irrigation system where all necessary facilities have been installed for the efficient application of water directly to the root zone of plants by means of applicators (orifices, emitters, porous tubing, perforated pipe, etc.) operated under low pressure. The applicators may be placed on or below the surface of the ground.

Flooding: The water is released from field ditches and allowed to flood over the land.

furrow: The water is applied to row crops in ditches made by tillage implements.

spinkler: The water issprayed over the soil surface through nozzles from a pressure system.

subirrigation: The water is applied in open ditches or tile lines until the water table is raised sufficiently to wet the soil.

wild-flooding: The water is released at high points in the field and distribution is uncontrolled.

Interstratification: Mixing of silicate layers within the structural framework of a given silicate clay.

Isomorphous substitution: The replacement of one atom by another of similar size in a crystal lattice without disrupting or changing the crystal structure of the mineral.

Isotopes: Two or more atoms of the same element that have different atomic masses because of different numbers of neutrons in the nucleus.

J

Jevenile: New or young animal or plant or soil.

Joule: The SI energy unit defined as a force of one newton applied over a distance of one meter, 1 joule = 0.239 calorie.

K

Kame: A conical hill or ridge of sand or gravel deposited in contact with glacial ice.

Kainite [KMg $(SO_4)Cl.3H_2O$] A mineral used as a source for potassic fertilizers and potassic salts. It is an impure mixture of potassium and magnesium chloride, sometimes containing sulphate of magnesia. It contains about 12 per cent K_2O. It is quite soluble in water and is non-hygroscopic.

Kalium: The Latin word for alkali. In chemistry it stands for potassium and the symbol 'K' used for potassium is derived from kalium.

Kaolinite: An aluminosilicate mineral of the 1:1 crystal lattice group, that is, consisting of single silicon tetrahedral sheets alternating with single aluminium octahedral sheets.

Kinetic energy: Energy resulting from motion. In equation form, KE= ½ mass x (velocity squared).

Kjedahl nitrogen: The amount of nitrogen contained in organic material as determined by a method based on the digestion of the sample in a sulphuric acid base reagent, which converts the nitrogenous organic material to carbon dioxide, water and ammonia. Subsequently the ammonia is quantified.

L

Labile: Descriptive of a substance in soil that readily undergoes transformation or is readily available to plants.

Lacustrine deposit: Material deposited in lake water and later exposed either by lowering of the water level or by the elevation of the land.

Lambert's Law: The law states that light absorbed by homogeneous transparent media is proportional to the thickness of the media and intensity of light. Mathematically, Lambert's law is described as:

$$\text{Log}\ \frac{Io}{It} = KH$$

where,

Io: Intensity of incident light
It: Intensity of transmitted light
H: Thickness of media
K: Constant.

Land: A broad term embodying the total natural environmental of the areas of the Earth not covered by water. In addition to soil, its attributes include other physical conditions such as mineral deposits and water supply; location in relation to centers of commerce, populations, and other land, the size of the individual tracts or holdings, and existing plant cover, works of improvement, and the like.

Land capability classification: A grouping of kinds of soil into special units, subclasses, and classes according to their capability for intensive use and the treatments required for sustained use. One such system has been prepared by the USDA Soil Conservation Service.

Land classification: The arrangement of land units into various categories based upon the properties of the land or its suitability for some particular purpose.

Landslide: A rapid downhill movement of soil and rock under the force of gravity. A term also used to described the resulting landform.

Land-use planning: The development of plans for the uses of land that, over long periods, will best serve the general welfare, together with the formulation of ways and means for achieving such uses.

Latent heat: A property of a material. The amount of heat involved per unit mass when the material undergoes a change of phase (*e.g.*, the heat of melting or the heat of vaporization).

Laterite: An iron-rich subsoil layer found in some highly weathered humid tropical soils that, when exposed and allowed to dry, becomes very hard and will not soften when rewetted. When erosion removes the overlying layers, the laterite is exposed and a virtual pavement results.

Layer silicates: Another term for aluminosilicate minerals that are platelike in form and composed of layers of atoms. Mica and the aluminosilicate clay minerals are layer silicates.

Leaching: Removal of plant nutrients in solution by the passage of water through soil. This is one of the ways in which plant nutrients are lost from the soil. Among the major nutrients, nitrogen is lost in large quantities by leaching.

Leaf area index (LAI): Area of leaves per unit of land on which the plants are growing.

Legume Inoculation: Treatment of legume seed with rhizobium culture. There is a specific symbiotic relationship between different species of rhizobium and various legume crops. For example, a rhizobium specie that will live symbiotically with soyabean will not do so with alfalfa. It is, therefore, necessary to use only specific cultures for different crops.

Legume: A pod-bearing member of the leguninosae family, one of the most important and widely distributed plant families. Includes many valuable food and forage species, such as peas, beans, peanuts, clovers, alfalfas, sweet clovers, lespedezas, vetches, and kudzu. Nearly all legumes are associated with nitrogen-fixing organisms.

Lichen: Symbiosis between fungi and algae or bluegreen bacteria, commonly forming a flat, speading growth on surfaces of rocks and tree trunks.

Liebig's law: In 1862, Justus Von Liebing (A German Chemist) stated that the growth and reproduction of an organism are determined by the nutrient substance (oxygen, carbon dioxide, calcium, etc) that is available in minimum quantity, the limiting factor.

Ligand exchange: A class of surface reactions on minerals in which anions from solution take the place of some of the anions (ligands) normally bound to cations in the mineral.

Light soil: A term used for sandy or coarse-textured soil. Leaching of nutrients is more in light soils, as such it is advisable to apply nitrogenous fertilizers in small doses. Wherever practicable, green manuring should be done to improve the physical condition of a light soil.

Lignin: The complex organic constituent of woody fibers in plant tissue that, along with cellulose, cements the cells together and provides strength. Lignins resist microbial attack and after some modification may be come part of the soil organic matter.

Lime (agricultural): In strict chemical terms, calcium oxide. In practical terms, a material containing the carbonates, oxides and/or hydroxides of calcium and/or magnesium used to neutralize soil and acidity.

Lime: Generally, the term lime, or a agricultural lime, is applied to ground limestone, hydrated lime or burned lime. These materials are used as amendments to reduce the acidity of acid soils. In strict chemical terminology, lime refers to calcium oxide (CaO), but by an extension of the meaning, it is now used for all limestone-derived materials applied to neutralize acid soils.

Lime requirement: The amount of standard ground limestone required to bring a 6 inch layer of an acre (about two million pounds) of acid soil to some specific lesser degree of acidity. In common practice, lime requirements are given in tons per acre of nearly pure limestone.

Limestone: The term refers to rocks consisting chiefly of calcium carbonate ($CaCO_3$) or calcium and magnesium carbonate $CaMg(CO_3)_2$. When magnesium carbonate is present in a large percentage, limestone is called dolomite limestone.

Liming: Addition of ground limestone (calcium carbonate), hydrated lime (calcium hydroxide) or burned lime (calcium oxide) to acid soils to permit growth of deeply rooted legumes and many other valuable crops, and to create soil conditions favourable for the utilization of plant nutrients.

Liquid fertilizers: Commercial fertilizers in liquid form. Such fertilizers are chiefly anhydrous ammonia, aqueous solution of nitrogen, and some mixed fertilizers. Liquid fertilizers are applied to the soil, through irrigation water, starter solutions, or with the help of special equipment. Use of liquid fertilizers is common in the United States of America.

Liquid limit: In engineering, the water content corresponding to the limit between a soil's liquid and plastic states of consistency.

Lithosequence: Two or more soils in which the soil forming factor that varies the most is the parent material. Other soil-forming factors are constant or vary much less than parent material.

Loam: The textural class name for soil having a moderate amount of sand, silt, and clay. Loam soils contain 7-27 per cent clay, 28-50 per cent silt, and 23-52 per cent sand.

Loamy: Intermediate in texture and properties between fine-textured and coarse-textured soils. Includes all textural classes with the words loam or loamy as a part of the class name, such as clay loam or loamy sand.

Localized placement of fertilizers: This refers to the application of fertilizers into the soil close to the seed or plant. This method is specifically adopted with phosphatic fertilizers for three reasons, namely;

1. Restricted contact of fertilizers with soil lessens fixation of phosphate,
2. Necessary plant food is placed within easy reach of plant roots, and
3. Application of fertilizers in a band along the rows does not readily furnish nutrients to weeds growing between the rows.

Loess: Material transported and deposited by wind and consisting of predominantly silt-sized particles.

London forces: Weak, close-range attraction between atoms, molecules, or colloid particles due to transitory dipole-dipole interactions between their electrons.

Longwave radation: In meteorology, radiation in the infrared and radio wavelengths, emitted at the earth's fsurface and partly absorbed in the stomosphere. Distinct from the sun's shortwave radiation.

Low-analysis fertilizer materials: The fertilizers containing a low percentage of plant nutrients. Such fertilizers are Chilean nitrate containing 16 per cent nitrogen and single superphosphate containing 16 per cent nitrogen and single superphosphate containing 16 to 18 per cent P_2O_5.

Luxury consumption: The intake by a plant of an essential nutrient in amounts exceeding what is needs. For example, if potassium is abundant in the soil, alfalfa may take in more than it requires.

Lysimeter: A device for measuring perolation and leaching and evapotranspiration losses from a column of soil under controlled conditions.

M

Macronutrient: A nutrient required by plants in relatively large amounts. There are three macronutrients namely, nitrogen, phosphorus and potassium which are also called major nutrients or macro elements. Other nutrients which are normally required in lesser but still considerable amounts like calcium, magnesium and sulphur are called secondary nutrients. Some authors group secondary nutrients along with macronutrients.

Magnesium sulphate: ($MgSO_4 7H_2O$) is commonly used to supply magnesium to the soil through fertilizer mixtures or foliar sprays.

Manure: The excreta of animals–dung and urine, with straw or other materials used as the absorbent. The decomposed manure is called farmyard manure or farm manure or barnyard manure. The average composition of well-rotted farmyard manure is 0.5 per cent nitrogen, 0.3 per cent P_2O_5 and 0.5 per cent K_2O.

Marl: Soft and unconsolidated calcium carbonate, usually mixed with varying amounts of clay or other impurities.

Marsh: Periodically wet or continually flooded area with the surface not deeply submerged. Covered dominantly with sedges, cattails, rushes, or other hydrophytic plants. Subclasses include freshwater and saltwater marshes.

Marsh spot of pea: Disease caused in peas due to deficiency of manganese.

Mature soil: A soil with well developed soil horizons produced by the natural processes of soil formation and essentially in equibrium with its present environment.

Maximum water-holding capacity: The average moisture content of a disturbed sample of soil, 1 cm hgh, which is at equilibrium with a water table at its lower surface.

Medium texture: Intermediate between fine-textured and coarse textured (soils). It includes the following textural classes: very fine sandy loam, loam, silt loam, and silt.

Mellow Soil: A very soft, very friable, prorous soil without any tendency toward hardness or harshness.

Metamorphic rock: A rock that has been greatly altered from its previous condition through the combined action of heat and pressure. For example, marble is a metamorphic rock produced from limestone, gneiss is produced from granite, and slate from shale.

Metamorphic rock: Rocks which have undergone change in structure and mineral composition due to very high pressure and temperature conditions, *e.g.* when limestone is altered to marble.

Methane, CH_4: An odorless, colorless gas commonly produced under anaerobic conditions. When released to the upper atmosphere, methane contributes to global warming.

Micas: Primary aluminosilicate minerals in which two silica tetrahedral sheets alternate with one alumina/ magnesia octahedral sheet with entrapped potassium atoms fitting between sheets. They separate readily into thin sheets or flakes.

Micron (µ): A unit of length denoting a millionth part of a metre.

$$1\ \mu = 10^{-6} \text{ metre.}$$

In instrumental analysis it is used to measure wavelength of radiation.

Microfauna: That part of the animal population which consists of individuals too small to be clearly distinguished without the use of a microscope. Includes protozoans and nematodes.

Microflora: That part of the plant population which consists of individuals too small to be clearly distinguished without the use of a microscope. Includes actinomycetes, algae, bacteria, and fungi.

Micronutrients: The essential plant nutrients required in minute quantities. These nutrients are seven in number, namely, iron, manganese, boron, molybdenum, copper, zinc and chlorine. Micronutrients are also called 'minor elements' or 'trace elements'.

Micro-organisms: The most primitive plant and animal life whose structure is very simple. Their size being small, they are often found in soil and have direct or indirect bearing on soil formation and soil fertility. They are divided into two main groups:

1. Micro flora, *e.g.* bacteria, actinomycetes, fungi and algae.
2. Micro fauna, *e.g.* protozoa.

Microrelief: Small-scale local differences in topography, including mounds, swales, or pits that are only a meter or so in diameter and with elevation differences of up to 2 m.

Mineral soil: A soil consisting predominantly of, and having its properties determined predominantly by mineral matter. Usually contains <20 per cent organic matter, but may contain an organic surface layer upto 30 cm thick.

Mineralization: The conversion of an element from an organic form to an inorganic state as a result of microbial decomposition.

Mitscherlich's growth law: It states that increase in the yield of a crop as a result of increasing a single growth factor is proportional to the decrement from the maximum yield obtainable by increasing the particular growth factor.

Mixed fertilizers: Mixed fertilizers consist of individual or straight fertilizer materials blended together to permit application in the field in one operation. Mixed fertilizers supply two or three major plant nutrients. The percentage of these nutrients are expressed as fertilizer grade, like 10-5-5. See 'fertilizer grade' for further information.

Moderately coarse texture: Consisting predominantly of coarse particles. In soil textural classification, it includes all the sandy loams except the very fine sandy loam.

Moderately fine texture: Consisting predominantly of intermediate-sized (soil) particles or with relatively small amounts of fine or coarse particles. In soil textureal classification, it includes clay loam, sandy loam, sandy clay loam, and silty clay loam.

Moisture equivalent: The weight percentage of water retained by a previously saturated sample of soil 1 cm in thickness after it has been subjected to a centirifugal force of 1000 times gravity for 30 min.

Molar solution: One mole (gram molecular weight) of a substance dissolved in solvent and the fiinal volume made to one litre of solution, *e.g.* the gram molecular weight of sodium dydroxide is 40. Therefore a molar solution is prepared by dissolving 40 g of sodium hydroxide in water and making up the final volume of the solution (by adding water) to one litre.

Molar concentration refers to number of moles in one litre solution, *e.g.* 2 molar concentration of sodium hydroxide solution means 2 moles (40 x 2 = 80 g) of sodium hydroxide present in one litre of solution.

Mole drain: It is unlined drain formed by pulling a bullet-shaped cylinder through the soil.

Montmorillonite: An aluminosilicate clay mineral in the smectite group with a 2:1 expanding crystal lattice, with two silicon tetrahedral sheets enclosing an aluminum octahedral sheet. Isomorphous substition of magnesium for some of the aluminum has occurred in the octahedral sheet. Considerable expansion may be caused by water moving between silica sheets of contiguous layers.

Mor raw: It is a type of forest humus layer of unincorporated organic material, usually matted or compacted or both, distinct from the mineral soil, unless the latter has been blackened by washing in organic matter.

Moraine: An accumulation of drift, with an initial topographic expression of its own, built within a glaciated region chiefly by the direct action of glacial ice. Examples are ground, lateral, recessional, and terminal moraines.

Morphology, soil: The constitution of the soil including the texture, structure, consistence, color, and other physical, chemical and biological properties of the various soil horizons that make up the soil profile.

Most profitable rate or MPR or optimum dose: Refers to the amount of fertilizer that produces the greatest profit per acre. If, more than the optimum dose is given, the profit will fall and less than the optimum is used, some profit will be sacrificed.

Motting: Spots or blotches of different color or shades of color interspersed with the dominant color.

Mucigel: The gelatinous material at the surface of roots grown in unsterilized soil.

Muck: Highly decomposed organic material in which the original plant parts are not recognizable. Contains more mineral matter and is usually darker in color than peat.

Muck soil: (1) A soil containing 20-50 per cent organic matter. (2) An organic soil in which the organic matter is well decomposed.

Mull: A humus-rich layer of forested soils consisting of mixed organic and mineral matter. A mull blends into the upper mineral layers without an abrupt change in soil characteristics.

Multiple nutrient materials: The fertilizers containing more than one nutrient element. When fertilizers supply two major plant nutrients they are called binary fertilizers. Ammonium phosphate and nitrophosphate are binary fertilizers. Fertilizers supplying three major plant nutrients are termed ternary fertilizer. Ammonium potassium phosphate is a ternary fertilizer.

Muriate of potash: It is potassium chloride, of 90 to 99 per cent grade. It contains nearly 60 per cent K_2O. It is sold in granular as well as powder form.

Mycorrhiza: A mycorrhiza is an infected root system arising from the root lets of a seed plant. The word mycorrhiza (my-koe-rye-zee), derived from Greek meaning "fungus root" Mycorrhizae are fungi that form symbiotic association of a fungus with the roots of a higher plants.

N

Necrosis: Death associated with discoloration and dehydration of all or parts of plant organs, such as leaves.

Neem coated urea: Neem has a nitrification inhibition property, hence is used in India for coating of urea granules to produce slow release fertilizer. The technique as developed at the Indian Agriculture Research Institute for coating urea with neem cake is as follows: About 100 kg urea is mixed in a drum with a solution of 1 kg coaltar in 2 litres of kerosene. To this, 20 kg of powdered neem cake is added and thoroughly mixed.

Nematodes: Very small worms abundant in many soils and important because some of them attack and destroy plant roots.

Neutral soil: A soil in which the surface layer, at least to normal plow depth, is neither acid nor alkaline in reaction. In practice this means the soil is within the pH range of 6.6–7.3.

Night soil: Night soil is human excreta, solid and liquid, In India, it is directly applied to the soil to a limited extent but converted mainly as town compost. In cities which have sewage facilities, sewage water and sludge are used directly to raise crops. On an average nigh-soil contains 5.5 per cent nitrogen, 4.0 per cent phosphorus (P_2O_5) and 2.0 per cent potash (K_2O) on oven dry basis.

Niter (KNO_3): Potassium bearing mineral also known as potassium nitrate or saltpeter. It contains about 46 per cent K_2O.

Nitrate of soda: A commercial nitrogenous fertilizer carrying 16 per cent nitrogen in nitrate form. Nitrate of soda is imported from Chile hence the name Chilean nitrate. In other countries, it is manufactured synthetically.

Nitrification: The formation of nitrates and nitrites from ammonia (or ammonium compounds) as in soils, by micro-organisms.

$$NH_3 \xrightarrow{\textit{Nitrosomonas}} NO_2 + 3\,H+$$

$$NH_3 \xrightarrow{\textit{Nitrobactor}} NO_3$$

Nitrification inhibitors: Chemicals which inhibits the nitrification process, used extensively to check mineralization of nitrogen and thereby control the release of nitrogen. Some of the important nitrification inhibitors are N-serve, AM, DD, Thiourea and ST.

Nitrogen cyle: The sequence of chemical and biological changes undergone by nitrogen as it moves from the atmosphere into water, soil, and living organisms, and upon death of these organisms (plants and animals) is recycled through a part or all of the entire process.

Nitrogen fixation: The assimilation of free nitrogen from the soil air by soil micro-organisms and the formation of nitrogen compounds that eventually become available to plants. The nitrogen fixing organisms associated with legumes are called symbiotic; those not definitely associated with higher plants are non-symbiotic.

Nitrophosphate: This fertilizer contains 20 per cent phosphorus. Nitrophosphate is manufactured by treating rock phosphate with nitric acid and neutralizing this solution with gaseous ammonia. The nitrogen is present in ammonium nitrate form, and P_2O_5 partly in water-soluble and partly in citrate-soluble form.

Nucleic acids: Complex compounds found in plant and animal cells may be combined with proteins as nucleoproteins.

Nutrient deficiency symptoms: When any essential plant nutrient is seriously lacking in the soil, plants growing on it show certain colour development in leaves and certain changes in growth. These symptoms varies from crop to crop and with the degree of deficiency. On the basis of deficiency symptoms, fertilizer recommendations can be done.

Nutrient: Any mineral element that functions in plant, animal and other organism, metabolisms, whether or not its action is specific.

O

O horizon: Organic horizon of mineral soils.

Oilcakes: When oil is extracted from oilseeds, the remaining solid portion is the oilcake. Oilcakes are of two types, edible oilcakes which can be safely fed to livestock, and non-edible oilcakes which are not fit for feeding to livestock. Oilcakes are added to the soil as concentrated organic manures. They supply organic matter and all the three major plant nutrients (N, P_2O_5, and K_2O) but mostly they supply nitrogen.

Open-formula mixed fertilizers: Fertilizer mixtures of which the composition and the complete make up are disclosed by manufacturers. Thus the fertilizer grade, the straight fertilizers and the filler used are disclosed on each bag. Cultivators and extension workers are in a better position to know the quantity and quality of nutrients contained in such a fertilizer mixture.

Organic colloids: In soils, these present as humus which is a temporary intermediate product left after considerable decomposition of plant and animal residues. Organic colloids having high surface area with higher charge density and higher cation exchange capacity.

Organic farming: According to the U.S. Department of Agriculture "a production system which avoids or largely excludes the use of synthetically compounded fertilizers, pesticides, growth regulators and livestock feed additives. To the maximum extent feasible, organic forming systems rely upon crop rotations, crop residеues, animal manures, legumes, green manures, off-farm organic waste, mechanical cultivation, mineral bearing rocks and aspects of biological pest control to maintain soil productivity and tilth, to supply plant nutrients and to control insects, weeds and other pests".

Organic fertilizer: By product from the processing of animal or vegetable substances that contain sufficient plant nutrients to be of value as fertilizers.

Organic soil: A soil that contains at least 20 per cent organic matter (by weight) if the clay content is low and at least 30 per cent if the clay content is as high as 60 per cent.

Organic soil materials (As used in Soil taxonomy in the United States). (1) Saturated with water for prolonged periods unless artificially drained and having 18 per cent or more organic carbon (by weight) if the mineral fraction is more than 60 per cent clay, more than 12 per cent, organic carbon if the mineral fraction has no clay, or between 12 and 18 per cent carbon if the clay content of the mineral fraction is between 0 and 60 per cent. (2) Never saturated with water for more than a few days and having more than 20 per cent organic carbon. Histosols develop on these organic soil materials.

Organic matter: The terms organic means "living", therefore, organic matter is any material derived from living organism like plants and animals. The organic matter content in soil is an important indicator of nutrient status especially of nitrogen content in soils. Generally, the status of soil organic carbon content less than 0.5 per cent is termed as low and more than 0.75 per cent as high.

Ortstein: An indurated layer in the B horizon of spodosols in which the cementing material consists of illuviated sesquioxides (mostly iron) and organic matter.

Osmotic pressure: Pressure exerted in living bodies as a result of unequal concentrations of salts on both sides of a cell wall or membrane. Water moves from the area having the lower salt concentration through the membrane into the area having the higher salt concentration and, therefore, exerts additional pressure on the side with higher salt concentration.

Outwash plain: A deposit of coarse-textured materials (*e.g.*, sands, gravels) left by streams of melt water flowing from receding glaciers.

Oven-dry soil: Soil that has been dried at 105°C until it reaches constant weight.

Oxidation ditch: An artificial open channel for partial digestion of liquid organic wastes in which the wastes are circulated and aerated by a mechanical device.

Oxisols: A soil order which is highly weathered dominant in oxides of iron and aluminium but depleted of silica. Oxisols are reddish, yellowish or greyish soils of tropical and sub-tropical regions and low in fertility.

Pedon: The smallest volume that can be called "a soil". It has three dimensions. It extends downward to the depth of plant roots or to the lower limit of the genetic soil horizons. Its lateral cross section is roughly hexagonal and ranges from 1 to 10 m^2 in size depending on the variability in the horizons.

Pedology: It is branch of soil science which deals with the survey, genesis and classification of soils for land use planning.

Petrology: It is science of rocks which form the units of the earth's crust.

Peneplain: A once high, rugged area that has been reduced by erosion to a lower, gently rolling surface resembling a plain.

Penetrability: The ease with which a probe can be pushed into the soil. May be expressed in units of distance speed, force, or work depending on the type of penetrometer used.

Percolation, soil water: The downward movement of water through soil. Especially, the downward flow of water in saturated or nearly saturated soil at hydraulic gradients of the order of 1.0 or less.

Permafrost: (1) Permanently frozen material underlying the solum. (2) A perennially frozen soil horizon.

Permeability, soil: The ease with which gases, liquids, or plant roots penetrate or pass through a bulk mass of soil or a layer of soil.

pH: A term used to indicate the degree of acidity or alkalinity, Technically, pH is the common logarithm of the reciprocal of the hydrogen ion concentration of a solution. A pH of 7.0 indicates precise neutrality, higher values indicate increasing alkalinity, and lower values indicate increasing acidity.

Phase, soil: A subdivision of a soil series or other unit of classification having characteristics that affect the use and management of the soil but do not vary sufficiently to differentiate it as a separate series. Included are such characteristics as degree of slope, degree of erosion, and content of stones.

pH-dependent charge: That portion of the total charge of the soil particles that is affected by, and varies with, changes in pH.

P

Pans: Horizons or layers, in soils, that are strongly compacted, indurated, or very high in clay content.

Parent material: The unconsolidated and more or less chemically weathered mineral or organic matter from which the solemn of soils is developed by pedogenic processes.

Particle density: The mass per unit volume of the soil particles. In technical work, usually expressed as mega gram per cubic meter (Mg/m^3) grams per cubic centimeter (g/cm^3).

Particle size: The effective diameter of a particle measured by sedimentation, sieving, or micro-metric methods.

Particle-size analysis: Determination of the various amounts of the different separates in a soil sample, usually by sedimentation, sieving, micrometry, or combinations of these methods.

Particle-size distribution: The amounts of the various soil separates in a soil sample, usually expressed as weight percentages.

Pascal: An SI unit of pressure equal to one newton per square meter.

Peat: Unconsolidated soil material consisting largely of undecomposed, or only slightly decomposed, organic matter accumulated under conditions of excessive moisture.

Ped: A unit of soil structure such as an aggregate, crumb, prism, block, or granule, formed by natural processes (in contrast to a clod, which is formed artificially).

Phosphobacterium: A microbial culture containing bacteria which increases the availability of applied and native soil phosphorus. Several species of bacteria are effective for increasing phosphorus availability but *Bacillus magatherium var phosphaticum* is most commonly used. In Russia and several Eastern European countries phosphobacterium culture is sold commercially and is used to inoculate the soil for increasing phosphorus availability.

Phosphoric acid: A term that refers to the phosphorus content of a fertilizer, expressed as phosphoric acid (P_2O_5).

Photomap: A mosaic map made from aerial photographs to which place names, marginal data, and other map information have been added.

Phyllosphere: The leaf surface.

Physical properties (of soils): Those characteristics, processes, or reactions of a soil that are caused by physical forces and that can be described by or expressed in, physical terms or equations. Examples of physical properties are bulk density, water holding capacity, hydraulic conductivity, porosity, pore-size distribution, and so on.

Plant tissue test: This refers to the rapid chemical testing of plant tissue, which is analogous to chemical testing and diagnosis in medicine. The basis of rapid chemical testing of plant tissue is the colorimetric determination of the levels of nitrate, phosphorus and potassium in the sap of fresh plant tissue. Sometimes, tests are also carried out for magnesium, calcium, manganese and zinc. Plant tissue tests, though only qualitative, are highly useful guides to interpretation of the relative supply of nutrients actually being taken by the plant. The underlying assumption is that an adequate supply of the elements is indicated by abundance in the plant sap.

Plastic soil: A soil capable of being molded or deformed continuously and permanently, by relatively moderate pressure, into various shapes.

Platy: Consisting of soil aggregates that are developed predominantly along the horizontal axes, laminated, flaky.

Plinthite (brick): A highly weathered mixture of sesquioxides or iron and aluminum with quartz and other diluents that occurs as red mottles and that changes irreversibly to hardpan upon alternate wetting and drying.

Plough-sole placement: With the plough-sole method, the fertilizer is placed in a continuous band on the bottom of a furrow in the process of ploughing. Each band is covered as the next furrow is turned. This method is advocated in areas where the soil becomes quite dry for a few centimeters below the surface during the growing season.

Plow layer: The soil ordinarily moved when land is plowed, equivalent to surface soil.

Plow pan: A subsurface soil layer having a higher bulk density and lower total porosity than layers above or below it, as a result of pressure applied by normal plowing and other tillage operations.

Plowing: A primary broad base tillage operation that is performed to shatter soil uniformly with partial to complete inversion.

Polynutrient fertilizer: A fertilizer containing more than one major plant nutrient. It is synonymous for multiple nutrient material and complex fertilizers.

Polypedon (as used in soil taxonomy)**:**Two or more contiguous pedons, all of which are within the defined limits of a single soil series, commonly referred to as a soil individual.

Pore size distribution: The volume of the various sizes of pores in a soil. Expressed as percentages of the bulk volume (soil plus pore space).

Porosity, soil: The volume percentage of the total soil bulk not occupied by solid particles.

Potash (potassium oxide)**:** The potassium content of fertilizer is expressed as K_2O.

Potassium chloride: Commercial potassium chloride is a potash salt that contains not less than 48 per cent potash, chiefly as chloride, Normally, potassium chloride or muriate of potash sold in India supplies 60 per cent K_2O.

Potassium sulphate: Commercial potassium sulphate contains not less than 48 per cent potash (K_2O) chiefly as sulphate, and not more than 20.5 per cent as chloride.

Precision agriculture: It is the application of modern information technologies and makes uses of information system and planning software to provide, process and analyze multi-source data of high spatial and temporal resolution for decision making and operation in the management of crop production.

Primary mineral: A mineral that has not been altered chemically since deposition and crystallization from molten lava.

Primary plant foods: The primary plant foods are nitrogen (N)., phosphoric acid (P_2O_5) and potash (K_2O). These nutrients are also called major plant nutrients.

Prismatic soil structure: A soil structure type with prism-like aggregates that have a vertical axis much longer than the horizontal axes.

Productivity, soil: The capacity of a soil for producing a specified plant or sequence of plants under a specified system of management. Productivity emphasizes the capacity of soil to produce crops and should be expressed in terms of yields.

Profile, soil: A vertical section of the soil through all its horizons and extending into the parent material.

Protein: Any of a group of nitrogen-containing organic compounds formed by the polymerization of a large number of amino acid molecules and that, upon hydrolysis, yield these amino acids. They are essential parts of living matter and are one of the essential food substances of animals.

Puddled soil: Dense, massive soil artificially compacted when wet and having no aggregated structure. The condition commonly results from the tillage of a clayey soil when it is wet.

Putrefaction: A process of degradation by which protein rich material is decomposed under anaerobic conditions thereby releasing foul smelling gases. The final products in putrefaction are ammonia, amines, carbon dioxide, organic acids, hydrogen sulphide etc.

Q

Quartz (SiO_2): Hard crystalline silica found extensively in igneous and sedimentary rocks. The sand used as filler in fertilizers has quartz as the main constituent.

Quantitative chemical analysis: The chemical testing of a substance to determine its chemical nature. It identifies chemical substance in a product.

Quick soil tests: These are simple and rapid chemical tests of soils designed to give an approximation of the nutrients available to plants. Interpretation of results depends upon previous standardization of soil tests with responses to fertilizers. It varies for different kinds of soil.

R

Radiation: Processes in which energy is sent out as waves and particles through space from atoms and molecules as they rotate, vibrate, and undergo internal change.

Rangeland: Land used for free-grazing livestock.

Reagent: Any chemical compound used to produce a chemical reaction in laboratory analysis to detect and identify specific constituents of a substance being examined.

Recalcitrant: The property of a material to stub-bornly resist decomposition. A chemical such as a pesticide that is not decomposed by microorganisms is a recalcitrant pesticide.

Reclamation: In regard to soil, the treatment of soil to correct waterlogging or severe excesses such as salinity or sodicity.

Reduction: Addition of electron(s) to an atom or molecule. Opposite of *axidation*.

Reemitted energy: Radiation of previously absorbed energy.

Reflected energy: Radiant energy that is thrown back from an object, with no change except direction.

Regolith: The unconsolidated mantle of weathered rock and soil material on the earth's surface, loose earth materials above solid rock. (Approximately equivalent to the term "soil" as used by many engineers).

Relative humidity: The concentration of water in the air or soil atmosphere relative to the maximum concentration it can hold at the given temperature.

Remote sensing: Identifying and observing objects without having any contact with the object. It is feasible through the use of aerial photographs or setellite imagery.

Reradiation: Radiant emission of energy previously absorbed.

Residual effect of fertilizers: This refers to the amount of fertilizer that remains in the soil after one or more cropping seasons. This is particularly true of phosphatic fertilizers.

Residual effect of manure: This refers to the residual beneficial effect of application of farmyard manure on the succeeding crops. This beneficial effect is due to improvement in the physical condition of the soil, and also due to the unutilized plant nutrients. It is estimated that only one third of the nitrogen present in farmyard manure is utilized by the first crop. Similarly, about two third of the phosphate is effective but most of the potash is available for the first crop.

Residual sodium carbonate (RSC): It is a parameter to assess the quality of irrigation water can be expressed by following formula:

$$RSC = (CO_3^{2-} + HCO_3^{-}) - (Ca^{2+} + Mg^{2+})$$

All concentrations are expressed in me/L.

Residual material: Unconsolidated and partly weathered mineral materials accumulated by disintegration of consolidated rock in place.

Residual parent material: Unconsolidated and partly weathered mineral materials accumulated by disintegration of consolidated rock in place.

Residue conservation: Leaving straw, stubble, trash–crop residues– to rot on or in the ground instead of removing or burning them.

Resistance: A measure of the difficulty with which a material conducts heat, ions, electricity, or water. Reciprocal of *conductance*.

Respiration: The set of metabolic processes in which an organism obtains energy from the oxidation of sugars to CO_2 and water, usually with O_2 as the oxidizing agent.

Response curve: A graph relating a plant's quantitative response-changes in growth, yield, or any desired attribute-to different levels o a treatment such as fertilizer application.

Rhizobia: Bacteria capable of living symbiotically with higher plants, usually in nodules on the roots of legumes, from which they receive their energy, and capable of converting atmospheric nitrogen to combined organic forms; hence, the term *symbiotic nitrogen-fixing bacteria.* (Derived from the generic name *Rhizobium).*

Rhizosphere: The term was introduced by L. Hiltner (1904) to describe that portion of the soil which is in close contact with roots and is influenced by the root system with regard to microbial activity and other related phenomena.

Rill: A small, intermittent water course with steep sides, usually only a few centimeters deep and hence no obstacle to tillage operations.

Ripening: The process by which reclaimed land develops characteristics suitable for farming. It includes drainage, cultivation, and neutralizing acidity.

Riprap: Broken rock, cobbles or boulders placed on earth surfaces, such as the face of a dam or the bank of a stream, for protection against the action of water (waves); also applied to brush or pole mattresses, or brush and stone, or other similar materials used for soil erosion control.

Rock: The material that forms the essential part of the Earth's solid crust, including loose incoherent masses such as sand and gravel, as well as solid masses of granite and limestone.

Rock phosphate: Rock containing phos-phorus such as tricalcium phosphate. Rock phosphate is treated with strong acids to make superphosphate. Finely ground rock phosphate is used in strongly acid soils to supply P_2O_5. The finer the rock phosphate, more effective it will be on acid soils.

Root cap: Cells at the tip of the root protecting the growing cells as the root pushes through the soil.

Root density: Number mass or length of roots within a given soil volume.

Root distribution: Arrangement of roots in space within the soil, especially the vertical root density profile.

Root exudate: A mixture of organic acids, sugars, and other soluble plant components that escape from roots.

Root hair: A hairlike structure produced from a root epidermal cell. Probably helps obtain water and immobile nutrients.

Rotational slide: Mass wasting in which the mass rotates about a point rather than sliding horizontally.

Runoff: The portion of the precipitation on an area that is discharged from the area through stream channels. That which is lost without entering the soil is called surface runoff and that which enters the soil before reaching the stream is called *groundwater runoff or seepage flow* from groundwater. (In soil science "runoff" usually refers to the water lost by surface flow; in geology and hydraulics "runoff" usually includes both surface and subsurface flow).

S

Saline: Containing large concentrations of soluble salts. Operationally defined as the electrical conductivity of a saturation extract of > 4 deciSiemens per meter (dS m^{-1})

Saline soil: A saline soil contains enough soluble salts so distributed in the soil that they interfere with growth of most crop plants. Ordinarily the saline soil is only slightly alkaline in reaction (pH 7.4 to 8.5) and contains very little absorbed sodium. Saline soils are often recognized by the presence of white salt crusts. As such, a saline soil is sometimes referred to as white alkali soil.

Salinization: The process of accumulation of salts in soil.

Saltation: Particle movement in water or wind where particles skip or bounce along the stream bed or soil surface.

Salt index: It is one of criteria for evaluating the quality of an irrigation water. It can be expressed as follows:

$$SI = (\text{Total Na-24.5}) - [\text{Total Ca} - \text{Ca in } CaCO_3) \times 4.85]$$

Sand: Soil of particle size ranging from 0.02 to 2 mm in diameter formed as a result of the weathering of rocks. Sand is one of the fillers commonly used in fertilizer mixtures.

Saturated: Generally, occupying all of a capacity. With respect to water, the condition of a soil when all pores are filled with water. With respect to a particular cation or group of cations, the condition in which cations of the specified kind occupy all the exchange capacity.

Saturation extract: The solution extracted from a saturated soil paste.

Saturation percentage: The water content of a saturated soil paste, expressed as a dry weight percentage.

Savanna mineral: A mineral resulting from the decomposition of a primary mineral or from the repreipitation of the products of decomposition of a primary mineral.

Seal: A thin, dense layer that develops when the soil is wet.

Second bottom: The first terrace above the normal flood-plain of a stream.

Secondary plant nutrients: The secondary plant nutrients are calcium, magnesium and sulphur. These nutrients are called secondary because these are not applied as straightcommercial fertilizer but are applied to the soil, indirectly while adding N, P_2O_5 and K_2O in the form of commercial fertilizers. Thus for manufacturers of fertilizers containing major plant nutrients, calcium, magnesium and sulphur are of secondary importance.

Section-range system: A legal survey system used to accurately locate a parcel of land.

Sedimentary rock: A rock fo4med from materials deposited from suspension or precipitated from solution and usually being more or less consolidated. The principal sedimentary rock are sandstones, shales, limestones, and conglomerates.

Sedimentation: A technique for determining particle size distribution. Particles fall through water at a rate proportional to their size.

Seedbed: The soil prepared to promote the germination of seed and the growth of seedlings.

Self-mulching soil: A soil in which the surface layer becomes so well aggregated that it does not crust and seal under the impact of rain but instead serves as a surface mulch upon drying.

Semiarid: Term applied to regions or climates where moisture s more

Sensible heat: Ordinary everyday heat, sensed by feel or by thermometer.

Separate, soil: One of the individual-sized groups of mineral soil particles–sand, silt, or clay.

Septic tank: An underground tank used in the deposition of domestic wastes. Organic matter decomposes in the tank, and the effluent is drained into the surrounding soil.

Sesquioxides: The general term for the aluminum and iron oxides in soil.

Seventh Approximation: The first widely distributed version of the present U.S. soil classification system.

Sewage sludge: Settled sewage solids combined with varying amounts of water and dissolved materials, removed from sewage by screening, sedimentation, chemical precipitation, or bacterial digestion.

Shear: Force, as of a tillage implement, acting at right angles to the direction of movement.

Shear strength: The resistance of a soil to forces acting at right angles to the soil body. A plow produces shearing forces.

Sheet (mineralogy): A flat array of more than one atomic thickness and composed of one or more levels of linked coordination polyhedra. A sheet is thicker than a plane and thinner than a layer. Example, tetrahhedral sheet, octahedral sheet.

Sheet erosion: Water erosion that removes a uniform layer of soil from the land surface.

Sheet flow: A thin, relatively uniform water runoff (a few millimeters deep), in which the water is not concentrated in channels.

Shelterbelt: A wind barrier of living trees and shrubs established and maintained for protection of farm fields. Syn. Windbreak.

Shifting cultivation: A farming system in which land is cleared, the debris burned, and crops grown for 2-3 years. When the farmer moves on to another plot, the land is then left idle for 5-15 years, then the burning and planting process is repeated.

Shortwave radiation: In meteorology, high-energy radiation of the ultraviolet, visible, and near infrared wavelengths, sunshine. Unlike longwave radiation, it passes freely through air.

Shrinkage limit: The maximum water content at which a reduction in water content will not cause a decrease in the soil's volume.

Side dressing: The application of fertilizer alongside row crop plants, usually on the soil surface. Nitrogen materials are most commonly side-dressed.

Silica/ alumina ratio: The molecules of silicon dioxide (SiO_2) per molecule of aluminum oxide (Al_2O_3) plus ferric oxide (Fe_2O_3) in clay minerals or in soils.

Silicates: Those mineals in which silicon and oxygen are the major elemental constituents.

Silt: 1) A soil separate consisting of particles between 0.05 and 0.002 mm in equivalent diameter. 2) A soil textural class.

Silting: The deposition of water-borne sediments in stream channels, lakes, reservoirs, or on floodplains, usually resulting from a decrease in the velocity of the water.

Sink: Somewhere or something for the disposal of wastewater.

Site index: A quantitative evaluation of the productivity of a soil for forest growth under the existing or specified environment.

Size separate: Individual mineral particles less than 2.0 mm in diameter, ranging between specified size limits. Sand, silt, and clay are common soil size separates.

Slag: A product of smelting, containing mostly silicates, the substances not sought to be produced as matte or metal and having a lower specific gravity.

Slake: A process of aggregate breakdown caused by internal pressures generated as water enters aggregates and air escapes.

Slash spots: Small areas in a field field that are slich when wet because of a high content of alkali or exchangeable sodium.

Slope: The degree of deviation of a surface from horizontal, measured in a numerical ratio, percent, or degrees.

Sludge: The solid portion of sewage. The sludge is obtained by treating sewage by different methods. Accordingly, sludges of different types are formed. Various types of sludges are settled sludge, digested sludge, activated sludge, digested activatedsludge and chemically precipitated sludge, On an average, sludge contains 1.5 to 3.5 per cent nitrogen, 0.75 to 4.0 per cent P_2O_5 and 0.3 to 0.6 per K_2O.

Slump: Collapse of soil and rock due to gravity.

Smectite: A group of silicate clays having a 2:1 type lattice structure with sufficent isomorphous substitution in either or both the tetrahedral and octahedral sheets to give a high interlayer negative charge and high cation exchange capacity and to permit significant interlayer expansion and consequent shrinking and swelling of the clay. Montmorillonite, beidellite, and saponite are in the smectite group.

Sodic soil: A soil containing sufficient exchangeable sodium to interfere with the growth of most crop plants and containing appreciable quantities of soluble salts. The exchangeable sodium adsorption ratio is > 13, the conductivity of the saturation extract is >4 dS/m (at 25°C), and the pH is usually 8.5 or less in the saturated soil.

Sodium (Na): It is one of the most common alkali metals. Its atomic number is 11 and atomic weight 22.9 per cent. It forms a number of chemical compounds used very widely in industry.

It is found in plant ash but is not yet accepted as an essential element. However, it is known to have influence on plant cell osmotic pressure and metabolism. From the angle of agriculture, sodium is useful for the following purposes:

1. It is an essential element for certain plants like sugar beet.
2. It acts as a substitute for potassium in certain plant metabolic functions.
3. It acts as an agent to maintain a high degree of availability of phosphates applied to the soil as superphosphate.

Its excess in soil is harmful for growth of plants. It is a dominanating cation in alkali soils.

Sodium adsorption ratio (SAR): The concentration of Na divided by the square root of the sum of Ca and Mg concentrations, both expressed as molarities and measured in the saturation extract.

$$SAR = \frac{Na+}{\sqrt{\frac{Ca^{2+} + Mg^{2+}}{2}}}$$

Soil acidity factors: A set of related factors often inhibiting plant growth in acidic soils (deficiency of Ca^{2+} and excess of H^{+}, Al^{3+}, and Mn^{2+}).

Soil air: The soil atmosphere; the gaseous phase of the soil, being that volume not occupied by soil or liquid.

Soil alkalinity: The degree or intensity of alkalinity of a soil, expressed by a value > 7.0 on the pH scale.

Soil amendment: Any material, such as lime, gypsum, sawdust, or synthetic conditioner, that is worked into the soil to make it more amenable to plant growth.

Soil association: A group of defined and named taxonomic soil units occurring together in an individual and characteristic pattern over a geographic region, comparable to plant associations in many ways.

Soil classification (Soil Taxonomy): The systematic arrangement of soils into groups or categories on the basis of their characteristics.

Soil complex: A mapping unit used in detailed soil surveys where two or more defined taxonomic units are so intimately intermixed geographically that it is undesirable or impractical, because of the scale being used, to separate them. A more intimate mixing of smaller areas of individual taxonomic units than that described under *soil association*.

Soil conditioners: Are chemicals which are added to maintain physical condition of the soil. Such chemicals are polyvinytites polyacrylates, cellulose gums, lignin derivatives and silicates.

Soil conservation: A combination of all management and land-use methods that safeguard the soil against depletion or deterioriation caused by nature and / or humans.

Soil correlation: The process of defining, mapping, naming, and classifying the kinds of soils in a specific soil syrvey area, the purpose being to ensure that soils are adequately defined, accurately mapped and uniformly named.

Soil density: A measure of the relative amount of pores and solid particles. The mass per unit volume.

Soil drainage class: An interpretive grouping of soils based on the level of the water table during the growing season and the rate of water flow through soil.

Soil fabric: The pattern resulting from the arrangement of solids and pores.

Soil fertility: Soil fertility refers to the inherent capacity of a soil to supply nutrients to plants in adequate amount and in suitable proportion.

Soil forming factors: The five interrelated natural factors that are active in the formation of soil: parent material, climate, organisms, topography, and time.

Soil forming process: The biological, chemical and physical processes that act under the influence of the soil forming factors to create soils.

Soil genesis: The mode of origin of the soil, with special reference to the processes or soil-forming factors responsible for the development of the solum, or true soil, from the unconsolidated parent material.

Soil geography: A subspecialization of physical geography concerned with the areal distributions of soil types.

Soil individual: The pedon; a three-dimensional body large enough to contain all the properties necessary to describe a soil completely.

Soil management: The sum total of all tillage operations, cropping practices, fertilizer, lime, and other treatments conducted on or appliedto a soil for the production of plants.

Soil map: A map showing the distribution of soil types or other soil mapping units in relation to the promiment physical and cultural features of the Earth's surface.

Soil matrix: Like soil fabric, the combination of solids and pores in a soil.

Soil mechanics and engineering: A subspecialization of soil science concerned with the effect of forces on the soil and the application of engineering principles to problems involving the soil.

Soil monolith: A vertical section of a soil profile removed from the soil and mounted for display or study.

Soil morphology: The physical constitution, particularly the structural properties, of a soil profile as exhibited by the kinds, thicknesses, and arrangement of the horizons in the profile, and by the texture, structure, consistence, and porosity of each horizon.

Soil order: In soil classification, the most general level of classification. All soils fit into 12 orders.

Soil organic matter: The organic fraction of the soil that includes plant and animal residues at various stages of decomposition, cells and tissues of soil organisms, and substances synthesized by the soil population. Commonly determined as the amount of organic material contained in a soil sample passed through a 2 mm sieve.

Soil pH: It refers to reaction of the soil as expressed in terms of pH. Soil having pH 7 is neutral and with less than pH 7 is acidic. If the soil pH is more than 7, it is alkaline in reaction.

Soil potential: The usefulness of a site for a specific purpose using available technology at a cost expressed in economic, social, or environmental units of value.

Soil productivity: Productivity is the present capacity of a soil to produce crop yield under a defined set of management practices.

It is measured in terms of the yield in relation to the input of production factors.

Soil profile: A vertical section of the soil from the surface through all its horizons, including C horizons.

Soil reaction: The degree of acidity of alkalinity of a soil mass, expressed in either pH value or in words

Soil Reaction		*pH*
Extremely acid	-	Below 4.5
Very strongly acid	-	4.5–5.0
Strongly acid	-	5.1–5.5
Medium acid	-	5.6–6.0
Slightly acid	-	6.1–6.5
Neutral	-	6.6–7.4
Mildly alkaline	-	7.4–7.8
Moderately alkaline	-	7.9–8.4
Strongly alkaline	-	8.5–9.0
Very strongly alkaline	-	9.1 and higher

Soil salinity: The amount of soluble salts in a soil, expressed in terms of percentage, milligrams per kilogram, parts per million (ppm), or other convenient ratios.

Soil solution: The aqueous liquid phase of the soil and its solutes consisting of ions dissociated from the surfaces of the soil particles and of other soluble materials.

Soil structure: The combination or arrangement of primary soil particles into secondary particles, units, or peds. These secondary units may be but usually are not, arranged in the profile in such a manner as to give a distinctive characteristic pattern. The secondary units are characterized and classified on the basis of size, shape, and degree of distinctness into classes, types, and grades, respectively.

Soil structure classes: A grouping of soil structural units or peds on the basis of size from the very fine to very coarse.

Soil structure grades: A grouping or classification of soil structure on the basis of inter- and intra-aggregate adhesion, cohesion, or stability within the profile. Four grades of structure, designated from 0 to 3, are recognized.

0: *Structureless–no observable aggregation.*

1: Weakly durable peds.

2: moderately durable peds.

3: *Strong, durable peds.*

Soil structure types: A classification of soil stucture based on the shape of the aggregates or peds and their arrangement in the profile, including platy, prismatic, columnar, blocky, subangular blocky, granulated, and crumb.

Soil survey: The systematic examination, description, classification, and mapping of soils in an area. Soil surveys are classified according to the kind and intensity of field examination.

Soil taxonomy: The U.S. Department of Agriculture system of soil classification.

Soil temperature classes: Classes are based on mean annual soil temperature and on differences between summer and winter temperatures at a depth of 50 cm.

1. Soils with 5°C and greater difference between summer and winter temperatures are classed on the basis of mean annual temperatures.

   Frigid: < 8°C mean annual temperature.

   Mesic: 8-15 °C mean annual temperature.

   Thermic: 15-22 °C mean annual temperature.

   Hyperthermic: > 22°C mean annual temperature.

2. Soils with < 5°C difference between summer and winter temperatures are classed on the basis of mean annual temperatures.

   Isofrigid: < 8°C mean annual temperature.

   Isomesic: 8-15 °C mean annual temperature.

   Isothermic: 15-22 °C mean annual temperature.

   Isohyperthermic: <22°C mean annual temperature.

Soil-testing: This refers to chemical tests of the soil that can be made rapidly and with low cost, as compared to conventional methods of soil chemical analysis which are more accurate but more time-consuming and expensive. Soil-testing covers both rapid analysis in the field and in the laboratory.

Soil textural class: A grouping of soil textural units based on the relative proportions of the various soil separates (sand, silt, and clay). These textural classes, listed from the coarsest to the finest in texture, are sand, loamy sand, sandy loam, loam, silt loam, silt, sandy clay loam, clay loam, silty clay loam, sandy clay, silty clay, and clay. There are several subclasses of the sand, loamy sand, and sandy loam, classes based on the dominant particle size of the sand fraction (*e.g.*, loamy fine sand, coarse sandy loam).

Soil texture: The relative proportions of the various soil separates in a soil.

Soil water potential (total): A measure of the difference between the free energy state of soil water and that of pure water. Technically it is defined as "that amount of work that must be done per unit quantity of pure water in order to transport reversibly and isothermically an infinitesimal quantity of water from a pool of pure water, at a specified elevation and at atmospheric pressure to the soil water (at the point under consideration)." This total potential consists of the following potentials.

Matric potential: That portion of the total soil water potential due to the attractive forces between water and soil solids as represented through adsorption and capillarity. It will always be negative.

*Osmotic potential:*That portion of the total soil water potential due to the presence of solutes in soil water. It will generally be negative.

Gravitational potential: That portion of the total soil water potential due to differences in elevation of the reference pool of pure water and that of the soil water. Since the soil water elevation is usually chosen to be higher than that of the reference pool, the gravitational potential is usually positive.

Solid waste: Waste other than sewage, mostly solid. Household garbage is a familiar form of solid waste.

Solum (plural **sola):** The upper and most weathered part of the soil profile; the A, E, and B horizons.

Solute: A material dissolved in a solvent to form a solution.

Specific gravity: The ratio of the weight or mass of a given volume of a substance to that of an equal volume of a reference substance. Water is the reference for liquids and solids.

Specific surface: The solid particle surface area per unit mass or volume of the solid particles.

Spectral signature: In remote sensing, an objects spectral reflectance pattern. The wavelength and intensity of reflectance.

Splash erosion: The detachment and airborne movement of small soil particles by raindrop impact.

Spoil (overburden): Material overlying sought after resources that is removed and piled to gain access to the minerals, coal, or other underground materials of value. Also, the material remaining after rock has been crushed to remove valuable minerals.

Starter solutions: The term refers to solutions of fertilizers, generally consisting of N-P_2O_5-K_2O in the ratio of 1:2:1 and 1:1:2. These solutions are applied to young vegetable plants at the time of transplanting to help the plants to establish.

Stele: The central conductive tissue of a root or some kinds of stem.

Stereo pair: Two photographs of the same object taken from slightly different positions that when viewed together give the viewer a three dimensional image.

Stomates (or stomata): The controllable openings in the epidermis of leaves and other parts of a plant shoot.

Straight fertilizers: Are the fertilizers containing only one plant nutrient *e.g.* ammonium sulphate, superphosphate, muriate of potash etc.

Stratified: Arranged in or composed of strata or layers.

Stress: A force exerted on a body, especially one that strains or deforms its shape. An adverse condition imposed on an organism.

Strip cropping: The practice of growing crops that require different types of tillage, such as row and sod, in alternate strips along contours or across the prevailing direction of wind.

Strip mining: A process in which rock and top soil strata overlying ore or fuel deposits are scraped away to expose the desired deposits.

Stubble mulch: The stubble of crops or crop residues left essentially in place on the land as a surface cover before and during the preparation of the seedbed and at least partly during the growing of a succeeding crop.

Subangular blocky: A structure type whose aggregates are nearly equidimensional, square, or blocklike with rounded edges.

Subsoil: That part of the soil below the plow layer.

Subsoiling: Breaking of compact subsoils, without inverting them, with a special knife-like instrument (chisel), which is pulled through the soil at depths usually of 30-60 cm and at spacings usually of 1-2 m.

Substrata::Layers of material underlying the soil horizons–for example, sediment or rock.

Substrate: A substratum. More often, the reactant or starting compound for a biochemical reaction or a material that provides food for microbes.

Subsurface diagnostic horizons: One or more soil horizons below the surface horizon with specific properties as described in the soil taxonomy (*e.g.* the argillic horizon).

Subtractive trial: A fertility trial in which deficiencies are detected by observing plant response to the elimination of individual elements from a complete mixture.

Sugar: A simple carbohydrate, consisting of one or two monomer units, each usually containing five or six Catoms, Rapidly used food sources for microbes in soil. Examples are glucose, fructose, and sucrose.

Sulphur coated urea: Urea granules or prills uniformly coated with a layer of sulphur used as slow release fertilizer. The average composition of sulphur coated urea is:

Urea	80-85 per cent
Sulphur	13-16 per cent
Wax	2 per cent

Rate of release of nitrogen from sulphur coated urea is governed by thickness of coating, placement of fertilizer, use of micobicide like coaltar oil, soil temperature, redox potential etc.

Summer fallow: A cropping system that involves management of uncropped land during the summer to control weeds and store moisture in the soil for the growth of a later crop.

Superphosphate: This term is most commonly used to denote ordinary superphosphate which contains nearly 16 percent available phosphoric acid. This fertiliseris made by treating rock phosphate with concentrated sulphuric acid.

Surface soil: The uppermost part of the soil, ordinarily moved in tillage, or its equivalent in uncultivated soils. Ranges in depth from 7 to 25 cm. Frequently designated as the "plow layer" the "Ap layer", or the "Ap horizon".

Surface chelation: A chelation reaction that holds a cation at the surface.

Surface tension: The force required per unit length to separate or pull apart a liquid surface.

Suspension: A fluid, usually liquid, bearing small particles held suspended but not dissolved.

Sustainable agriculture: According to Consultative Group on International Agricultural Research (CGIAR), it is the successful management of resources to satisfy the changing needs, while maintaining or enhancing the quality of environment and conserving natural resources.

Symbiosis: The living together in intimate association of two dissimilar organisms, the cohabitation being mutually beneficial.

Synthesis: Combination of simple molecules to form another substance *e.g.*, the union of carbon dioxide and waer under the action of sunlight in photosynthesis. Adjective from synthesis is synthetic, *e.g.* synthetic ammonia.

T

Talus: Fragments of rock and other soil material accumulated by gravity at the foot of cliffs or steep slopes.

Taxonomy, soil: The science of classification of soils; laws and principles governing the classifying the soil.

Technical classification: A soil classification of inferred rather than observed properties.

Temperature gradient: The rate at which temperature changes with distance, in soil, usually vertical distance.

Tensile strength: The resistance of a soil to forces pulling from opposite directions.

Tensiometer: A device for measuring the negative pressure (or tension) of water in soil in situ; a porous, permeable ceramic cup connected through a tube to a manometer or vacuum guage.

Terrace: (1) A level, usually narrow, plain bordering a river, lake, or the sea. Rivers sometimes are bordered by terraces at different levels. (2) A raised, more or less level or horizontal strip of earth usually constructed on or nearly on a contour and designed to make the land suitable for tillage and to prevent accelerated erosion by diverting water from undesirable channels of concentration; sometimes called diversion terrace.

Tetrahedron: A four sided arrangement of atoms with a cation, such as Si, in the center surrounded by four oxygen atoms or hydroxyls.

Texture: The relative proportion of the various soil separates–sand, silt, and clay–that make up the soil texture classes as described by the textural triangle.

Thermal analysis (differential thermal analysis): A method of analyzing a soil sample for constituents, based on a differential rate of heating of the unknown and standard samples when a uniform source of heat is applied.

Thermophilic organisms: Organisms that grow readily at temperatures above 45°C.

Thiobacillus: Sulphur oxidizing chemoautotrophic bacteria that convert inorganic sulphur to sulphate.

Tile, drain: Pipe made of burned clay, concrete, or ceramic material, in short lengths, usually laid with open joints to collect and carry excess water from the soil.

Till: (1) Unstratified glacial drift deposited directly by the ice and consisting of clay, sand, gravel, and boulders intermingled in any proportion. (2) To plow and prepare for seeding; to seed or cultivate the soil.

Tillage: The mechanical manipulation of soil for any purpose; but in agriculture it is usually restricted to the modifying of soil conditions for crop production.

Tillage, conservation: Any tillage sequence that reduces loss of soil or water relative to conventional tillage, including the following systems.

Minimum tillage: The minimum soil manipulation necessary for crop production or meeting tillage requirements under the existing soil and climatic conditions.

Mulch tillage: Tillage or preparation of the soil in such a way that plant residues or other materials are left to cover the surface; also called *mulch farming, trash farming, stubble mulch tillage, plowless farming.*

No-tillage system: A procedure whereby a crop is planted directly into a seedbed not tilled since harvest of the previous crop; also zero tillage.

Plow-planting: The plowing and planting of land in a single trip over the field by drawing both plowing and planting tools with the same power source.

Ridge till: Planting on ridges formed by cultivation during the previous growing period.

Sod planting: A method of planting in sod with little or no tillage.

Strip till: Planting is done ina narrow strip that has been tilled and mixed, leaving the remainder of the soil surface undisturbed.

Subsurface tillage: Tillage with a special sweep like plow or blade that is drawn beneath the surface, cutting plant roots and loosening the soil without inverting it or without incorporating residues of the surface cover.

Wheel track planting: A practice of planting in which the seed is planted in tracks formed by wheels rolling immediately ahead of the planter.

Tillage, conventional: The combined primary and secondary tillage operations normally performed in preparing a seedbed for a given crop grown in a given geographic area.

Tillage, primary: Tillage that contributes to the major soil manipulation, commonly with a plow.

Tillage, rotary: An operation using a power-driven rotary tillage tool to loosen and mix soil.

Tillage secondary: Any tillage operations following primary tillage designed to prepare a satisfactory seedbed for planting.

Tilth: The physical condition of soil as related to its ease of tillage, fitness as a seedbed, and its impedance to seedling emergence and root penetration.

Top dressing: An application of fertilizer to a soil after the crop stand has been established.

Toposequence: A sequence of related soils that differ, one from the other, primarily because of *topography* as a soil formation factor.

Topsoil: (1) The layer of soil moved in cultivation. (2) Presumably fertile soil material used to top-dress roadbanks, gardens, and lawns.

Tortuosity: A term describing the winding, twisting, and crooked nature of soil pores. A factor contributing to the rate of water movement in soil.

Trace elements: An old term used for the nutrients required in small quantities. Such nutrients are called 'micronutrients' or 'minor elements'.

Transitional horizon: A horizon with properties intermediate between the horizons above and below.

Transitional slide: A break causing a mass to slide along the face of the supporting mass under the force of gravity.

Transmitted energy: In remote sensing, energy that passes through an object.

Transpiration: Evaporation from leaves, the flow of water through plants from soil to atmosphere.

Transported parent material: The opposite of *residual parent material,* unconsolidated material carried by wind or water that, once stabilized, acts as the starting material for soil.

Trial: A term often used to describe a field or container experiment designed to determine nutrient deficiencies.

Triple superphosphate: A fertilizer which contains about 45 to 46 per cent available phosphoric acid. This differs from ordinary superphosphate in that it contains very little calcium sulphate. In the fertilizer trade, the product is also called treble superphosphate or multiple super-phosphate.

Truncated: Having lost all or part of the upper soil horizon or horizons.

Tuff: Volcanic ash usually more or less stratified and in various states of consolidation.

Tundra: A level or undulating treeless plain characteristic of arctic regions.

Turgulent transfer: Diffusion-like movement of heat, gases, and solutes greatly accelerated by irregular mixing motion–turbulence–in the fluid medium.

Turgor: The normal state of turgidity in living cells produced by pressure of the cell contents on the cell walls.

Turgor pressure: Pressure exerted on a cell wall by the cell's constituents (and vice versa).

U

Ultisols: Strongly weathered soils formed in warm, humid regions under forest vegetation.

Undifferentiated group: A soil map unit consisting of two or more similar soil units not in a regular geographic association. The soil units have the same or very similar use and management.

Unit cell: The smallest deinable repeating structural unit in a crystalline material.

Universal soil loss equation (USLE)**:** An equation for predicting the average annual soil loss per unit area per year, A = RKLSPC, where R is the climatic erosivity factor (rainfall plus runoff), K is the soil erodibility factor, L is the length of slope, S is the percent slope, C is the cropping and management factor, and P is the soil erosion practice factor.

Unsaturated flow: The movement of water in a soil that is not filled to capacity with water.

Urease: A crystalizabale protein enzyme that activates hydrolysis of urea.

$$\underset{\text{Urea}}{NH_2CO.NH_2} + \underset{\text{Water}}{2H_2O} \longrightarrow \underset{\text{Amm.Carbonate}}{(NH_4)_2CO_3}$$

Its richest source is soyabean. It is found extensively in jack beans and numerous fungi. It is marketed as white tablets or powder. It is soluble in dilute alkali. Its isoelectric point is pH

5.5. It is activated by the presence of metals. Urease is found in soils in various degrees. If its concentration is high, urea hydrolysis in soils takes place very rapidly. Urease enzyme taken from soyabean or jack bean is also used to determine urea in mixed fertilizers.

Uronite: A sugar with a COOH group.

V

Van der Waal's forces: Combined polar and London attractions between molecules in close proximity.

Variable charge: An electrical charge on clay or organic matter that changes with changes in soil pH.

Varnish, desert: A glossy sheen or coating on stones and gravel in arid regions.

Vascular: Having to do with veins.

Vermiculite: A 2:1-type silicate clay usually formed from mica that has a high net negative charge stemming mostly from extensive isomorphous substitution of aluminum for silicon in the tetrahedral sheet.

Vertical photo: In remote sensing, a photo taken from directly above an object rather than at an angle to the object.

Vertisols: Soils high in clay that form deep cracks at least one centimeter wide during dry season.

Vesicular arbuscular mycorrhiza (VAM): A common endomycorrhizal association produced by phycomycetous fungi of the genus *Endogone* and characterized by the development of two types of fungal structures: a) within root cells small structures known as arbuscles and b) between root cells storage organs known as vesicles. Host range includes many agricultural and horticultural crops.

Vesicular pore: A soil pore not connected to other pores.

Virgin soil: A soil that has not been significantly disturbed from its natural environment.

Visible spectrum: Electromagnetic radiations which are within the range of 400-700 nm and visible to the naked eye. The beam formed due to these is called the visible spectrum. It consists of seven colours corresponding to the following wavelength:

Colour of Light	*Range of Wavelength (in nm)*
Red	747-700
Orange	585-647
Yellow	575-585
Green	491-575
Blue	424-491
Violet	400-424

Void ratio: A measurement of porosity. The ratio of void volume to soil bulk volume.

Volcanic ash: Fine particles of rock blown into the air by a volcano. Settles, often in layers, to become soil parent material. Often consists of short-range-order minerals.

Volumetric analysis: A group of methods of quantitative analysis based on accurate measurement of the volume of a reagent of known concentration used in the analysis. These are extensively used in fertilizer analysis of three types: Neutralization, oxidation reduction and complex formation.

W

Washed in zone: Formed by clay separating from the top few tenths of a millimeter of soil and being translocated into the next few tenths to clog the pores.

Water erosion: The natural wearing away of the earth's surface by rainfall and surface runoff.

Waterlogged: Soils saturated with water.

Water potential: The tendency of soil water to move, the sum of gravity, pressure, matric and solute components.

Water profile: The pattern of vertical variation in water content or potential with depth in the soil.

Water retention curve or soil water characteristic curve: The graphical relationship between soil water content (by mass or colume) and the soil water matric potential (the energy required to remove the water).

Water stable aggregate: A soil aggregate stable to the action of water such as falling drops, or agitation as in wet-sieving analysis.

Water table: The upper surface of groundwater or that level below which the soil is saturated with water.

Water table, perched: The surface of a local zone of saturation held above the main body of groundwater by an impermeable layer of stratum, usually clay, and separated from the main body of groundwater by an unsaturated zone.

Water use efficiency: Dry matter or harvested portion of crop produced per unit of water consumed.

Weathering: All physical and chemical changes produced in rocks, at or near the Earth's surface, by atmospheric agents.

Wetting front: The advancing boundary between dry soil and wetted soil during infiltration.

Wilting point (permanent wilting point)**:** The moisture content of soil, on an oven-dry basis, at which plants wilt and fail to recover their turgidity when placed in a dark humid atmosphere.

Windbreak: Planting of trees, shrubs, or other vegetation perpendicular, or nearly so to the principal wind direction to protect soils, crops, homesteads, etc., from wind and snow.

X

X-ray diffraction: An analytical method that involves the exposing of samples to X-rays. The radiation is reflected from the sample in response to structure of the crystals that compose the sample. It is useful in identification of elements in materials and coatings.

Xeric: Refers to a dry environment *e.g.* desert.

Xerophytes: Plants that grow in or on extremely dry soils or soil materials.

Xenon: A nobal gas, which found in atmosphere in trace amount.

Xylem: One of two main kinds of conductive tissue in plants. Xylem has tubelike cells that conduct water and ions rapidly from roots to leaves. Wood is xylem tissue.

Y

Yellow cake: Uranium oxide (U_3O_8) that results from the refining of uranium ore. The purified material contain 93.3 percent uranium-238 and 0.7 percent uranium-235. The term yellow cake is applied because of the colour and texture of the material.

Yield: The individuals or biomass removed when the plant population is harvested.

Yield potential: The total production capacity of a crop.

Z

Zero order reaction: A chemical reaction in which the rate of reaction is independent of reactants concentration.

Zero tillage: It is a system which refers to planting crops with minimum of soil disturbance. In this seed are placed directly into narrow slits 2–4 cm wide and 4–7 cm deep made with a drill fitted with chisel "inverted T" or double disc openers without land preparation. In this soil macro and micro fauna builds and maintain an open-pore structure of the soil.

Zeolite: Natural hydrated silicate of calcium and aluminium used in water softening process by the ion exchange method.

Zone of depletion: Top layer of A horizon of a soil, characterised by the removal of soluble and insoluble soil components through leaching and eluviation by gravitational water.

Zinc (Zn): An essential plant micronutrient found deficient in a variety of cultivated soils all over the world. Its atomic number in 30 and atomic weight is 65.37. Total zinc content in soils varies from 5–50 ppm. In some soils its deficiency is attributed to phosphate application. Plants contain zinc in the range of 1-50 ppm.

Appendices

Appendix-I

Commonly Used SI Prefixes

Order of Magnitude	*Prefix*	*Symbol*
10^{15}	Peta	P
10^{12}	Tera	T
10^{9}	Giga	G
10^{6}	Mega	M
10^{3}	Kilo	k
10^{2}	Hecto	h
10^{1}	deca	da
10^{-1}	deci	d
10^{-2}	centi	c
10^{-3}	milli	m
10^{-6}	micro	μ
10^{-9}	nano	n
10^{-12}	pico	p

Appendix-II

Factor for Converting Nonmetric to SI-metric Unit

Nonmetric Unit		*SI-metric Unit*
Length		
inches	x 25.4	= millimeters
feet	x 0.305	= meters
miles	x 1.61	= kilometers
Area		
square inches	x 645	= square millimeters
square feet	x 0.0929	= square meters
acres	x 0.405	= hectares
square mile	x 2.54	= square kilometers
Volume		
cubic inches	x 1.64×10^{-5}	= cubic meters
cubic feet	x 0.028	= cubic meters
gallons	x 3.78	= liters
pints	x 0.473	= liters
Mass		
ounces	x 28.4	= grams
pounds	x 454	= grams
tons	x 907	= kilograms
Yield		
Pound/ acre	x 1.12	= kilograms / hectare
Tons/ acre	x 2.24	= megagrams/ hectare.
Other units		
temp. (°F-32)	x 0.556	= temperature (°C)
horse power	x 7.46	= watts
kilocalories	x 1.163	= watt-hours
kilocalories	x 4.190	= joules

Appendix-III

Important Gravimetric Conversion Factors

Multiply the weight of	*By this Factor*	*To Obtain Weight of*
Al	1.8895	Al_2O_3
Al_2O_3	0.52913	Al
Ba	1.6994	$BaSO_4$
$BaSO_4$	0.58845	Ba
$BaSO_4$	0.13737	S
$BaSO_4$	0.41154	SO_4
Ca	2.4973	$CaSO_4$
Ca	4.2959	$CaSO_{4.}2H_2O$
$CaCO_3$	0.4004	Ca
$CaSO_{4.}2H_2O$	0.23277	Ca
CaO	0.715	Ca
CaC_2C_4 (oxalate)	0.3128	Ca
$CaSO_{4.}$ $2H_2O$	0.5579	SO_4
$CaCO_3$	0.43971	CO_2
AgCl	0.24737	Cl
Cl	4.0426	AgCl
KCl	0.47557	Cl
Cl	2.1027	KCl
NaCl	0.60664	Cl
Cl	1.6484	NaCl
Cu	3.9296	$CuSO_4.2H_2O$
$CuSO_4$ $5H_2O$	0.25448	Cu
$FeSO_4$ $(NH_4)_2$ $SO_4.6H_2O$	0.14242	Fe

Contd...

Appendix-III Contd...

Multiply the weight of	*By this Factor*	*To Obtain Weight of*
Fe	1.4297	Fe_2O_3
Fe_2O_3	0.69944	Fe
$FeSO_4 \cdot 7\,H_2O$	0.20088	Fe
$FeSO_4 \cdot 7\,H_2O$	0.28720	Fe_2O_3
MgO	0.60317	Mg
$Mg_2\,P_2O_7$	0.2185	Mg
$MgSO_4 \cdot 7H_2O$	0.09866	Mg
MgP_2O_7	0.36228	MgO
N	1.2159	NH_3
N	4.7168	$(NH_4)_2\,SO_4$
N	4.4266	NO_3
NH_3	0.82245	N
NH_3	3.6407	NO_3
$NaNO_3$	0.1648	N
$(NH_4)_2\,SO_4$	0.21201	N
NO_3	0.22591	N
NO_3	0.27467	NH_3
NO_2	0.30447	N
P	2.2914	P_2O_5
P	3.0661	PO_4
PO_4	0.32613	P
PO_4	0.7473	P
P_2O_5	0.43642	P
P_2O_5	2.18	$Ca(PO_4)_2$
P_2O_5	1.3381	PO_4
$Ca(PO_4)_2$	0.45762	P
$Ca(PO_4)_2$	0.1997	P
$Ca(H_2PO_4)2H_2O$	0.2457	P
KH_2PO_4	0.2276	P
K	1.2046	K_2O
KCl	0.52443	K

Contd...

Appendix-III Contd...

Multiply the weight of	*By this Factor*	*To Obtain Weight of*
K_2O	0.83015	K
K_2SO_4	0.44874	K
$KC1O_4$	0.28219	K
AgCl	0.2473	Cl
$AgNO_3$	0.84371	AgCl
$AgNO_3$	0.2087	Cl
Na	1.5422	Cl
Na_2CO_3	1.2855	$CaSO_4$
NaCl	0.6066	Cl
NaCl	0.39336	Na.
$NaHCO_3$	0.27367	Na
Na_2SO_4	0.32371	Na.
S	5.3695	$CaSO_4.2H_2O$
S	7.2795	$BaSO_4$
S	2.995	SO_4
$CaSO_4.2H_2O$	0.1862	S
$CaSO_4.2H_2O$	0.5566	SO_4
SO_4	1.7812	$CaSO_4.2H_2O$
SO_4	0.3217	S
$BaSO_4$	0.13737	S
$BaSO_4$	0.41155	SO_4
ZnO	0.8033	Zn
Zn	1.2447	ZnO
$Zn\ SO_4.\ 7H_2O$	0.22736	Zn
Zn	4.3982	$ZnSO_4.\ 7H_2O$

Appendix-IV

Nutrients (N, P_2O_5 and K_2O) content of some organic waste materials, FYM, compost, cakes and residues

Class	*Nutrient Source*	*Nutrient Content (%)*		
		N	*P_2O_5*	*K_2O*
Animal wastes	Sheep dung	0.65	0.5	0.03
	Night soil	1.2-1.5	0.8	0.5
	Cattle dung	0.3-0.4	0.10-0.15	0.15-0.20
	Cattle urine	0.80	0.01-0.2	0.5-0.70
	Human urine	1.0-1.2	0.1-0.2	0.2-0.3
	Leather waste	7.0	0.1	0.2
	Hair and wool waste	12.3	0.1	0.3
FYM and compost	Farmyard manure	0.5-1.0	0.15-0.20	0.5-0.6
	Poultry manure	2.87	2.90	2.35
	Town compost	1.5-2.0	1.0	1.5
	Rural compost	0.5-1.0	0.2	0.5
	Water hyacinth compost	2.0	1.0	2.3
Oil cakes	Cotton seed	3.9	1.8	1.6
	Groundnut	4.5	1.7	1.5
	Castor	5.5-5.8	1.8	1.0
	Coconut	3.0-3.2	1.8	1.7
	Karanj *(Pongamia pinnata)*	3.9-4.0	0.9-1.0	1.3
	Neem *(Azadirachta indica)*	5.2	1.0	1.4
	Mahua *(B.latifolia)*	2.5-2.6	0.8	1.8
	Rapeseed	5.1	1.8	1.0
	Linseed	5.5	1.4	1.2
	Safflower	4.8	1.4	1.2
	Sesame	6.2	2.0	1.2

Contd...

Appendix-IV Contd...

Class	*Nutrient Source*	*Nutrient Content (%)*		
		N	P_2O_5	K_2O
Animal meals	Blood meal	10-12	1.2	1.0
	Fish meal	4-10	3-9	1-8
	Meat meal	10.5	2.5	0.5
	Horn and hoof	13.0	0.3-1.5	-
	Raw bone	3-4	20-25	-
	Steamed bone	1-2	25-30	1.8

Appendix-V

Level of Nutrient Element in Plants

Deficient	When an essential element is at low concentration that severely limits yield and produces more or less distinct deficiency symptoms. Extreme deficiency will lead to death of the plant.
Insufficient	When the level of an essential plant nutrient is below that required for optimum yields or when there is an imbalance with another nutrient. Symptoms of this condition are seldom evident.
Toxic	When the concentration of either essential or other elements is sufficiently high to reduce plant growth severely. Severe toxicity will result in death of plants.
Excessive	When the concentration of an essential plant nutrient is sufficiently high to result in a corresponding shortage of another nutrient.

Appendix-VI

Some Common Term Used to Describe Nutrient Deficiency Symptoms

Bronzing	Development of bronze copper colour on the tissues.
Chlorosis	Loss of chlorophyll, resulting in loss of green colour, result paleness, yellow tissue.
Decline	Onset of general weakness as indicated by loss of vigour, poor growth and low productivity.
Die-back	Collapse of the growing tip affecting the youngest leaves.
Firing	Burning of tissue accompanied with dark brown or reddish brown colour.
Lesion	A localised wound of the tissue accompanied with loss of normal colour.
Necrosis	Death of tissue.
Scorching	Burning of the tissue accompanied with the light brown colour.

Appendix-VII

Disease Caused by the Deficiency of Nutrients

Sl.No.	*Name of Diseases*	*Nutrient Responsible*
1.	Khaira disease of rice	Zn
2.	White bud of maize	Zn
3.	Little leaf of cotton	Zn
4.	Mottle leaf/frenching in citrus	Zn
5.	Fern leaf in potato	Zn
6.	Grey speack of oat/cereals	Mn
7.	Marsh spot of field peas	Mn
8.	Speckled yellow of sugarbeet	Mn
9.	Streak disease of sugarcane	Mn
10.	Yellow spot of citrus	Mo
11.	Whiptail in cauliflower	Mo
12.	Blacknened heart of sugarbeet	B
13.	Corking of apple	B
14.	Gummosis of apple	B
15.	Crinckle leaf of cotton	Mn toxicity
16.	Molydenosis of cattle	Mo toxicity in forages

Appendix-VIII

Rhizobium Species and their Cross Inoculation Group

Rhizobium sps.	*Cross Inoculation Group*	*Legumes*
R. trifolii	Clover group	Trifolium
R.meliloti	Alfalfa group	Melilotus
		Medicago
		Trigonella
R. phaseoli	Bean group	Phaseolus
R. lupini	Lupine group	Lupinus,
		Ornithopus
R. leguminosarum	Pea group	Pisum, Vicia, Lens
R. japonicum	Soybean group	Glyciine
Rhizobium sp.	Cowpea group	Vigna, Arachis.
(Miscellany)	(Miscellany)	

Appendix-IX

Nutrient Susceptibility

Sl.No.	*Nutrient*	*Most Susceptible Crops*
1.	Nitrogen	Maize, wheat, rice, mustard
2.	Phosphorus	Maize, barley, lettuce, tomato
3.	Potassium	Potato, clover, lucerne, cotton, bean
4.	Calcium	Lucerne, legumes
5.	Magnesium	Potato, cauliflower
6.	Sulphur	Lucerne, clover, raya
7.	Zinc	Maize, rice, onion, citrus, apple.
8.	Iron	Sorghum, rice, soybean, citrus
9.	Manganese	Beans, oat, peas, wheat, citrus, maize, reddish, apple
10.	Copper	Wheat, oat, alfalfa, apple, citrus, barley, maize, lettuce
11.	Boron	Alfalfa, cauliflower, sugarbeet, turnip, apple, roses, peach
12.	Molybdenum	Cauliflower, clover, lettuce, spinach, citrus, oat
13.	Chlorine	Sugarbeet, barley, maize, spinach, tomato
14.	Cobalt	Legume

Appendix-X

Multiple Deficiencies and Nutrient Interactions

Antagonistic Effects

— Excess of Ca may induce P deficiency

— Excess of Ca & Mg may depress K uptake

— Excess of Ca may reduce Mg uptake if ratio widen than 7:1

— Higher concentration of K & NH_4 may restrict Mg uptake

— Excess application of N, K & Ca control B toxicity

— Higher availability of Fe and SO_4 accentuate Mo deficiency

— Higher concentration of Cu, Mn and NH_4-N depresses Mo uptake

— High P availability induce Zn deficiency

— Excess accumulation of P shows symptoms resembling Zn deficiency

— Increasing use of N, P, K may aggravate Cu deficiency

— Higher concentration of Zn & Al dipresses Cu absorption

— Excess of P and Mo may cause Fe deficiency

— Nitrate-N application may cause Fe deficiency

— High pH, CO_3 & HCO_3 may induce Fe deficiency

Synergistic Effects

— Application of Cl containing and acid forming N fertilizers will boost Mn uptake

— Application of N-usually enhance micronutrient uptake and utilization

— Mg may induce P-uptake

— Application of NH_4-N improve P uptake

— N & P fertilization improve K uptake

— Higher availability of Mg and NO_3-N boost Mo uptake

Appendix-XI

General Descriptions of the Soil Orders

Alfisols	Soils with medium-to-light-colored A horizons and with significant clay accumulation in their B horizons. Most Alfisols formed under forest vegetation.
Andisols	Soils with relatively high contents of glass and/or extractable aluminum and iron. Most Andisols are formed in relatively young volcanic ash materials.
Aridisols	Soils of arid regions. Aridisols are light colored and most are alkaline in reaction.
Entisols	Very young soils that have little or no horizon differentiation.
Histosols	Soils dominated by organic materials. Histosols form in wet and/or cold conditions.
Inceptisols	Soils in an early stage of development that lack significant clay accumulations. Most Inceptisols formed under forest vegetation in humid climaties.
Mollisols	Soils with thick, dark-colored A horizons. Most Mollisols formed under grass vegetation in temperate climates.
Oxisols	Very highly weathered soils. Most Oxisols occur in the tropics and have low natural fertility.
Spodosols	Very strongly leached soils of cool, humid areas. Spodosols have bright colors, high acidity, and low fertility.
Ultisols	Strongly weathered soils formed in warm, humid regions under forest vegetation. Ultisols are redder and less fertile thn Alfisols.
Vertisols	Soils high in clay that form deep cracks at least 1 cm wide during dry seasons.
Gallisols	Newly identified order.

www.ingramcontent.com/pod-product-compliance
Ingram Content Group UK Ltd.
Pitfield, Milton Keynes, MK11 3LW, UK
UKHW020207290926
15470UKWH00001B/52

9 789351 240006